Statistics and Computing

Series Editor

Wolfgang Karl Härdle, Humboldt-Universität zu Berlin, Berlin, Germany

Statistics and Computing (SC) includes monographs and advanced texts on statistical computing and statistical packages.

Giovanni Cerulli

Fundamentals of Supervised Machine Learning

With Applications in Python, R, and Stata

 Springer

Giovanni Cerulli 🆔
IRCRES-CNR, Research Institute for
Sustainable Economic Growth
National Research Council of Italy
Rome, Italy

ISSN 1431-8784   ISSN 2197-1706 (electronic)
Statistics and Computing
ISBN 978-3-031-41336-0   ISBN 978-3-031-41337-7 (eBook)
https://doi.org/10.1007/978-3-031-41337-7

This Springer imprint is published by the registered company Springer Nature Switzerland AG
The registered company address is: Gewerbestrasse 11, 6330 Cham, Switzerland

Paper in this product is recyclable.

To the memory of my grandmother Esterina and aunt Letizia who silently but constantly accompanied the writing of this book. For the good, the wisdom, and the knowledge they pass on to me. For their priceless and everlasting love.

Preface

Scope of the Book

This is my second book published with Springer. The first, *Econometrics of Program Evaluation: Theory and Applications* (second edition), deals with causal inference for policy impact assessment and has a strong inferential character. This second book, dealing with supervised Machine Learning, largely complements the first in many regards, in that it privileges predictive analytics over inferential analysis. Inference and prediction are however both essential for data science and analysis, as they shed light on two different, although increasingly linked, modes of scientific investigation.

This book presents the fundamental theoretical notions of supervised Machine Learning along with a wide range of applications with Python, R, and Stata software. The book is dedicated to Ph.D. students, academics, and practitioners in various fields of study (social sciences, medicine, epidemiology, as well as hard sciences) who intend to learn the fundamentals of supervised Machine Learning to apply it to concrete case studies.

The book assumes the reader to have a good understanding of basic statistics (both descriptive and inferential), the meaning and the writing of algorithms, and a working knowledge of Python, R, or Stata software. Mathematics is used only when strictly necessary, and a large focus is paid to graphical explanations instead of analytic proofs.

Why should the reader consider this book as a valuable source of knowledge for learning theoretical and applied supervised Machine Learning? Here are three compelling reasons:

1. *Comprehensive coverage of theoretical foundations.* The book starts by presenting the fundamental theoretical notions of supervised Machine Learning. It covers key concepts such as regression, classification, ensemble methods, and evaluation metrics, providing a solid foundation for understanding the principles and techniques behind supervised learning. This theoretical grounding ensures that readers gain a deep understanding of the subject matter before moving on to practical applications.

2. *Diverse applications and case studies.* The book goes beyond theory and offers a wide range of applications using three popular software packages: Python, R, and Stata. By including multiple software options, the book caters to the diverse needs and preferences of readers. Python, known for its versatility and extensive libraries like scikit-learn, allows readers to implement Machine Learning algorithms and explore advanced techniques. R, a widely used language in statistics and data analysis, provides readers with additional tools and packages—such as CARET—specifically tailored for statistical learning modeling and results' visualization. Stata, a statistical software widely used in social sciences, medicine, and epidemiology, offers readers an alternative perspective with its unique features and capabilities. By including all three software packages, this book ensures that readers can leverage the software they are most comfortable with or choose the one best suited to their specific domain.

3. *Differentiating factors among other books.* While there are indeed several good books available on Statistical Learning and software, this book stands out by addressing the specific needs of Ph.D. students, academics, and practitioners in various fields of study. It bridges the gap between theory and application, making it particularly relevant for those who wish to apply supervised Machine Learning techniques to concrete case studies in their respective disciplines. The inclusion of multiple software packages further enhances the book's value, as readers can gain exposure to different tools and methodologies and develop a versatile skill set.

In summary, this book offers a comprehensive introduction to supervised Machine Learning, catering to the needs of Ph.D. students, academics, and practitioners in diverse fields of study. Its inclusion of Python, R, and Stata software packages allows readers to choose the tool that aligns with their preferences and domain-specific requirements. By combining theoretical foundations with practical applications and addressing the specific needs of its target audience, this book provides a unique and valuable resource for anyone looking to master supervised Machine Learning for real-world scenarios.

Every chapter presents an initial theoretical part, where the basics of the methodologies are explained, followed by an applicative part, where the methods are applied to real-world datasets. Each chapter is self-contained, but the reader is invited to consider reading the two introductory chapters before going through the study of chapters dealing with single methods.

For ease of reproducibility, the Python, R, and Stata codes used in the book for carrying out the applications, along with the related datasets, are available at this GitHub link: https://github.com/GioCer73/Book_FSML_Ed1.

Organization of the Book

The book is organized into eight chapters for a structured and cohesive learning experience.

Chapter 1. The first chapter introduces the rationale and ontology of Machine Learning. Its purpose is to pave the way for the subsequent chapters and facilitate the understanding of the material covered.

Chapter 2. This chapter focuses on the statistics of Machine Learning, which serves as a crucial foundation for comprehending the following chapters. Understanding this section will ensure a smooth transition into the presentation and application of individual Machine Learning methods.

Chapter 3. Model selection and regularization are the main topics discussed in this chapter. Special emphasis is placed on lasso and elastic-net regression and classification, subset selection models, and the lasso's role in inferential analysis.

Chapter 4. The fourth chapter delves into discriminant analysis, nearest neighbor methods, and support vector machines. Although primarily utilized for classification purposes, it is worth noting that regression analysis can also be performed using the nearest neighbor and support vector machine methods.

Chapter 5. In this chapter, tree modeling is explored in both classification and regression scenarios. After outlining the tree-building algorithm, the chapter introduces three ensemble methods: bagging, random forests, and boosting machines, which are valuable for prediction and feature-importance detection.

Chapter 6. The sixth chapter provides an introduction to artificial neural networks. It covers fully connected neural networks, explaining their construction logic and demonstrating software applications for intelligent tasks such as image recognition.

Chapter 7. Building upon the foundation laid in the previous chapter, the seventh chapter delves into deep learning modeling. This section focuses on special artificial neural networks that leverage specific data ordering to enhance predictability and computation efficiency. Two deep learning architectures are presented and discussed: convolutional neural networks and recurrent neural networks.

Chapter 8. The concluding chapter serves as a primer on sentiment analysis. This approach involves developing a predictive mapping between human textual documents and the corresponding human polar sentiment associated with those documents. The chapter explores textual feature engineering and its application in predictive analytics.

By structuring the book into these eight chapters, readers can systematically progress from understanding the fundamentals to exploring advanced topics and applications in Machine Learning.

Rome, Italy   Giovanni Cerulli

Acknowledgments

Writing a book is a challenging and arduous experience. In this regard, writing this book made no exception. I want to thank all my colleagues at IRCRES-CNR for their support, in particular, Emanuela Reale (our current director), Edoardo Lorenzetti (responsible for the research unit of Rome, where I work), Antonio Zinilli, and Lucio Morettini.

A special thank goes to Timberlake Ltd and Tstat Srl for the opportunity they gave me to run several courses on Machine Learning with Stata and Python.

I want to thank my lovely wife Rossella, and my sweet daughters Marta and Emma, for sustaining me during the time I was writing this book.

Another special thank goes to my parents, Felice and Barbara, my parents-in-law, Maria and Franco, and to my brother Riccardo.

Some fellow colleagues read this book. I want to thank them all.

All the errors are mine.

January 2023   Giovanni Cerulli

Contents

List of Figures

List of Tables

Chapter 1
The Basics of Machine Learning

1.1 Introduction

This chapter offers a general introduction to supervised Machine Learning (ML) and constitutes the basics to get through the next and subsequent chapters of this book.

It starts by providing an introduction to the basics of Machine Learning, discussing its definition, rationale, and usefulness. We point out that Machine Learning is a transformative technology within the field of artificial intelligence that enables computers to learn and make predictions or decisions without explicit programming. The rationale behind Machine Learning is the need to extract meaningful insights from the vast amount of available data. Traditional analytical approaches are often inadequate for handling such data, while Machine Learning algorithms excel in identifying patterns and relationships within them.

The usefulness of Machine Learning can be demonstrated across various domains of application. For example, in healthcare, Machine Learning assists in diagnosing diseases, predicting patient outcomes, and personalizing treatment plans. In finance, it aids in fraud detection, credit scoring, and investment strategies. In e-commerce, Machine Learning powers recommendation systems to enhance customer experiences. Of course, a large set of other examples of Machine Learning applications can be provided.

The chapter goes on by highlighting the shift from symbolic AI to statistical learning, with Machine Learning leveraging statistical models and algorithms to identify patterns and make predictions. Symbolic AI relied on explicit rules and logical reasoning, which struggled with complex real-world scenarios. Machine Learning, on the other hand, stands out in handling complex and ambiguous situations by analyzing data and making probabilistic decisions. In this regard, the chapter presents a comparison between symbolic AI and Machine Learning using the example of numeral recognition. Symbolic AI requires explicit rules for recognizing numerals, while Machine Learning algorithms can learn to recognize numerals by analyzing labeled examples and identifying common features and patterns. This illustrates the flexibility and scalability of Machine Learning.

G. Cerulli, *Fundamentals of Supervised Machine Learning*, Statistics and Computing, https://doi.org/10.1007/978-3-031-41337-7_1

The chapter finally discusses the challenge of non-identifiability of the mapping between the features and the target variable in supervised Machine Learning, known as the *curse of dimensionality*. We stress that the conditional expectation of the target given the features (i.e., the mapping) is generally not identified by data, and this depends on the presence of numerically continuous features that generate an inability to point-wisely identify the mapping. This raises the need for modeling and thus the use of various Machine Learning fitting algorithms.

The chapter ends with some concluding remarks.

1.2   Machine Learning: Definition, Rationale, Usefulness

Machine Learning (ML) (also known as statistical learning) has emerged as a leading data science approach in many fields of human activities, including business, engineering, medicine, advertisement, and scientific research. Placing itself in the intersection between statistics, computer science, and artificial intelligence, ML's main objective is to turn information into valuable knowledge by "letting the data speak", limiting the model's prior assumptions, and promoting a model-free philosophy. Relying on algorithms and computational techniques, more than on analytic solutions, ML targets Big Data and complexity reduction, although sometimes at the expense of the results' interpretability (Hastie et al., 2009; Varian, 2014).

In the literature, the two terms, *Machine Learning* and *statistical learning*, are used interchangeably. The term Machine Learning, however, was initially coined and used by engineers and computer scientists. Historically, this term was popularized by Samuel (1959) in his famous article on the use of experience-based learning machines for playing the game of checkers as opposed to the symbolic/knowledge-based learning machines relying on hard-wired programming rules. Samuel proved that *teaching* a machine increasingly complex rules to carry out intelligent tasks (as in the case of expert systems) was a much less efficient strategy than letting the machine learn from experience, that is, from data. As a consequence, he made the point that statistical learning is at the basis of a better ability of machines to learn from past (stored) events. We may say, in this sense, that Machine Learning is based on statistical learning. Of course, the use of either term also reflects the scientific community of developers, with engineers preferring mainly to use the label "Machine" Learning, and statisticians the label "statistical" learning.

Machine Learning is the branch of artificial intelligence mainly focused on statistical prediction (Boden, 2018). The literature distinguishes between supervised, unsupervised, and reinforcement learning, referring to a setting where the outcome variable is known (supervised) or unknown (unsupervised), with reinforcement learning representing the bridge to artificial intelligence and robotics applications.

Figure 1.1, shows the classical ML taxonomy. We can see that, in statistical terms, supervised learning coincides with a regression or classification setup, where regression represents the case in which the outcome variable is numerical, and classification the case in which it is categorical. In contrast, unsupervised learning deals

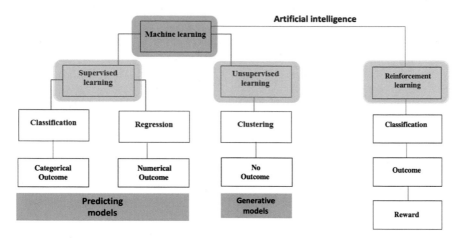

Fig. 1.1 Machine Learning (ML) standard taxonomy

with unlabeled data, and it is mainly concerned with generating a brand-new target variable via proper clustering algorithms or latent factor models. Unsupervised learning can thus be encompassed within statistical cluster analysis. Supervised learning focuses mainly on prediction, while unsupervised learning encompasses many types of the so-called generative models; models where an outcome is generated from raw data. Reinforcement learning represents a world apart, although it can use many techniques drawn from supervised and unsupervised learning. Reinforcement learning is based on the triplet *state-action-reward*, where action entails taking a decision on the part of an agent (a robot, for example), based on a reward function and a given environmental setting representing the state-of-the-world. According to the sequence of rewards, some decisions may be reinforced, while others weakened. In this way, one can describe the behaviors of artificial agents such as robots, exhibiting the ability to carry out specific intelligent tasks (such as climbing a mountain by avoiding barriers, or picking up and moving specific objects toward specific directions/places).

Strictly related to ML, deep learning (DL) is part of a broader family of Machine Learning methods based on various architectures of artificial neural networks (ANNs) embedded within representation learning (LeCun et al., 2015; Gulli et al., 2019). A deep neural network is an ANN with multiple layers between the input and output layers. There are different types of neural networks, but they largely consist of the same components that are neurons, synapses, weights, biases, and functions. The functioning of these components resembles that of the human brain. As we will see in the book, deep learning models can take different forms (i.e., architectures) that can accommodate different settings of data ordering (either sequential/temporal or spatial). Like ML, deep learning models can be supervised and unsupervised and can be trained like any other ML algorithm.

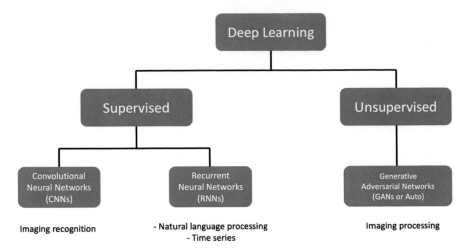

Fig. 1.2 Deep learning (DL) taxonomy

Figure 1.2 illustrates a simplified taxonomy of the most popular deep learning models, where we distinguish between supervised and unsupervised learning too. Among the supervised deep learning, we have two popular types of ANN architectures, that is, convolutional neural networks (CNNs), and recurrent neural networks (RNNs). CNNs are popular for imaging recognition and thus well suited to exploiting the spatial ordering of the data. RNNs are extensively used with sequential data, and largely exploited for natural language processing (speech recognition, for example), and time series forecasting. The main family of unsupervised deep learning models is the generative adversarial networks (GANs), used mainly for imaging processing (as, for example, to re-generate old unreadable documents or videos). GANs are able to generate new synthetic instances of data by extracting only salient components by getting rid of noisy signals.

In a nutshell, we can say that ML methods are techniques mainly suitable for carrying out predictive purposes. But there is more than this. Indeed, ML methods can pursue other purposes, such as feature-importance detection, signal-from-noise extraction, model-free clustering, model-free classification, and correct specification via optimal model selection.

In terms of applications, ML is used for many tasks, such as identifying risk factors for certain specific diseases; predicting genetic predisposition to specific syndromes; predicting strokes or mental disorders based on social and clinical measurements; customizing email spam detection systems; forecasting stock market prices; allowing for cars to become self-driving; classifying pixels of land-satellite images; setting the relationship between some socio-economic variables and some of their determinants; carrying out pattern recognition of handwritten symbols; optimizing the tuning of devices for automated languages translators or vocal recognition systems; generating new text from pre-existing documents; and so forth.

It is worth stressing again that, in this chapter, as well as throughout the entire book, I will focus only on supervised learning. Interested readers can learn about unsupervised learning and reinforcement learning from other specialized publications.

1.3   From Symbolic AI to Statistical Learning

Artificial intelligence (AI) is intelligence exhibited by artifacts (e.g., machines), as opposed to the natural intelligence (NI) displayed by animals, including humans. AI endows machines with human-like cognitive functions, such as learning and problem-solving. In a very general way, we can define intelligence as the ability to perceive an environment, elaborate on it, and take actions/decisions that maximize the chance of achieving given goals. Figure 1.3 highlights this process.

AI is the product of a tremendous increase of three essential ingredients to carrying out an intelligent task: computation capacity, availability of huge data, and algorithm development.

Just to give an idea of the massive growth in computer power that has taken place in recent decades, we can observe that in the early 70s of the last century, a supercomputer's CPU contained around 1,000 transistors. Nowadays, a single CPU can contain more than 50 billion transistors.

Computation development has been accompanied by an unprecedented availability of large masses of data (the so-called Big Data), collecting information from enterprises, social media, VoIP (Voice-over-IP), and the Internet of things, as well as data from various types of sensors. Big data have increased along three dimensions (the so-called "three-Vs"), which are velocity, volume, and variety.

Algorithm development has also experienced comparable development, partly dependent on the growth in computer power and data availability. Figure 1.4 illustrates a tentative timeline development of artificial intelligence paradigms and methods (Nilsson, 2010).

Although tentative, this figure shows the evolution of the AI paradigms over the years starting from the early 50s of the last century to date.

The initial and dominant AI paradigm was the so-called "Symbolic AI" (Herrero, 2013). It refers to that branch of artificial intelligence attempting to explicitly represent human learning in a declarative form, i.e., based on facts and rules. Allen Newell and Herbert A. Simon were the pioneer in Symbolic AI. Following the famous Thomas Hobbes statement according to which "thinking is the manipulation

Fig. 1.3 Process carrying out an intelligent task

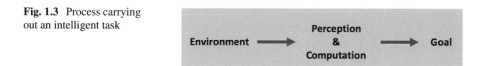

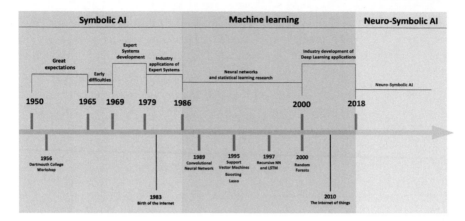

Fig. 1.4 Timeline development of artificial intelligence paradigms and methods

of symbols and reasoning is computation", developers of Symbolic AI aimed at building *general problem solvers* (GPS), leading to hard-wired rule-based reasoning systems like the *Expert Systems*, virtual machines able to solve specific intelligent tasks such as finding an object within a finite set of different objects, or automating computations for producing specific industrial products (Nikolopoulos, 1997).

For almost 40 years, Symbolic AI was the dominant paradigm in AI research. At the core of this paradigm, is placed the concept of "symbol", defined as a "perceptible something that stands for something else". Symbols are alphabet letters, numerals, road signs, music signs, and so forth. According to this paradigm, new knowledge comes up from manipulating symbols via hard-wired logical rules, and to do this, computers have to be *explicitly* programmed.

Although the expert system turned out to be useful for carrying out numerous intelligent tasks, with application in industry and public administration, more complex intelligence tasks, general pattern recognition, and unspecialized procedural tasks were not successfully carried out using the Symbolic AI paradigm. This awareness became more evident in the early '80s of the last century, the years of the so-called "fall" of Symbolic AI (Singh, 2019). What was wrong with Symbolic AI?

In a nutshell, the main problem with Symbolic AI was that it aimed at practicing intelligence by "teaching a machine logical rules" instead of "letting the machine learn from experience". This critique was raised by Arthur Samuel in 1959, who opposed and pioneered what he called the Machine Learning (ML) paradigm, defined as the "field of study that gives computers the ability to learn without being explicitly programmed".

At the basis of the fall of Symbolic AI, there was the inability to solve the so-called "learning-by-experience paradox". Let's do a simple example. Figure 1.5 shows two young girls, Marta and Emma Cerulli, when they were six years old. Both used to speak fluent Italian at that age. They were able to understand and make themselves understood by other Italian people. Curiously, Marta and Emma were never taught

Fig. 1.5 Marta and Emma Cerulli when they were six years old. Examples of intelligent systems able to learn from experience. Photographs courtesy of Giovanni Cerulli

a single Italian grammar rule. They did not even know about the existence of grammar rules. Yet, they were able to recognize words and a large number of symbols, understand their meaning, and process them accordingly. This was possible because Marta and Emma were able to learn Italian from experience.

If human beings, as intelligent systems, are able to learn from experience, let machines learn from data, namely, from past events' frequencies/experiences. As such, ML can be better understood as a set of computational methods that use experiences (events stored into "datasets") to accomplish specific intelligent tasks, by generating accurate predictions.

The statistical learning paradigm hinges on the idea of translating a cognitive task into a statistical problem. But what kind of statistical problem? Basically, statistical learning starts with the collection of information (i.e., the experiences) from the past by storing it into an object $\mathcal{D}$, the *dataset*. A dataset is a collection of information about N cases, over which we measure an outcome (or target variable) y, and a set of p features (or predictors) $\mathbf{x} = (x_1, \ldots, x_p)$:

$$\mathcal{D} := \{(y_i, \mathbf{x}_i),\ i = 1, \ldots, N\} \tag{1.1}$$

The main task of (supervised) statistical learning is to learn from $\mathcal{D}$ – in the most predictive way—a *mapping* of the features $\mathbf{x}$ onto the target y. This mapping is represented by the algorithm f as follows:

$$(x_1, \ldots, x_p) \xrightarrow{f} y \tag{1.2}$$

What is the fundamental statistical object shaping the mapping represented in Eq. (1.2)? In other words, among all the infinite ways to map $\mathbf{x}$ onto y, which is the best to adopt? To respond to this question, we have to introduce the concept of prediction error and of a loss function defined over it. For the sake of exposition, by considering the case of a numerical target variable y (the case in which y is categorical follows a similar logic), we look for a function $f(\mathbf{x})$ that minimizes a specific loss which is an increasing function of the prediction error, simply defined as $e = [y - f(\mathbf{x})]$. By conditioning on $\mathbf{x}$, i.e., by holding $\mathbf{x}$ constant, and considering squared errors

Table 1.1 A one-to-one word translation of the statistical and the Machine Learning jargon

Statistics	Machine learning
Statistical model	Learner
Estimation sample	Training dataset
Out-of-sample observations	Test dataset
Estimation method	Algorithm
Observation	Instance (or sample)
Predictor	Feature
Dependent variable	Target

(thus giving positive and negative errors the same weight), we can define as our predictive loss the so-called *conditional mean squared error* defined as:

$$\text{MSE}(\mathbf{x}) = E[(y - f(\mathbf{x}))^2 | \mathbf{x}] \tag{1.3}$$

that can be written as:

$$\text{MSE}(\mathbf{x}) = E[y^2 - 2yf(\mathbf{x}) + f^2(\mathbf{x}) | \mathbf{x}]$$
$$= E(y^2 | \mathbf{x}) - 2E(y|\mathbf{x})f(\mathbf{x}) + f^2(\mathbf{x}) \tag{1.4}$$

where the expectation is taken over the distribution of y. We can minimize the $\text{MSE}(\mathbf{x})$ over $f(\mathbf{x})$, by equalizing to zero its first derivative over $f(\mathbf{x})$:

$$\frac{\partial \text{MSE}(\mathbf{x})}{\partial f(\mathbf{x})} = -2E(y|\mathbf{x}) + 2f(\mathbf{x}) = 0 \tag{1.5}$$

which leads to discover that the function of $\mathbf{x}$ which minimizes it is the conditional expectation of y given $\mathbf{x}$, i.e.:

$$f(\mathbf{x}) = E(y|\mathbf{x}) \tag{1.6}$$

In general, the conditional expectation of y given $\mathbf{x}$ is not identified by observation due to the so-called *curse of dimensionality* (also known as *sparseness problem*). We will come back to this essential point later, after showing first an example in which an intelligent task, numeral recognition, can be problematic for Symbolic AI while easily handled by a statistical learning approach. This example is paradigmatic of the success of AI applications based on statistical learning.

Before going into this example, however, it is worth stressing that statistical and ML scientists use sometimes different words to mean the same concepts. Table 1.1 sets out a one-to-one word translation of the statistical and the Machine Learning jargon. It is useful, as depending on the reference literature, we can encounter different words to indicate the same concepts. This scheme can prevent possible confusion.

Fig. 1.6 Examples of two numerals to be recognized by a machine

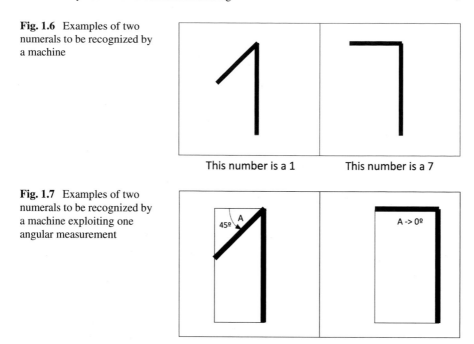

This number is a 1   This number is a 7

Fig. 1.7 Examples of two numerals to be recognized by a machine exploiting one angular measurement

1.3.1 Numeral Recognition: Symbolic AI Versus Machine Learning

People are generally good at recognizing symbols like numerals. However, a trickier task is making a machine able to recognize whether a numeral is, let's say, a one or a seven. Consider Fig. 1.6, where two numerals are displayed, that is, a one and a seven written in a way easily understandable to humans. How can a machine recognize that the numeral on the left is a one and that the numeral on the right is a seven? The response to this question is not banal.

An AI scientist rooted in Symbolic AI could follow the strategy highlighted in Fig. 1.7: first, insert the numbers into rectangles; second, draw the angle A; third, apply the following rule to teach a machine how to detect the numeral:

- if $(A \leq 45°)$: it is a 7;
- else if $(A > 45°)$: it is a 1.

This rule is very likely to provide correct recognition.

It is true that the numeral seven is hardly written as shown in Fig. 1.7. Numeral seven is more often written using a steeper vertical bar, as in Fig. 1.8. In this case, to accomplish the same task, we can exploit angle B, and a new predictive rule is:

- if $(A \leq 45°)$ and $(B > 45°)$: it is a 7;
- else if $(A > 45°)$ and $(B \leq 45°)$: it is a 1.

Fig. 1.8 Examples of two
numerals to be recognized by
a machine exploiting two
angular measurements.
Numeral seven is written
with a steeper vertical bar

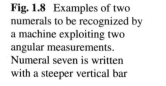

Fig. 1.9 Prediction of two
numerals to be recognized by
a machine exploiting two
angular measurements

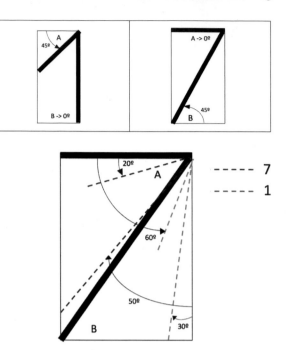

This rule is likely to be predictive of whether a numeral is a one or a seven. Figure 1.9 shows an example in which such a prediction rule works well.

When numerals are written in a schematic way, as in the previous examples, Symbolic AI seems to be successful to predict which number one is writing. However, what about more complex (but also usual) situations when one has to deal with handwritten numerals? Figure 1.10 shows a series of numbers handwritten by humans, where it is immediately visible that the same number is written in many different ways. Can we design rule-based symbolic algorithms able to let a machine recognize handwritten numerals as displayed in Fig. 1.10? The response is self-evident and negative, as even very sophisticated rules based on geometry, mathematics, and logic would be largely unable to let a machine predict the correct numeral. From a Machine Learning perspective, on the contrary, this task is relatively simple, whenever we apply the "learning from experience" principle.

To better catch the ML functioning for this specific task, consider two handwritten numerals, a one and a three, and let the machine recognize them. ML re-organizes this task as a statistical mapping problem where numerals are transformed into a dataset $\mathcal{D}$ as in Eq. (1.1). ML thus generates a dataset able to map $\mathbf{x}$ (the usual ways people write numerals one and three by hand) onto y (the true numeral). The trick to accomplish this recognition task is to decompose the numeral into adjacent equally spaced pieces (called "pixels") and use them as predictors ($\mathbf{x}$) for the numeral y. Figure 1.11 shows how to implement data representation of images representing two handwritten numerals. Every single square represents one pixel, where we consider

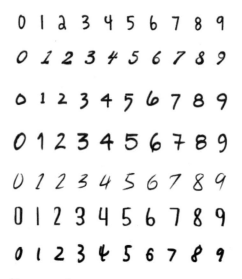

Fig. 1.10 Some handwritten numerals

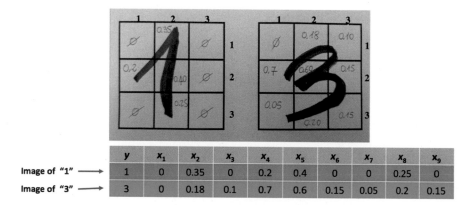

	y	x_1	x_2	x_3	x_4	x_5	x_6	x_7	x_8	x_9
Image of "1" →	1	0	0.35	0	0.2	0.4	0	0	0.25	0
Image of "3" →	3	0	0.18	0.1	0.7	0.6	0.15	0.05	0.2	0.15

Fig. 1.11 Data representation of the two handwritten numerals

as image resolution a total of 3×3 pixels. The number inside each pixel is the percentage of the pixel area covered by the ink. Every pixel is thus transformed into a variable. For example, pixel 1 corresponds to x_1, pixel 2 corresponds to x_2, and so on. We have thus transformed two handwritten numerals provided as images into a dataset $\mathcal{D}$.

If we were able to collect a large number of numeral images, we could produce a mapping between the way the numeral has been written by a human (**x**) and every numeral y by means of a specific mapping algorithm (i.e., a learner) that "trains" this mapping over the collected dataset (called, in fact, *training* dataset). In this way, we can predict over new unlabeled images what numerals they represent. Figure 1.12 sets out a training dataset made of many handwritten images of 1s and 3s, and two

	y	x_1	x_2	x_3	x_4	x_5	x_6	x_7	x_8	x_9
Image of "1" ⟶	1	0	0.35	0	0.2	0.4	0	0	0.25	0
Image of "3" ⟶	3	0	0.18	0.1	0.7	0.6	0.15	0.05	0.2	0.15
Image of "1" ⟶	1	0	0.30	0	0.1	0.4	0	0.01	0.22	0
Image of "1" ⟶	1	0	0.35	0.01	0.3	0.4	0	0	0.23	0
Image of "3" ⟶	3	0	0.16	0.1	0.65	0.6	0.16	0.09	0.3	0.14
Image of "3" ⟶	3	0	0.15	0.1	0.75	0.7	0.14	0.06	0.2	0.15
Image to predict ⟶	?	0	0.30	0	0.1	0.4	0	0.01	0.22	0
Image to predict ⟶	?	0	0.35	0	0.2	0.4	0	0	0.25	0

Fig. 1.12 Data representation of the two handwritten numerals

unlabeled images (put at the end of the table) that one can predict based on the cases collected in the training dataset.

This procedure is general, can be extended to any type of image recognition task (not only numerals), and can be summarized as follows:

- build a training dataset made of many images;
- train a learner, i.e. a mapping between y and $\mathbf{x} = (x_1, \ldots, x_p)$;
- as long as certain configurations of $\mathbf{x}$ are likelier to be associated to a specific y, a proper learner can learn how to link a new configuration (i.e., a new image) to a specific image label.

Figure 1.13 shows an example of *face completion*, a task aimed at predicting the lower half of a face by knowing the upper half using different learners for building the mapping between the upper half and the lower half of the face (Jaiswal, 2022). It is interesting to see that different learners can have different performances. Measuring learners' predictive performance is thus a fundamental task of any ML study, and we will give it a large relevance in this chapter and those that will follow.

1.4   Non-identifiability of the Mapping: The *Curse of Dimensionality*

We saw that the conditional expectation of y given $\mathbf{x}$ is the function of $\mathbf{x}$ that minimizes the conditional prediction error. The central role of the conditional expectation in statistical learning is ubiquitous, as whatever regression or classification analysis, as we will discuss more in-depth later, is ultimately estimating a conditional expectation.

Quite surprisingly, in general settings, this central statistical object is not identified by data. To better appreciate this apparent paradox, let's first start from a situation when $E(y|\mathbf{x})$ is fully identified by data. Figure 1.14 shows such a situation when only one feature x (i.e., gender) is considered, and this feature is discrete taking only two values, x_0 (male) and x_1 (female). By assuming that the outcome variable is

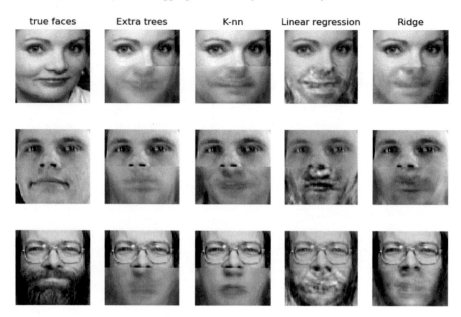

Fig. 1.13 Face completion. Predicting the lower half of a face by knowing the upper half. Olivetti faces dataset. The dataset contains a set of freely available face images taken between April 1992 and April 1994 at AT&T Laboratories Cambridge

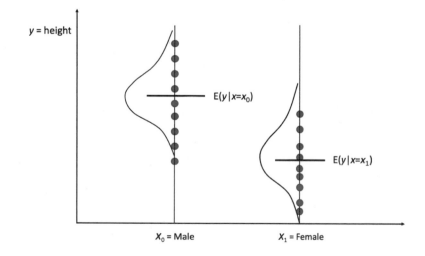

Fig. 1.14 Identification of the conditional mean with no sparse data

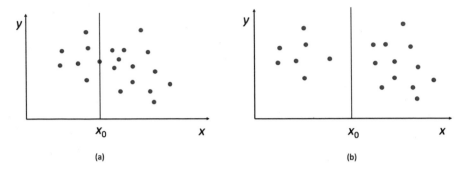

Fig. 1.15 Non identification of the conditional mean with sparse data

the height of people, we can see that the conditional expectation of the height given the gender is perfectly identified by data, as we can calculate, both for men and for women, an estimate of the average height. More precisely, we have a sample for men and a sample for women, and these samples are perfectly aligned with the two categories. Apart from the presence of a sampling error—generally decreasing with the availability of larger samples—the conditional expectation is fully identified by observation.

Look, however, at Fig. 1.15. In this case, the unique feature x is a continuous variable, thus taking an infinite number of values. In point x_0, we can have two situations: either no data are available (panel (b)), or only one or a few points are observable (panel (a)). In both cases, we no longer have perfect data alignment in any point of the support of x, as data are sparse. In both cases, the conditional expectation of y given x is not identifiable by observation, as the dimensionality of the support of x is too large (it is infinite, indeed), thus originating a sparseness (non-alignment) condition. This phenomenon is known as the *curse of dimensionality*, and it is at the basis of the inability to identify $E(y|\mathbf{x})$ through data.

When the conditional expectation's identification is prevented because of sparseness, we have to rely on some hypotheses about the nature of this function. We have two routes that we can hit: one is to assume that the unknown function has a parametric form; the other is to assume a nonparametric form.

Parametric models assume a functional form for $E(y|\mathbf{x})$—that is, a specific analytic formula. This family of models encompasses both linear and generalized linear models (probit, logit, poisson, etc.), as well as shrinkage models. For this family of models, the problem is to estimate some unknown parameters characterizing the assumed functional form of the conditional expectation, as when these parameters become known, $E(y|\mathbf{x})$ is automatically identified.

Nonparametric models assume no pre-existing functional form. Instead of relying on formulas, $E(y|\mathbf{x})$ is estimated using *computational procedures* that can have a global, local, or semi-global nature. Indeed, the nonparametric family can be split into three sub-families: global, semi-global, and local methods. Global methods estimate the conditional expectation in one shot using the entire support of the features. Local

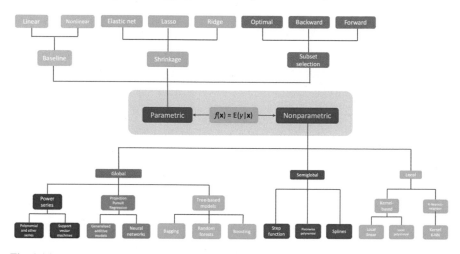

Fig. 1.16 Taxonomy of parametric and nonparametric statistical learning methods

methods estimate the conditional expectation pointwise. Semi-global methods represent a compromise between global and local approaches; they are interval estimators, splitting the support of the feature into intervals formed by setting specific nodes and fitting a global method within each interval. Figure 1.16 presents a tentative taxonomy to classify parametric and nonparametric statistical learning methods.

Because $E(y|\mathbf{x})$ is not identified by observation, whatever method in Fig. 1.16 can represent a valid alternative. How can we decide which method is better than another? It clearly depends on what purpose our analysis wants to pursue. In Machine Learning, predicting the target variable is so central that we can better define ML as a collection of modeling strategies (either parametric or nonparametric) to recover a reliable approximation of $E(y|\mathbf{x})$ using *prediction accuracy* as a leading principle. Therefore, some methods can be found to be better than others, provided that prediction is the only purpose of our analysis.

Notwithstanding the large variety of methods available for estimating the conditional expectation, its measurement will always come with some error. The statistical estimation of $E(y|\mathbf{x})$ is plagued by two possible errors one can commit: (1) *sampling* error, and (2) *specification* error. Big Data and ML can be used for attenuating these errors as shown in Fig. 1.17.

The sampling error is the error due to the fact that we consider a sample drawn from a larger population. As soon as the sample size increases, this error gets smaller and smaller. The availability of Big Data means large N (number of cases), as well as large p (number of features). Thus, by enlarging the sample size N, Big Data are likely to have a positive impact in reducing the sampling error (given, of course, no presence of specification error).

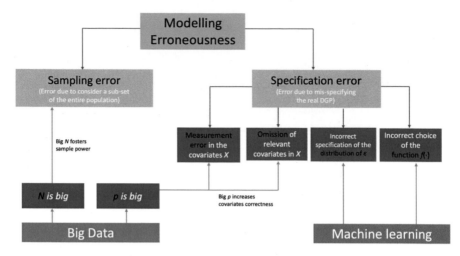

Fig. 1.17 How Machine Learning and big data impact sampling and specification errors

The specification error is a more subtle type of error, and can be decomposed into four main errors: (i) the measurement error in the features, (ii) the omission of relevant features, (iii) an incorrect specification of the distribution of y (generally driven by the type of distribution attributed to the residual term ϵ in the population equation $y = f(x_1, \ldots, x_p) + \epsilon$), and (iv) an incorrect specification of the mapping $f(\cdot)$.

A big p is likely to have a positive impact in reducing both the measurement error in the target and in the features, and the omission of relevant features. ML, on the contrary, is very likely to be useful for reducing the errors coming from assuming a wrong distribution for the error term, and an incorrect specification of the mapping.

1.5   Conclusions

This chapter has offered a general introduction to the rationale and ontology of Machine Learning (ML). It started by discussing the definition, rationale, and usefulness of ML in the scientific context. Then, it underscored the transition from symbolic AI to statistical learning and examined the *curse of dimensionality* as the main source of the non-identifiability of the mapping between the features and target variable in supervised Machine Learning. Next, the chapter provided a guiding taxonomy of Machine Learning methods and discussed some ontological aspects related to Machine Learning as a scientific paradigm.

References

Boden, M. A. (2018). *Artificial intelligence: A very short introduction.* Oxford, United Kingdom: OUP Oxford.

Gulli, A., Kapoor, A., & Pal, S. (2019). *Deep learning with TensorFlow 2 and Keras: Regression, ConvNets, GANs, RNNs, NLP, and more with TensorFlow 2 and the Keras API* (2nd ed.). Birmingham Mumbai: Packt Publishing.

Hastie, T., Tibshirani, R., & Friedman, J. (2009). *The elements of statistical learning: Data mining, inference, and prediction* (2nd ed.). Springer series in statistics. New York: Springer. https://www.springer.com/gp/book/9780387848570

Herrero, J. C. (2013). *Symbolic and connectionist artificial intelligence.* CreateSpace Independent Publishing Platform.

Jaiswal, A. (2022). Face recognition system using python. https://www.analyticsvidhya.com/blog/2022/04/face-recognition-system-using-python/

LeCun, Y., Bengio, Y., & Hinton, G. (2015). Deep learning. *Nature, 521*(7553), 436–444. Publisher: Nature Publishing Group.

Nikolopoulos. (1997). *Expert systems: Introduction to first and second generation and hybrid knowledge based systems* (1st ed.). CRC Press.

Nilsson, N. J. (2010). *Quest for artificial intelligence.* Cambridge; New York: Cambridge University Press.

Samuel, A. L. (1959). Some studies in machine learning using the game of checkers. *IBM Journal of Research and Development, 3*(3), 210–229. Conference Name: IBM Journal of Research and Development.

Singh, R. (2019). Rise and fall of symbolic AI. https://towardsdatascience.com/rise-and-fall-of-symbolic-ai-6b7abd2420f2

Varian, H. R. (2014). Big data: New tricks for econometrics. *Journal of Economic Perspectives, 28*(2), 3–28.

Chapter 2
The Statistics of Machine Learning

2.1 Introduction

This chapter introduces the statistical aspects of Machine Learning and discusses various concepts and considerations.

It sets out by discussing three fundamental trade-offs coming up in Machine Learning statistical modeling: prediction versus inference, flexibility versus interpretability, and goodness-of-fit versus overfitting.

Next, it goes on by describing the regression setting—which involves predicting continuous numerical values, and the classification setting—which involves predicting categorical labels and class probabilities.

Subsequently, an example is presented to illustrate the concepts of training and testing error in both parametric (linear model) and nonparametric (nearest neighbor) regression.

The chapter goes on by presenting different measures for evaluating model's goodness-of-fit models. This includes regression metrics for assessing performance in regression tasks and classification metrics for assessing performance in classification tasks. Strategies for handling imbalanced class distributions in classification tasks are also explored.

Next, the chapter focuses on how to estimate the test error and find the optimal values for hyper-parameters. Specifically, it discusses four methods: information criteria, the bootstrap, K-fold cross-validation (CV), and the plug-in approach.

After presenting learning modes and architecture for predictive ML models, the chapter concludes by addressing the limitations and challenges associated with statistical learning. In particular, data ordering and data sparseness are widely discussed.

Some final remarks end the chapter.

© The Author(s), under exclusive licence to Springer Nature Switzerland AG 2023   19
G. Cerulli, *Fundamentals of Supervised Machine Learning*, Statistics and Computing,
https://doi.org/10.1007/978-3-031-41337-7_2

2.2   ML Modeling Trade-Offs

2.2.1   Prediction Versus Inference

The discussion about how to cope with the sampling and specification errors in modeling the conditional expectation addressed in Chap. 1 is general, but it applies differently depending on the purpose of our modeling strategy. Supervised statistical modeling can in fact have two distinct objectives: *predictive* vs. *inferential*. The two modeling purposes can require different methods, based on different model specifications and estimations. In other words, a model that can be suitable for inference, could not be suitable for prediction and vice versa.

To catch this point, we need to distinguish between the *direct* and the *indirect* effect of the features **x** on the target y. Suppose to have the following structural equation of y:

$$y = h(\mathbf{x}) + \epsilon \tag{2.1}$$

where we assume that ϵ is statistically dependent on **x**. By taking the conditional expectation over **x** of Eq. (2.1), we have

$$E(y|\mathbf{x}) = h(\mathbf{x}) + E(\epsilon|\mathbf{x}) \tag{2.2}$$

Because any conditional expectation of **x** is a function of **x**, we can rewrite the previous equation as

$$E(y|\mathbf{x}) = h(\mathbf{x}) + g(\mathbf{x}) = f(\mathbf{x}) \tag{2.3}$$

The function $h(\mathbf{x})$ is the *direct* effect of **x** on y, while the function $g(\mathbf{x})$ is the *indirect* effect of **x** on y, that is, any change in y activated by the effect of **x** on the unobservable components contained in ϵ. By definition, any Machine Learning method estimates $f(\mathbf{x})$, but it is not able—unless differently structured—to identify separately $h(\mathbf{x})$ and $g(\mathbf{x})$. We say that ML methods have *correlative* power, but their results have no *causal* meaning as they are not suited to tell direct from indirect effects. In a nutshell, ML methods estimate total effects (direct + indirect), but cannot separate them.

A model able to separately identify $h(\mathbf{x})$ and $g(\mathbf{x})$ would allow for a causal interpretation of the results, but it is likely to be sub-optimal for prediction purposes. To clarify this point, consider a linear setting and a comparison between ordinary least squares (OLS) and instrumental-variables (IV) estimation procedures. Suppose a baseline equation of this type:

$$y = \alpha + \beta x + \gamma w + \epsilon \tag{2.4}$$

Assume that, with no prior knowledge of it

$$\epsilon = b * w + \eta \tag{2.5}$$

with η a pure uncorrelated error. By substitution, we have

$$y = \alpha + \beta x + (\gamma + b)w + \eta \tag{2.6}$$

As coefficient of w, OLS estimate $(\gamma + b)$, but are unable to separate the direct effect (γ) and the indirect effect (b). Provided that an instrumental variable z for w is available, IV estimation would be able to estimate γ, thus identifying also b by taking the difference with the OLS estimate of the parameter of w. In terms of prediction, however, the two approaches might behave very differently. In fact, while OLS entail a one-step procedure, it is very likely that the prediction accuracy (for instance the R-squared) will be higher than the accuracy of the IV estimation which entails a two-step approach, thus increasing the prediction variance via an increased variance of the estimated parameters. Generally, in common applications, IV prediction accuracy is much worse than the OLS's. This leads to the conclusion that a model that can be useful for estimating structural parameters, cannot be optimal for prediction, and vice versa. Therefore, there is a divide between models suitable for inference and models suitable for prediction as we will discuss more in-depth in the chapter dedicated to the Lasso and the causal Lasso.

2.2.2 Flexibility Versus Interpretability

Simply put, a model for $E(y|\mathbf{x})$ is said to be *flexible* (or *complex*) when it does not rely on too stringent assumptions about the data generating process of the target variable y. For example, in a polynomial regression, model flexibility increases with the number of the included polynomial terms. The larger the polynomial terms, the more likely the model is able to catch also hidden non-linear patterns, thus increasing its ability to fit the data.

Parametric models characterized by a larger number of free parameters to estimate are generally more flexible. Therefore, in a parametric world, model flexibility is equivalent to the *degrees of freedom* of the model, i.e. the number of free parameters to estimate with a specific estimation procedure (for example, least squares).

A model for $E(y|\mathbf{x})$ is said to be *interpretable* when results obtained by fitting this model are easy to be understood and communicated. For example, a linear regression model with a few parameters to estimate is highly interpretable compared to a polynomial function in which many interactions with different powers can come up. Parametric models characterized by a lower number of free parameters to estimate are generally more interpretable than models characterized by lots of parameters. Of course, less complex models will not be able to catch very hidden patterns in the data, being it the main cost we incur using models with smaller degrees of freedom.

For parametric modeling, flexibility/complexity can thus be measured via the number of model's degrees of freedom. What about nonparametric models? In a nonparametric world, because parameters are not available, we cannot rely on the notion of degrees of freedom to measure complexity. However, we can measure it as

Table 2.1 Flexibility measures for a parametric model like the ordinary least squares (OLS), and for a nonparametric model like the K-nearest neighbor (KNN)

	Flexibility range		Flexibility measure
	Min	Max	
OLS	$p = 0$	$p = N$	p
KNN	$K = N$	$K = 1$	$1/K$

the extent to which we use the initial amount of independent information (i.e., the sample) to produce independent pieces of new information.

To fix the ideas, take the case of the K-nearest neighbor (KNN) regression. In this case, if we used all the N initial pieces of information (i.e., the sample) to identify $E(y|\mathbf{x})$ at each single point, we would obtain as prediction estimate the mean of the target variable y. This is the case of setting $K = N$, which corresponds to the case in which, in a regression setting, we set $p = 0$ thus obtaining a regression of the type $y = \alpha + \epsilon$. Fitting the latter equation via OLS provides a constant prediction value equal to the sample mean of y. This is a degenerate model, in which the information brought about $\mathbf{x}$ becomes uninformative.

The same phenomenon occurs to the KNN regression when $K = N$, as the prediction at each point would be in this case the same and equal to the mean of y, as for the OLS. Therefore, the case in which $K = N$ corresponds to the minimum flexibility of the model, with flexibility increasing as long as K decreases and reaches its maximum at $K = 1$. Thus, the equivalent of the degree of freedom in a KNN regression is $\rho = 1/K$, ranging from $1/N$ and 1. Table 2.1 summarizes these arguments.

We can conclude that model flexibility (or complexity) measures the number of new independent information that one can generate using an initial set of N independent pieces of information by using combinations of this initial set.

2.2.3   Goodness-of-Fit Versus Overfitting

ML models are suitable for predicting either numerical variables (regression) or label classes (classification) as a function of p predictors (or features) that may be either quantitative and qualitative. As said above, from a statistical point of view, ML aims at fitting the mapping between the set of p predictors, $\mathbf{x}$, and the target outcome, y, with the aim of reducing as much as possible the prediction error. Measuring the *prediction error* (or, equivalently, the *prediction accuracy*) is different when y is numerical (regression) from when it is categorical (classification). In what follows, we will go through these two settings separately.

2.3 Regression Setting

When y is a numerical variable (regression setting), the prediction error can be defined as $e = y - \hat{f}(\mathbf{x})$, with $\hat{f}(\cdot)$ indicating prediction from a specific ML method. In a regression setting, ML scholars focus on minimizing the mean square error (MSE), defined as:

$$\text{MSE} = E[y - \hat{f}(\mathbf{x})]^2.$$

However, while the in-sample MSE (the so-called "training MSE") is generally affected by *overfitting* (thus going to zero as the model's degrees of freedom—or complexity—increases), the out-of-sample MSE (also known as "testing MSE") has the property to be a convex function of model complexity, and thus characterized by having an optimal level of complexity (Hastie et al. 2009).

It can be proved (see equation (2.46) in Hastie et al. (2009)) that the testing MSE, when evaluated at one out-of-sample observation (also called "new instance" in the ML jargon), can be decomposed into three components—*variance*, *squared bias*, and *irreducible error*—as follows:

$$E[(y - \hat{f}(\mathbf{x}_0)]^2 = Var(\hat{f}(\mathbf{x}_0)) + [Bias(\hat{f}(\mathbf{x}_0)]^2 + Var(\epsilon) \qquad (2.7)$$

where $\mathbf{x}_0$ is a new instance, that is, an observation that did not participate in producing the ML fit $\hat{f}(\cdot)$. Figure 2.1 shows a graphical representation of the pattern of the previous quantities as functions of model complexity. It is immediate to see that, as long as model complexity increases, the bias decreases while the variance increases monotonically. Because of this, the test-MSE sets out a parabola-shaped pattern which allows us to minimize it at a specific level of model complexity. This is the optimal model tuning, whenever complexity is measured by a specific hyper-parameter λ. In the figure, the irreducible error variance represents a constant lower bound of the testing MSE. It is not possible to overcome this minimum testing MSE, as it depends on the nature of the data-generating process (that is, the intrinsic unpredictability of the phenomenon under analysis).

An interesting issue to explore is why variance and bias have these different patterns as a function of the model's flexibility. In other words, we would like to explain the so-called *trade-off* between prediction variance and prediction bias, and what causes it. To fix ideas, take the case of a polynomial model where the main hyper-parameter (measuring model flexibility) is M, the number of polynomial terms to include in the model. Let's see first what happens to the prediction bias when we increase the hyper-parameter M.

To begin with, it is worth noticing that the sample size N is constant, whatever the value taken by M. This means that, while increasing values of M, the value of N remains unchanged. At a small value of M, the corresponding polynomial is rather parsimonious in terms of number of terms. This suggests that the polynomial will be hardly able to detect hidden non-linearity and unusual patterns in the data, implying a rather high fitting bias. As long as M increases, however, the polynomial becomes

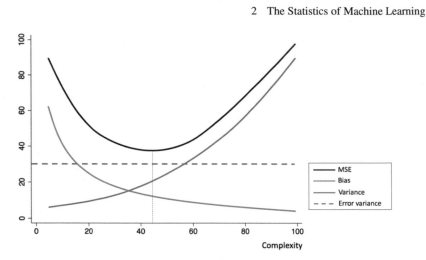

Fig. 2.1 Trade-off between bias and variance as functions of model complexity

increasingly able to fit the data, until it fits the data perfectly when M goes to infinity. This phenomenon—leading to perfect accuracy—is called model's *overfitting*.

Following this logic, one would be tempted to fit the largest polynomial possible in order to obtain the smallest prediction error possible (maximal accuracy). However, this reasoning does not take into account the constrain represented by fixed sample size. Indeed, because N is kept constant over increased values of M, larger polynomials require estimating a larger set of parameters using, however, the same amount of N independent pieces of information. This has a natural cost in the form of reduced precision in estimating this larger set of parameters. Such reduced precision corresponds to a larger variance of the point estimates of the parameters (commonly estimated by least squares). Metaphorically, it is like trying to cook as many cakes as possible using the same amount of flour, compared to when one wants to cook a smaller number of cakes. Of course, increasing the number of cakes to cook will make them less similar to proper cakes, as they will contain an insufficient amount of flour and probably too much water inside. The quality of the cakes will worsen gradually with the number of additional cakes one wishes to cook.

Because the prediction variance depends on the variance of the parameter estimates, also the prediction variance will increase with an increasing level of M. This reasoning leads to explain the prediction bias and variance patterns of Fig. 2.1, where we plotted the trade-off between them as function of model complexity. Prediction bias and prediction variance have an opposite behavior as functions of model flexibility, while the MSE, exhibits the parabola-shape visible in the figure.

Observe that, in general, the minimum of the MSE does not coincide with the point at which the bias and the variance meet each other.

2.4 Classification Setting

In the classification setting, the response variable y is categorical (i.e., qualitative). Our goal is to build a classifier $C(\mathbf{x})$ assigning a class label from J possible classes to an unlabeled observation $\mathbf{x}$. Examples of classification tasks include detecting whether an email is a spam (bad email) or a ham (good email); classifying the type of cancer among different cancer types; fraud detection, anomaly detection; image recognition; and so on.

What is the best way to build a classifier? We saw that the conditional expectation was the function of the predictors $\mathbf{x}$ that minimizes the conditional MSE. Unfortunately, in the classification setting, the MSE is meaningless as in this case we have class labels and not numerical values. For classification purposes, the correct objective function to minimize is the mean classification error (MCE) defined as:

$$\text{MCE}(C) = E[I(y \neq C(\mathbf{x}))] \tag{2.8}$$

where $I(\cdot)$ is an index function (taking value 1 if its argument is a true statement and 0 otherwise), and $C(\mathbf{x})$ a classifier. Notice that the MCE depends on the specific classification rule adopted. Among all the possible classifying rules $C \in \mathcal{C}$, which is the one generating the minimum MCE? It can be proved that, for a categorical outcome y, taking values $y = 1, \ldots, j, \ldots, J$, the Bayes classification rule is the one minimizing the MCE. This rule is as follows:

$$C(\mathbf{x}) = j \text{ if } p_j(\mathbf{x}) = \max\{p_1(\mathbf{x}), p_2(\mathbf{x}), \ldots, p_J(\mathbf{x})\} \tag{2.9}$$

where

$$p_j(\mathbf{x}) = \text{Prob}(y = j|\mathbf{x}) \tag{2.10}$$

is the conditional probability to belong to class j, given $\mathbf{x}$. Classification rule (2.9) suggests to classify a new instance within the class having, for this instance, the largest probability. We can prove the optimality of the Bayes rule in the case of a two-class target variable (the multi-class proof is similar):

$$\begin{aligned}
\text{MCE}(C) &= E[I(y \neq C(\mathbf{x}))]\} \\
&= E_{\mathbf{x}}\{E_{y|\mathbf{x}}[I(C(\mathbf{x}) \neq y)|\mathbf{x}]\} \\
&= E_{\mathbf{x}}\{p(\mathbf{x}) \cdot I(C(\mathbf{x}) \neq 1) + (1 - p(\mathbf{x})) \cdot I(C(\mathbf{x}) \neq 0)\} \\
&= E_{\mathbf{x}}\{p(\mathbf{x}) \cdot I(C(\mathbf{x}) = 0) + (1 - p(\mathbf{x})) \cdot I(C(\mathbf{x}) = 1)\}
\end{aligned} \tag{2.11}$$

where the second equality follows from the law of iterated expectations (LIE). From the last equality, we have that:

$$\text{MCE}(C(\mathbf{x}) = 1) = 1 - p(\mathbf{x})$$

and

$$\text{MCE}(C(\mathbf{x}) = 0) = p(\mathbf{x})$$

implying that MCE(C) is minimized when:

$$\text{MCE}(C^\star) = E_\mathbf{x}[\min\{p(\mathbf{x}); 1 - p(\mathbf{x})\}]$$

For example, suppose that for a new instance, $\mathbf{x}^\star$, we have that $p(\mathbf{x}^\star) = 0.7$ and $1 - p(\mathbf{x}^\star) = 0.3$; then, $\mathbf{x}^\star$ is classified as "1" because, in this case, the MCE(C) is minimized. Therefore, classification rule (2.9) follows.

As in the case of the conditional expectation, the conditional probability is not identified by observation because of data sparseness (i.e., the curse of dimensionality). We can however model the conditional probability either parametrically and non parametrically. Popular parametric models for binary outcomes are the linear, Probit and Logit models, while in a multi-class setting, multinomial models are generally employed. Nonparametric estimates of the conditional probabilty can be carried out with a vast set of models that includes pretty all the methods set out in Fig. 1.16 and used for estimating the conditional expectation.

As in the case of the MSE, it can be proved (Hastie et al. 2009) that the training-MCE overfits the data when model complexity increases, while the test-MCE allows us to find the optimal model's complexity. Therefore, the graph of Fig. 2.1 can be likewise extended also to the case of the testing MCE.

2.5   Training and Testing Error in Parametric and Nonparametric Regression: an Example

In this section, I will show how to compute the training and testing errors in two cases–a parametric and a nonparametric regression. As for the parametric approach, I consider the linear regression estimated by least squares. As for the nonparametric approach, I consider a K-nearest neighbor (KNN) regression.

Figure 2.2 visualizes the training and testing errors in a parametric (linear) regression. This figure shows a left-hand panel and a right-hand panel. In the left-hand panel, unit i is included in the estimation of the line, implying that the error estimated over this unit is a training error. In the right-hand panel, unit i is excluded from the estimation of the line; as a consequence, the error estimated over this unit is a testing error. We can observe that the two errors are different, with the testing error larger than the training error. As said above in this chapter, this is a general pattern, not one arising only in this example. In this specific case, however, the reason why this happens is quite clear: unit i shows up at the very top of the cloud; thus, excluding it from estimation generates a change in the slope and position of the line, by dragging it downwards. This yields a substantial difference between in-sample and out-of-sample errors. Therefore, a mapping between x and y in which their scatter-

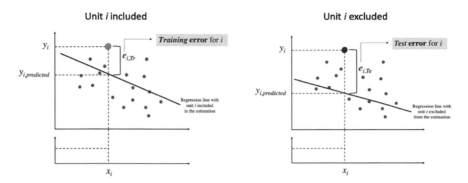

Fig. 2.2 Visualizing training and testing errors in a parametric regression (least squares)

plot is highly sparse, generally makes training prediction too optimistic. This event
has to do with the presence of larger noise in the data. When the noise in the data
is sufficiently sizable, with training estimation highly depending on data that are
either outliers (i.e., showing exceptional y), or characterized by some leverage (i.e.,
showing exceptional x), the testing error is generally much larger than the training
one.

A similar conclusion can be reached when considering a nonparametric setting,
as in the case of the nearest neighbor regression. Figure 2.3 illustrates this case. The
figure shows two sets of data: training and testing sets. The in-sample error is first
estimated on the training set (made of only two units), and then in the testing set
(made of two units as well). The testing predictions are estimated using the nearest
neighbor imputation; for example, because both units C and D have the same nearest
neighbor within the training set, which is unit B, the predicted y for these two units is
the y of unit B. We can thus compute the testing error, thus discovering that it is equal
to 925. By computing the training error (we let the reader do this as an exercise), we
find out that its value is equal to 25, implying a testing error that is 37 times larger
than the training error.

We can explain this huge difference by noticing again that the y and the x of
the units in the testing set are numerically very different from those of the units
belonging to the training set. As in the case of parametric regression, structural
differences between training and testing datasets lead to this kind of situation.

From both examples, we can learn an important lesson for data analysis and
prediction. As soon as the out-of-sample data come from a different *joint distribution*
than that of the training data, we cannot expect any learner to have a good predictive
performance. This is due to a very intuitive reason: we train the data over a sample
in which the y and the x come from a specific join distribution, but we predict out-of-
sample over data characterized by a different join distribution. This is a fundamental
limitation of any statistical learning's attempt to predict out-of-sample, an aspect that
we will cover more in-depth later on in this chapter.

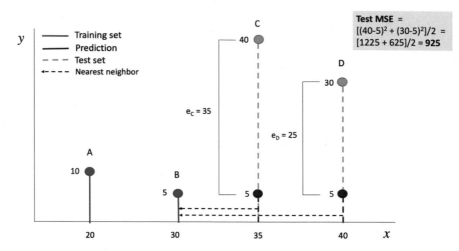

Fig. 2.3 Visualizing training and test error in a nonparametric regression (nearest neighbor)

2.6   Measures for Assessing the Goodness-of-Fit

In the previous sections, we stressed the centrality of the prediction performance for supervised ML models. Yet, we have not provided so far a way to measure in practice ML models' prediction ability. In what follows, we consider various metrics for assessing the goodness-of-fit of a generic ML method by distinguishing between a regression and a classification setting.

Regression Metrics

In the ML literature, various metrics to asses the regression goodness-of-fit of a given learner have been proposed over the years. The most popular however are the mean squared error (MSE), the coefficient of determination (R^2), and the mean absolute percentage error (MAPE). Below their descriptions and formulas.

Mean squared error (MSE). As seen above, the mean squared error is a risk index which corresponds to the expected value of the quadratic error (or loss). It is the average value over the sample of the prediction error. It is a dimensional measure, in that its size depends on the unit of measurement of the numerical variable y. Its formula is:

$$\text{MSE}(y, \hat{y}) = \frac{1}{N} \sum_{i=1}^{N} (y_i - \hat{y}_i)^2 \qquad (2.12)$$

Coefficient of determination (R^2). This index represents the proportion of the variance of y explained by the predictors included in the model. Its formula is:

$$R^2(y, \hat{y}) = 1 - \frac{\sum_{i=1}^{N}(y_i - \hat{y}_i)^2}{\sum_{i=1}^{N}(y_i - \bar{y})^2} \tag{2.13}$$

where $\bar{y} = \frac{1}{N}\sum_{i=1}^{N} y_i$. The largest score that the R^2 can take is 1, while it can take also negative values (as there are no limits to how bad a model can be in fitting the data). A model predicting for every unit the expected average value of y, thus disregarding the input features, would obtain a score equal to 0.

Mean absolute percentage error (MAPE). This metric is a relative metric, in that it considers relative errors and its value does not depend on the unit of measurement of the y. As such, it can be used to compare the goodness-of-fit for models that have different target variables. Its formula is:

$$\text{MAPE}(y, \hat{y}) = \frac{1}{N} \sum_{i=1}^{N} \frac{|y_i - \hat{y}_i|}{\max(\epsilon, |y_i|)} \tag{2.14}$$

where ϵ is an arbitrary small and positive number chosen to avoid undefined results when y is zero. Notice that an advantage of MAPE is that of providing a percentage measure of the prediction error by contrasting it with the absolute level of the target variable y.

Classification Metrics

Multi-label confusion matrix. A *confusion matrix* (or *error matrix*) is a table allowing for visualizing the performance of a classifier. Indicating by row the actual labels, and by columns the predicted labels, the confusion matrix shows all the possible crossings, with the diagonal displaying the frequencies of correct matches, and the off-diagonal the frequencies of the incorrect matches. Figure 2.4 shows an example of a confusion matrix for a three-class classification, where classes are labeled as A, B, and C. The diagonal of this matrix (in light gray color) shows the number of cases that have been correctly classified. All the other cells report the number of unmatched cases. For example, for class B, only 5 out of 18 actual B cases have been correctly classified within class B. The prediction accuracy for class B is around 27.8%, while the error rate is $100 - 27.8 = 72.2\%$. We can easily calculate the label-specific accuracy (and error rate), as well as a measure of the overall accuracy (overall error rate).

Accuracy. This metric can be used for multi-label classification (two or more classes). The accuracy takes value 1 if all the sample predicted labels match with the true labels; otherwise it takes values smaller than 1, and equal to 0 when no matches take place. Its formula is:

		Predicted			TOT
		A	B	C	
Actual	A	10	4	5	19
	B	8	5	5	18
	C	4	4	15	23
	TOT	22	13	25	60

Fig. 2.4 Example of a confusion matrix for a three-class classification, where classes are labeled as A, B, and C

	PREDICT COVID	PREDICT NO COVID
COVID	TP	FN
NO COVID	FP	TN

error

Fig. 2.5 Example of a two-class confusion matrix for the Covid-19 viral infection. TP = number of true positives; FP = number of false positives; FN = number of false negatives; TN = number of true negatives; $N = TP + FP + FN + TN$

$$\text{accuracy}(y, \hat{y}) = \frac{1}{N} \sum_{i=1}^{N} I(\hat{y}_i = y_i) \qquad (2.15)$$

Binary confusion matrix. A binary confusion matrix, is a confusion matrix for a two-class setting. Given the large popularity of two-class classification, the literature has put forward various accuracy indicators to deal with this special case. It is remarkably popular in medical studies, whenever a physician has to decide whether a patient has, or not has, contracted a given disease. A patient can be either *positive* (thus having the disease), or *negative* (thus not having the disease). Therefore, two kinds of errors can arise: an individual has the disease (it is positive), but she is classified as not having the disease, i.e. as negative (*false negative* error); or an individual does not have the disease, but she is classified has having it, i.e. as positive (*false positive* error). Of course, an individual can either be positive to the disease and correctly classified as positive (*true positive*), or negative and correctly classified as negative (*true negative*). Figure 2.5 shows these four cases using, as an example, the Covid-19 viral infection. We can thus define: TP = number of true positives; FP = number of false positives; FN = number of false negatives; TN = number of true negatives; $N = TP + FP + FN + TN$.

In medicine and related sciences, the two errors (false negative, and false positive) are not symmetric. Generally, mistakenly classifying someone having the disease as not having it (*false negative*), is more dangerous than mistakenly classifying someone not having the disease as having it (*false positive*). In the first case, in fact, we run the risk to leave the patient uncured, thus exposing her to dangerous (possibly fatal) consequences; in the second case, although we could give undue cures to someone who does not need them, we do not expose the patient to possible fatal consequences.

Based on this argument, we generally would like to give the false negative error a larger weight compared to the false positive one. For example, in choosing a classifier, we would be willing to select one providing zero false negatives. This can be reached, but at expense of increasing the number of false positives. It means that there is a *trade-off* between the two errors, which does not allow us to reduce the two errors contemporaneously. From a binary confusion matrix, we can extract information about the accuracy and the size of the two types of errors. Below we list and comment on some of the most popular metrics.

Precision. Among all the units that were predicted has positive (having the disease), precision measure the percentage of true positives (those actually having the disease). Its formula is:

$$precision = \frac{TP}{TP + FP},$$

The precision can be meant as the ability of a classifier not to label as positive a unit that is negative.

Recall (or **sensitivity**). Among all the actual positive cases (those having the disease), the recall (or sensitivity) measure the percentage of true positives (those actually having the disease). Its formula is:

$$recall = \frac{TP}{TP + FN}$$

The recall can be interpreted as the ability of a classifier to find all the positive units, or equivalently, to avoid false negatives.

Precision and recall cannot be increased simultaneously. As they are complements of the two errors discussed above, if one increases precision, the recall will be reduced, and vice versa. Therefore, there exists a precision/recall trade-off.

F1-score. This metric combines precision and recall using their harmonic mean. Its formula is:

$$F1\text{-score} = (1 + \beta^2)\frac{precision \times recall}{\beta^2 precision + recall}$$

where β is an importance weight usually taken as equal to 1.

When classes are unbalanced,[1] especially when positives are much lower than negative cases in the initial dataset, the accuracy can be too optimistic as a goodness-of-fit measure. The example in Fig. 2.6 higlights this situation. The number of true positive cases is 15, against 120 negatives. The accuracy is equal to 78%, while precision and recall are, on the contrary, equal to 20% and 33% respectively, more than three times lower than the accuracy. Nonetheless, a limitation of both precision and recall is that they only look at the relative size of the *TP* at expense of the size of *TN*. For this purpose, we can define another index called *specificity*.

---

[1] A more general way to deal with unbalanced data will be addressed later on in this chapter.

	Predicted positive	Predicted negative	
True positive	5	10	15
True negative	20	100	120
	25	110	135

Accuracy	0.78
Precision	0.20
Recall	0.33

Fig. 2.6 Example of imbalance data. Precision and recall are better goodness-of-fit measures than the accuracy when positives are much lower than negative cases in the initial dataset

Specificity. Among all the actual negative cases (not having the disease), the specificity measures the percentage of true negative cases. Thus, it looks at the other side of the accuracy, giving attention to *TN*. Its formula is:

$$\text{specificity} = TN/(TN + FP)$$

Receiver operator characteristics (ROC) curve. In Machine Learning, the ROC curve is used as a measure of a classifier (classification) accuracy over all possible classification probability thresholds within a two-class classification setting. It is thus a measure of global accuracy of a classifier. In an ideal world, we would like to maximize both sensitivity and specificity (that is, put both error types at zero). Unfortunately, this is not possible, as there exists a trade-off between sensitivity and specificity. If one increases sensitivity, then specificity decreases, and vice versa. A classification that is less sensitive (detecting fewer true positives), is also more specific (detecting fewer false positive).

Sensitivity (and specificity) are functions of the threshold probability $p^\star$ that one uses to classify. Suppose that $y \in \{1, 0\}$. Given $p^\star$, we can classify a unit with features $\mathbf{x}$ as "1" if $p(\mathbf{x}) = \text{Prob}(y = 1|\mathbf{x}) > p^\star$. By varying $p^\star$, we come up with different classifications and different sensitivity/specificity pairs. There is a one-to-one mapping between $p^\star$ and a sensitivity/specificity pair. The ROC curve plots this mapping. Figure 2.7 shows an example of ROC curve derived from a logistic classifier. In the ROC curve, sensitivity (i.e., the accuracy in detecting true positives, *TP/(TP+FN)* is plotted against (1 - specificity) (i.e., the error to mis-classify false positives, *FP/(FP+TN)*). As horizontal and vertical axes take values between 0 and 1, the total area of the square is $1 \times 1 = 1$. The *area under the curve* (AUC) takes values between 0 and 1 and is an overall measure of accuracy based on the sensitivity/specificity trade-off. A value of the AUC = 0.5 corresponds to the 45-degree line. This value corresponds to classifying by *random guessing*. The overall performance of a classifier, over all possible thresholds, is given by its AUC. The larger the AUC the better the classifier. Generally, we look for AUC larger than 90% or 95%, but it may depend on the context. Classifiers can be compared based on their AUC, with

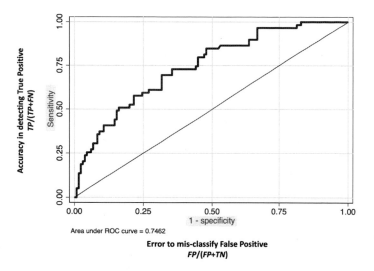

Fig. 2.7 Example of a receiver operator characteristics (ROC) curve

the best being the one having the largest AUC. The ideal classifier would have an AUC equal to 1.

In multi-class classification, we can still define the AUC using an averaging strategy. In the literature, two averaging strategies have been proposed: (1) the *one-vs-one* (OVO) approach, and (2) the *one-vs-all* (OVA) approach. The OVO strategy computes the average of the pairwise ROC AUC scores, while the OVA computes the average of the ROC AUC scores for each class against all the other classes. For example, if the y contains J categories, we average over $J(J-1)/2$ AUC scores in OVO, and over J AUC scores in OVA.

Cross-entropy (or deviance). The cross-entropy (sometimes refereed to as *log-loss*) is a goodness-of-fit measure used to evaluate the probability outputs of a classifier instead of its categorical predictions. Based on the logistic loss, it is predominantly used in binomial and multinomial logistic, and in neural networks classification. In the binary classification setting, with i.i.d. data, and y taking either value 1 or 0, the cross-entropy is equal to the sample negative log-likelihood of the classifier:

$$\text{cross-entropy}_B = -\frac{1}{N} \sum_{i=1}^{N} (y_i \log(p_i) + (1 - y_i) \log(1 - p_i))$$

where $p_i = \text{Prob}(y_i = 1)$. This formula extends naturally to the multi-class case. Specifically, for each of the J categories of the variable y, define the J dummy variables $y_{i,j}$ taking value 1 if unit i belongs to category j, and 0 otherwise. Then, the multi-class cross-entropy is equal to:

	Predicted positive	Predicted negative	Total
True positive	0	5	5
True negative	0	190	190
	0	195	195
Accuracy	0.97		

Fig. 2.8 Example of non-classification paradox

$$\text{cross-entropy}_M = -\frac{1}{N}\sum_{i=1}^{N}\sum_{j=1}^{J} y_{i,j}\log(p_{i,j})$$

where $p_{i,j} = \text{Prob}(y_{i,j} = 1)$. Observe that the cross-entropy is strictly related to the *deviance*. For a given model $\mathcal{M}$, the deviance is equal to:

$$\text{deviance} = -2\log\left(\frac{L_{\mathcal{M}}}{L_{\mathcal{S}}}\right) = -2(\log L_{\mathcal{M}} - \log L_{\mathcal{S}})$$

where $L_{\mathcal{M}}$ denotes the maximized likelihood of model $\mathcal{M}$, and $L_{\mathcal{S}}$ the maximized likelihood of the *saturated* model, which is a degenerate model having a separate parameter for each observation and thus providing a perfect fit. Because $L_{\mathcal{S}}$ is constant, whatever the model $\mathcal{M}$ is, the deviance of a model is basically equal to twice the negative log-likelihood of the current model plus a constant. As minimizing the cross-entropy and the deviance leads to the same estimation, they can be used interchangeably as goodness-of-fit measures.

2.6.1 Dealing with Unbalanced Classes for Classification

It can happen that a given classifier is *unable* to classify, although the accuracy is high (as high can be the AUC). This is known as the *non-classification paradox*. Figure 2.8 shows an example of this paradox.

In this example, we can see that whatever is the true status (either positive, or negative), every unit is predicted as negative. Therefore, no observations are predicted as positive. Actually, the classifier does not classify, but the accuracy is very large, equal to 97%. Cases like this makes classification useless, as we would like to classify observations within both classes. Why can a situation like this arise? Situations like this are typical when data are *unbalanced*, namely when one of the two classes has a much larger size compared to the other class. In a two-class setting, we generally distinguish between the majority and the minority class.

But why in the presence of a sizable majority class a classifier does not manage to classify? To respond to this question, we have to recall the logic through which we classify units within a specific class. As stressed earlier in this chapter, we classify an observation using its conditional probability to belong to the different classes. More precisely, a unit i is classified in class k if its probability to belong to class k is the largest among all the other conditional class probabilities (Bayes optimal classification rule). We can rewrite this conditional probability using the Bayes formula as follows:

$$\text{Prob}(y_i \in k | \mathbf{x}_i) = \frac{\text{Prob}(\mathbf{x}_i | y_i \in k) \cdot \text{Prob}(y_i \in k)}{\text{Prob}(\mathbf{x}_i)} \tag{2.16}$$

In this equation, the term $\text{Prob}(\mathbf{x}_i)$ does not count for classifying, as it does not depend on k. The terms that count for classifying are the density distribution of $\mathbf{x}_i$ within class k (i.e. $\text{Prob}(\mathbf{x}_i | y_i \in k)$), and the prior probability of unit i to belong to class k (i.e. $\text{Prob}(y_i \in k)$). Given the former, the probability to classify a unit i into class k increases as long as the prior probability increases. If the proportion of cases in the training set is unbalanced toward one class, it means that this class will have a larger prior probability and thus that units will be more likely classified within this class.

How can we deal with the presence of a large majority class in real applications? In the literature, the following strategies have been proposed:

1. class reweighting;
2. over-sampling;
3. under-sampling;
4. ROSE (random over-sampling examples);
5. SMOTE (synthetic minority oversampling technique).

Let's discuss them individually.

Class reweighting. This approach consists of giving the minority and majority observations a weight equal to the inverse of the class size:

$$\omega_1 = \frac{1}{N_1}, \, \omega_0 = \frac{1}{N_0}$$

If class "1" is the majority class and class "0" the minority one, it is clear that reweighting re-establish some balancing by giving more importance to the class with smaller size, and vice versa, less importance to the class with larger size. Notice that, in real applications, the employed ML classifiers must have a "weight" option, otherwise we cannot apply class reweighting. For a linear classifier, this approach coincides with Weighted Least Squares (WLS) estimation, with weights provided by ω_1 and ω_0.

Oversampling. Oversampling increases the number of the minority class by resampling to reach full balance. In its simplest version, oversampling randomly adds observations of the minority class to reach the same size of the majority class, thus obtaining a perfectly balanced dataset. The main idea is to resample with replacement from the minority class, until one reaches a perfect balance between the minority

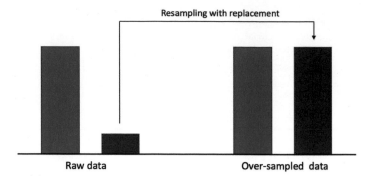

Fig. 2.9 Oversampling procedure

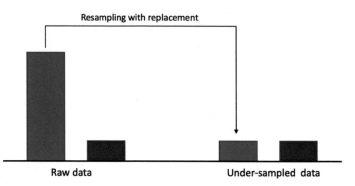

Fig. 2.10 Under-sampling procedure

and the majority classes. Figure 2.9 shows an intuitive graphical representation of the oversampling procedure.

Under-sampling. Under-sampling decreases the number of the majority class by resampling to reach full balance. In its simplest version, under-sampling randomly deletes observations of the majority class to reach the same size of the minority class, thus obtaining a perfectly balanced dataset. The main idea is to resample with replacement from the majority class, until one reaches a perfect balance between the minority and the majority classes. Figure 2.10 shows an intuitive graphical representation of the under-sampling procedure.

ROSE (Random oversampling examples). The ROSE algorithm oversamples the rare class by generating new *synthetic* data points. These new data are as similar as possible to the existing real ones. ROSE can also carry out an under-sampling of the majority class, possibly by cloning data with the same strategy. ROSE selects a random point from the minority set and generates a new point in the *neighborhood* of the selected point according to a kernel function. Just to have an intuitive idea, one can imagine to sample from a multivariate Normal distribution centered at the selected point, to then draw a new point from that distribution. ROSE however does

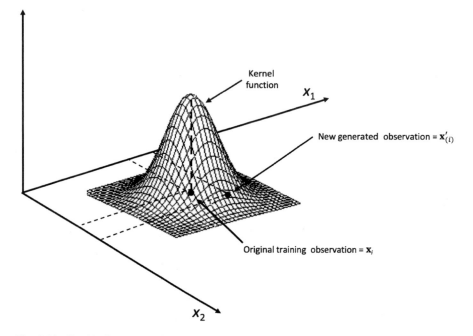

Fig. 2.11 Graphical representation of the ROSE (Random oversampling examples) procedure

not use a Normal distribution (parametric approach), but a kernel density estimation of the distribution of the features within the rare (or the prevalent) class. It can be showed that the ROSE algorithm is equivalent to obtaining a smoothed bootstrap resampling scheme.

Indicate by $P(\mathbf{x})$ the joint distribution the features $\mathbf{x}$ (multivariate probability density). Indicate by N_0 and N_1 the size of the two classes "0" and "1", respectively. The ROSE algorithm generates new artificial data by running this procedure:

1. select with probability $1/2$ the target y_j, with $j = \{0, 1\}$;
2. select $(\mathbf{x}_i, y_i)$ within the training set $\mathcal{T}_N$, such that $y_i = y_j$ with probability $p_i = 1/N_j$;
3. sample $\mathbf{x}$ from a kernel distribution $K_{\mathbf{H}_j}(\cdot, \mathbf{x}_i)$ centered at $\mathbf{x}_i$, which depends on a matrix of scale parameters $\mathbf{H}_j$;
4. obtain the new generated point as $(\mathbf{x}'_{(i)}, y_i)$, where $\mathbf{x}'_{(i)}$ is the sampled vector of features.

Figure 2.11 sets out a visual representation of the ROSE procedure. The matrix $\mathbf{H}_j$ controls for the variance-covariance structure of the estimated density function; as such, setting this matrix is equivalent to set a kernel local bandwidth which establishes how large is the probability mass of the the kernel density around the (central) point $\mathbf{x}_i$.

Fig. 2.12 Graphical
representation of the
SMOTE (Synthetic minority
oversampling technique)
procedure

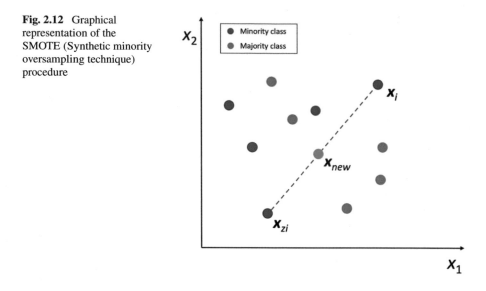

SMOTE (Synthetic minority oversampling technique). The SMOTE algorithms
generates artificial data using the *proximity* principle. It is similar to ROSE, but it uses
a nearest neighbor instead of a kernel proximity approach. The SMOTE algorithm
to generate new observations is as follows:

1. Consider an observation $\mathbf{x}_i$ in the minority class;
2. Consider the first K nearest neighbors of $\mathbf{x}_i$;
3. For each nearest neighbor $\mathbf{x}_{zi}$, a new sample $\mathbf{x}_{new}$ is generated as:

$$\mathbf{x}_{new} = \mathbf{x}_i + \lambda \times (\mathbf{x}_{zi} - \mathbf{x}_i) \tag{2.17}$$

where λ is a random parameter to draw from the range [0;1].

Figure 2.12 illustrates a graphical representation of the SMOTE algorithm. As we
can see, the new artificial point ($\mathbf{x}_{new}$) is generated along the line joining $\mathbf{x}_i$ and $\mathbf{x}_{zi}$.
In this case, the location of $\mathbf{x}_{new}$ is halfway, which corresponds to a value of λ equal
to 0.5 in Eq. (2.17).

2.7 Optimal Tuning of Hyper-Parameters

In the previous sections, we defined and discussed the central role played by the
test error related to any regression or classification model (i.e., any learner). In this
section, we show how to estimate the test error (either MSE or MCE) using the
available data. Importantly, the same procedures that allow us to estimate the test
error can be employed for the *optimal tuning* of the learners.

In Machine Learning, the optimal tuning of a learner consists of finding the optimal model complexity (or flexibility), whenever it is parameterized by a given set of G hyper-parameters $\lambda_1, \ldots, \lambda_G$. Usually, optimal tuning is carried out computationally, by defining a grid of potential optimal values for every hyper-parameter, to then look for the best ones using iterative methods. In a few cases, optimal tuning can be achieved without relying on intensive computational procedures. This is the case, as we will discuss, of the use of information criteria and plug-in approaches. In these cases, we either use adjustments of traditional in-sample goodness-of-fit methods (information criteria), or derive analytically the optimal value of the hyper-parameters (plug-in methods).

In the ML literature, we find basically four ways to tune a learner:

- Information criteria
- Bootstrap
- K-fold cross-validation
- Plug-in method

In what follows, we go through each of these methods by discussing both the applicability and pros and cons of each approach.

2.7.1 Information Criteria

Information criteria are based on goodness-of-fit formulas that adjust the training error by penalizing too complex models (i.e., models characterized by large degrees of freedom). Traditional information criteria comprise the Akaike criterion (AIC) and the Bayesian information criterion (BIC), and can be applied to both linear and non-linear models (probit, logit, poisson, etc.). Unfortunately, the information criteria are valid only for linear or generalized linear models, i.e. for parametric regression/classification. They cannot be computed for nonparametric methods like–for example–tree-based or nearest neighbor regression/classification.

Mallows's Cp. In the context of linear regression, the Mallows's Cp is an adjustment of the MSE, penalizing models characterized by too large degrees of freedom. Its formula is:

$$C_p = \frac{1}{n}(\text{RSS} + 2p\hat{\sigma}^2)$$

where the RSS is the residual sum of squares, p is the number of free parameters to estimate, and $\hat{\sigma}$ is an estimate of the error variance. The larger the model complexity, the higher the penalization and thus the value taken by the Mallows's Cp. Observe that models with lower Cp are generally preferred.

Akaike information criterion (AIC). The AIC is similar to the Mallows's Cp, but can be computed for any generalized linear model (GLM) estimated by maximum likelihood. Its formula is:

$$\text{AIC} = -2\log\hat{L} + 2p$$

where $\hat{L}$ is the value of the (estimated) maximized likelihood of the considered GLM model. In the case of the linear model with Gaussian errors, maximum likelihood estimation coincides with least squares, thereby Cp and AIC are equivalent. Notice that, in place of the RSS, the AIC uses the so-called *deviance*, that is, $-2\log\hat{L}$. For GLMs estimated by maximum likelihood, the deviance plays the same role that RSS plays for least squares estimation. Observe that models with lower AIC are generally preferred.

Bayesian information criterion (BIC). Closely related to the AIC, the BIC takes on this formula:

$$\text{BIC} = -2\log\hat{L} + 2p \cdot \log(N)$$

where the penalization term, compared to the AIC, depends also on the $\log(N)$. Since we have that $\log(N) > 2$ whenever $N > 7$, generally the BIC entails a larger penalization compared to the AIC. In model selection, as we will see later in this book, the BIC leads to select more parsimonious models (i.e., models characterized by lower degrees of freedom). Observe that models with lower BIC are generally preferred.

Adjusted R^2. The Adjusted R^2, as the name suggests, is a mathematical adjustment of the traditional coefficient of determination used as a goodness-of-fit measure in least squares regression. Its formula is:

$$\text{Adjusted } R^2 = 1 - \frac{\text{RSS}/(n-p-1)}{\text{TSS}/(n-1)}$$

where $\text{TSS} = \sum_{i=1}^{N}(y_i - \bar{y})^2$ is the total sum of squares. As clearly visible, this index penalizes too complex models. Observe that models with higher Adjusted R^2 are generally preferred.

2.7.2 Bootstrap

Information criteria are not available for nonparametric models. In this case, model complexity does not coincides with the number of free parameters to estimate. For nonparametric models, the testing error can be estimated via computational techniques, more specifically, by *resampling* methods. Among resampling approaches, a popular one if the *bootstrap*.

Developed and refined by Efron (1979), the bootstrap–i.e., resampling with replacement from the original sample–is a general approach for doing statistical inference computationally. The bootstrap has many uses in computational statistics, but it is predominantly employed to derive properties (standard errors, confidence intervals, and critical values) of the sampling distribution of estimators whose asymp-

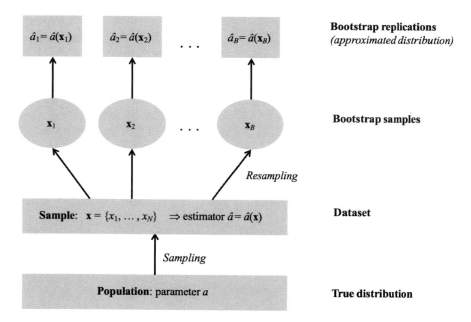

Fig. 2.13 Flow-diagram of the bootstrap procedure for computing the empirical distribution of a statistic for the population parameter a

totic analytic properties are unknown. It is similar to Monte Carlo techniques, but instead of fully specifying the Data Generating Process (DGP), the bootstrap uses directly information from the actual sample.

How does the bootstrap work? In a nutshell, the bootstrap consider the *sample* (i.e., the values of the independent and dependent variables in the dataset) as if it was the *population*, and the estimates of the sample as if they were the true values of the parameters. Then, it generates B different samples by resampling from the original sample, estimates the statistic of reference for every sample, and obtains the empirical distribution of the statistic. This empirical distribution is then used for testing purposes. Figure 2.13 illustrates the flow-diagram of the bootstrap procedure for computing the empirical distribution of a statistic for the population parameter a. From the original population, we draw randomly a sample $\mathbf{x}$, and estimate the statistic $\hat{a} = \hat{a}(\mathbf{x})$. Then, we draw B different samples with replacement from the original sample, thus obtaining B point estimates of the parameter a. Under reasonable assumptions, the bootstrap provides consistent estimation of the empirical distribution of the statistic $\hat{a}(\cdot)$.

The bootstrap could be in theory a practical solution to estimate the test error and optimally tune a learner, provided that the original dataset is used as validation dataset and the bootstrapped ones as training datasets. Unfortunately, the bootstrap has the limitation of generating observation overlaps between the test and the training datasets, as about two-thirds of the original observations appear in each bootstrap sample. This occurrence undermines its use to validate out-of-sample an ML

Fig. 2.14 Visualizing
bootstrap overlapping in a
sample with $N = 10$
observations and $B = 2$
bootstrap replications

Original dataset (Validation)	Boot 1 (training)	Val 1 (test) No overlap	Boot 2 (training)	Val 2 (test) No overlap
1	1	2	4	1
2	1	3	4	2
3	4	6	4	3
4	4	7	5	7
5	5		6	10
6	8		6	
7	9		6	
8	10		8	
9	10		8	
10	10		9	
	Train 1	Test 1	Train 2	Test 2

procedure. Figure 2.14 illustrates a simple example visualizing bootstrap overlapping in a sample with $N = 10$ observations and $B = 2$ bootstrap replications. The observations of the original sample are labeled: 1, 2, 3, 4, 5, 6, 7, 8, 9, 10. The first bootstrap produces the sample Boost 1, that we use as training dataset. If the original sample has to be the validation sample, it is clear that we have an overlap; specifically, units 1, 4, 5, 8, 9, 10 appear in both the training and testing samples. The non-overlapping sample is in the column with heading Val 1 and contains the following units: 2, 3, 6, and 7. We may use these units as validating observations. In conclusion, the model will be trained over Boost 1 and validated over Val 1, thus producing a first estimate of the prediction error (either MSE or MCE). By repeating the same procedure as many times as the number of the pre-set bootstrap samples, we obtain an estimate of the mean prediction error and its standard deviation.

2.7.3   K-fold Cross-validation (CV)

Cross-validation is the workhorse of the testing error estimation and hyper-parameters' optimal tuning in Machine Learning. Given its importance, we consider here two types of cross-validation, according to whether the data are cross-sectional, or time-series.

Cross-sectional CV

The idea of K-fold cross-validation is to randomly divide the initial dataset into K equal-sized portions called *folds*. This procedure suggests to leave out fold k and fit the model to the other $(K\text{-}1)$ folds (wholly combined) to then obtain predictions for

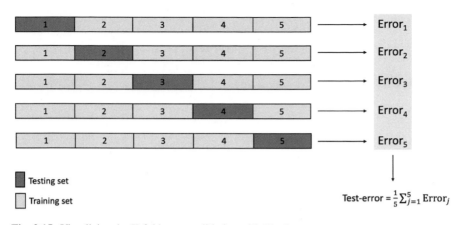

Fig. 2.15 Visualizing the K-fold cross-validation with $K = 5$

the left-out k-th fold. This is done in turn for each fold $k = 1, 2, \ldots, K$, and then the results are combined by averaging the K estimates of the error (either MSE or MCE). Figure 2.15 provides an example of the testing error estimation in the case of 5 folds.

In a regression setting, where y is numerical, the cross-validation procedure can be carried out as follows:

- Split randomly the initial dataset into K folds denoted as $G_1, G_2, \ldots, G_K$, where G_k refers to part k. Assume that there are n_k observations in fold k. If N is a multiple of K, then $n_k = N/K$.
- For each fold $k = 1, 2, \ldots, K$, compute:

$$\text{MSE}_k = \sum_{i \in C_k} (y_i - \hat{y}_i)^2 / n_k$$

where $\hat{y}_i$ is the fit for observation i, obtained from the dataset with fold k removed.
- Compute:

$$\text{CV}_{\text{MSE},K} = \sum_{i=1}^{K} \frac{n_k}{n} \text{MSE}_k$$

that is the average of all the out-of-sample MSEs obtained fold-by-fold.
- Compute:

$$\text{SE}_{\text{CV},\text{MSE},K} = \sqrt{\frac{\sum_{k=1}^{K} (\text{MSE}_k - \text{CV}_{\text{MSE},K})^2}{K - 1}}$$

that is the standard error of the cross-validation MSE.

Observe that, by setting $K = N$, we obtain the N-fold or *leave-one-out* cross-validation (LOOCV). The LOOCV is computationally costly, as we have to fit the learner as many times as the number of the training observations. However, in the case of least squares estimation (including, of course, polynomial fitting) the cost of carrying out the LOOCV is equal to a single model fit. Indeed, we can show that:

$$\text{CV}^{\text{LS}}_{\text{loocv}} = \frac{1}{N} \sum_{i=1}^{N} \left[\frac{y_i - \hat{y}_i}{1 - h_i} \right]^2$$

where h_i is the leverage of unit i. The leverage is a measure of *outlierness*, referred however to the predictors, not to the dependent variable. It measures how distant are the values of the features of an observation from the features of the other observations.

When $K = 1$, we obtain the 1-fold CV, also known as *validation* approach, entailing only one split generating one training set and one testing set. This approach can be used either for estimating the test error, or for tuning the hyper-parameters. Of course, as it based on only one split, the validation approach can be plagued by large estimation variance.

Importantly, as $\text{CV}_{\text{MSE},K}$ is an estimation of the true test error, estimating its standard error can be useful to provide accuracy's confidence interval, thereby providing a measure of test accuracy's uncertainty. In this sense, the ultimate assessment of the prediction ability of a learner, should not be based only on the point estimation of the test error but accompanied by some confidence measure expressing the degree of uncertainty in gauging learners' prediction ability.

For ML classification purposes, the K-fold cross-validation procedure follows a similar protocol, except for considering the MCE in place of the MSE. This protocol entails the following steps:

- Split randomly the initial dataset into K folds denoted as $G_1, G_2, \ldots, G_K$, where G_k refers to part k. Assume that there are n_k observations in fold k. If N is a multiple of K, then $n_k = N/K$.
- For each fold $k = 1, 2, \ldots, K$, compute:

$$\text{MCE}_k = \sum_{i \in G_k} I(y_i \neq \hat{y}_i)/n_k$$

  where $\hat{y}_i$ is the predicted label for observation i, obtained from the dataset with fold k removed.
- Compute:

$$\text{CV}_{\text{MCE},K} = \sum_{i=1}^{K} \frac{n_k}{n} \text{MCE}_k$$

  that is the average of all the out-of-sample MCEs obtained fold-by-fold.
- Compute:

$$\text{SE}_{\text{CV,MCE},K} = \sqrt{\frac{\sum_{k=1}^{K} (\text{MCE}_k - \text{CV}_{\text{MCE},K})^2}{K - 1}}$$

  that is the standard error of the cross-validation MCE.

Moreover, in the case of binary classification, other accuracy measures can be used such as the AUC (area under the ROC curve) and the F1-score (the harmonic mean of *recall* and *precision*).

Time-series CV

Time-series data have a natural sequential ordering. Thus, K-fold cross-validation cannot be carried out as in cross-sections, as randomizing over the observations does not make sense when a temporal/sequential ordering is in place. We have to comply with this issue. By and large, we have two kinds of cross-validations that we can perform in a time-series context: (i) with train varying window; and (ii) with train fixed window.

In general, time-series CV requires to fix these parameters:

1. size of the (initial) training horizon: n_{train}. We can start with the minimum sample size that allows to estimate our model;
2. size of the testing horizon: n_{test}. We choose the forecasting horizon that matters for our purposes;
3. size of the training step-ahead: h. We can use this parameter to reduce the computational burden.

Figure 2.16 sets out how CV is carried out with time-series data in the varying window set-up. For simplicity, assume that $h = 1$. We see that the training dataset becomes increasingly large, by augmenting of one unit in the subsequent step (provided that $h = 1$). The testing set size remains the same, equal to n_{test}. Observe that the

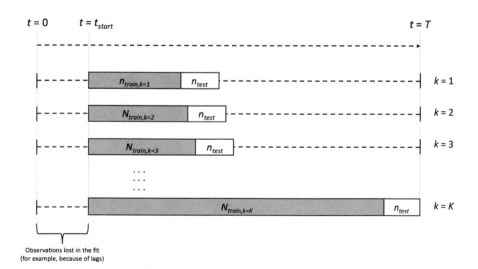

Fig. 2.16   Visualizing the cross-validation for time-series data. Varying window setting

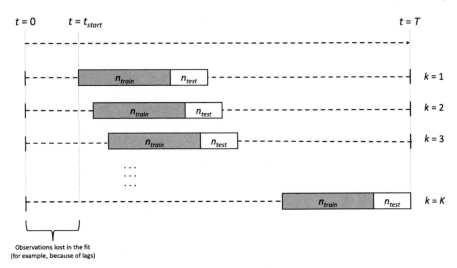

Fig. 2.17   Visualizing the cross-validation for time-series data. Fixed window setting

starting time of the CV cannot in general coincide with the initial time of the time-series. This may happen because the model to fit may contain lags of the outcome variable, thus losing one observation for every lag included in the model.

Figure 2.17 provides a schematic representation of CV carried out in a time-series context with fixed training window. In this case, both the size of the training set and that of the testing set remain fixed over the iterations. This allows for some balance between the two groups of observations.

Figure 2.18 illustrates an example of cross-validation for time-series data, where an auto-regressive structure of order 2–i.e. an AR(2)–is considered. We run cross-validation in a varying window setting. We obtain four MSEs, assuming a training horizon of 3 times, a testing horizon of 2 times, and one step ahead. The final MSE is the simple average of the four MSEs obtained fold-by-fold.

The Optimal Tuning Scheme

As a general approach, Fig. 2.19 shows the cross-validation optimal hyper-parameters' tuning scheme for a generic ML method. This scheme shows how ML practitioners tune in practice the hyper-parameters of their learners starting from the initial (original) dataset. The procedure entails the following steps:

1. split the original dataset into a training and a testing sample;
2. focusing on the training sample, define a grid of values for the hyper-parameter h, $\mathcal{H} = \{h_1, h_2, \ldots, h_H\}$;
3. for each $h \in \mathcal{H}$:

    a. split the training dataset into K folds;

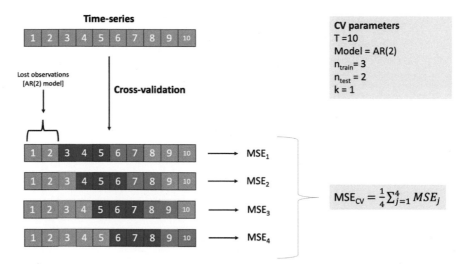

Fig. 2.18 Example of cross-validation for time-series data, where an auto-regressive structure of order 2–AR(2)–is considered. Varying window setting

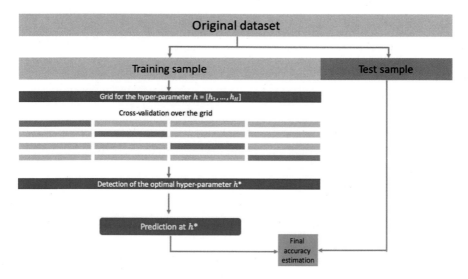

Fig. 2.19 Cross-validation optimal hyper-parameter's tuning scheme for a generic ML method

b. estimate the CV error (MSE(h) or MCE(h));
4. select the optimal hyper-parameter $h^\star$ as the value in the grid providing the minimum CV error;
5. fit the model at $h = h^\star$, and compute optimal predictions;
6. using the left-out initial test sample, estimate the out-of-sample accuracy of the learner.

This procedure is very general, and can be applied to any learner, both in a classification and regression setting. We will apply this procedure several times in the next chapters of this book.

2.7.4 Plug-in Approach

The plug-in approach uses analytic estimations of the optimal hyper-parameters. For some special nonparametric models, in fact, one can analytically minimize the (integrated) MSE or MCE, and find a formula for specific hyper-parameters, such as the *bandwidth* for local nonparametric regression or for kernel density estimates. These hyper-parameters' formulas involve the presence of unknown nested parameters that must be replaced in turn by some estimates. For example, for kernel nonparametric local regressions, such as the *local linear* regression, it is possible to obtain a formula for the optimal bandwidth (Silverman 1986). This formula, however, depends on the first and second derivatives of the unknown conditional mean function, whose estimates require in turn to carry out an optimal tuning of additional bandwidths. This makes the use of the plug-in approach rather complicated in many applications.

For many Machine Learning methods, the analytic minimization of the integrated MSE or MCE is even unfeasible, thus preventing the use of any plug-in approach. In more parametric ML settings such as, for example, the elastic-net regression, plug-in hyper-parameters' optimal tuning is still feasible. An example, as we will see later on in this book, is the *rigorous penalization* parameter for the Lasso regression, which uses a plug-in approach. Overall, however, tuning hyper-parameters via cross-validation is a more general and appealing approach, as it can be used for whatever ML model one wants to fit, independently of the possibility to carry out an analytic minimization of the integrated MSE or MCE.

2.8 Learning Modes and Architecture

By generalizing, we can define a learner L_j as a mapping from the set $[\mathbf{x}, \boldsymbol{\theta}_j, \boldsymbol{\lambda}_j, f_j(\cdot)]$ to an outcome y, where $\mathbf{x}$ is the vector of features, $\boldsymbol{\theta}_j$ a vector of estimation parameters, $\boldsymbol{\lambda}_j$ a vector of tuning parameters, and $f_j(\cdot)$ an algorithm taking as inputs $\mathbf{x}$, $\boldsymbol{\theta}_j$, and $\boldsymbol{\lambda}_j$. Differently from the members of the Generalized Linear Models (GLM) family (linear, probit or multinomial regressions are classical examples), that are

Table 2.2 Main Machine Learning methods and associated tuning hyper-parameters. GLM: Generalized Linear Models

ML method	Parameter 1	Parameter 2	Parameter 3
Linear Models and GLM	N of covariates		
Lasso	Penalization coefficient		
Elastic-Net	Penalization coefficient	Elastic parameter	
Nearest Neighbor	N of neighbors		
Neural Network	N of hidden layers	N of neurons	L2 penalization
Trees	N of leaves/depth		
Boosting	Learning parameter	N of sequential trees	N of leaves/depth
Random Forest	N of features for splitting	N of bootstraps	N of leaves/depth
Bagging	Tree-depth	N of bootstraps	
Support Vector Machine	C	Γ	
Kernel regression	Bandwidth	Kernel function	
Piecewise regression	N of knots		
Series regression	N of series terms		

highly parametric and are not characterized by tuning parameters, Machine Learning models–such as local-kernel, nearest neighbor, or tree-based regressions–may be highly nonparametric and characterized by one or more hyper-parameters λ_j which may be optimally chosen to minimize the *test prediction error*, i.e. the out-of-sample predicting accuracy of the learner, as stressed in the previous sections.

Table 2.2, however, lists the most popular Machine Learning algorithms proposed in the literature along with the most relevant associated tuning parameters. Most of them will be presented in detail over the next chapters of this book.

A combined use of these methods can produce a computational architecture (i.e., a virtual learning machine) enabling to increase statistical prediction accuracy and its estimated precision (van der Laan et al. 2007). Figure 2.20 presents the learning architecture proposed by Cerulli (2021). This framework is made of three linked learning processes (or modes): (i) the learning over the tuning parameter λ, (ii) the learning over the algorithm $f(\cdot)$, and (iii) the learning over new additional information. The departure is in point 1, from where we set off assuming the availability of a dataset $[\mathbf{x}, y]$.

The first learning process aims at selecting the optimal tuning parameter(s) for a given algorithm $f_j(\cdot)$. As seen above, ML scholars typically do it using K-fold cross-validation to draw test accuracy (or, equivalently, test error) measures and related standard deviations.

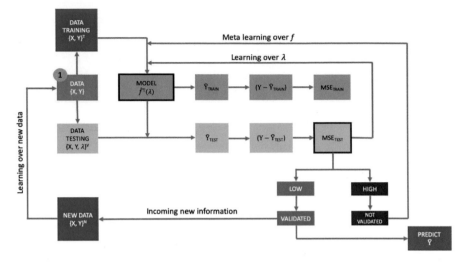

Fig. 2.20 The meta-learning machine architecture. This figure has been drawn from Cerulli (2021)

At the optimal λ_j, one can recover the largest possible prediction accuracy for the learner $f_j(\cdot)$. Further prediction improvements can be achieved only by learning from other learners, namely, by exploring other $f_j(\cdot)$, with $j = 1, \ldots, M$ (where M is the number of learners at hand).

Figure 2.20 shows the training estimation procedure that corresponds to the light blue sequence of boxes leading to the MSE_{TRAIN} which is, *de facto*, a dead-end node, being the training error plagued by *overfitting*.

Conversely, the yellow sequence leads to the MSE_{TEST}, which is informative to take correct decisions about the predicting quality of the current learner. At this node, the analyst can compare the current MSE_{TEST} with a benchmark one (possibly, pre-fixed), and conclude whether to predict using the current learner, or explore alternative learners in the hope of increasing predictive performance. If the level of the current prediction error is too high, the architecture would suggest to explore other learners.

In the ML literature, learning over learners is called *meta learning*, and entails an exploration of the out-of-sample performance of alternative algorithms $f_j(\cdot)$ with the goal of identifying one behaving better than those already explored (Laan and Rose 2011). For each new $f_j(\cdot)$, this architecture finds an optimal tuning parameter and a new estimated accuracy (along with its standard deviation). The analyst can either explore the entire bundle of alternatives and finally pick-up the best one, or decide to select the first learner whose accuracy is larger than the benchmark. Either case is automatically run by this virtual machine.

The third final learning process concerns the availability of new information, via additional data collection. This induces a reiteration of the initial process whose final outcome can lead to choose a different algorithm and tuning parameter(s), depending on the nature of the incoming information. This process is called "online learning" (Bottou 1999; Parisi et al. 2019).

As a final step, one may combine predictions of single optimal learners into one single super-prediction (*ensemble learning*). What is the advantage of this procedure? The prediction error of a learner depends on the sum of (squared) prediction bias and prediction variance. As the sum of i.i.d. random variables has smaller variance than that of the single elements of the sum, the benefit of aggregating learners is that of obtaining a predictor with smaller variance than the individual learners (Zhou 2012). However, as the bias of the ensemble predictor is not guaranteed to be the smallest among the biases of the individual learners, its estimated error might not be smaller than that of a single learner.[2] As a consequence, using an ensemble method could not lead to accuracy improvements although, in applications, the opposite case occurs rather often (Kumar 2020).

2.9 Limitations and Failures of Statistical Learning

In Sect. 1.3 we discussed the fall of symbolic artificial intelligence (symbolic AI), and the rise of AI based on statistical learning (Machine Learning). By a few illustrative examples, we saw how plenty of AI tasks, such as numeral or face recognition, can be fairly well carried out by ML algorithms based on a proper statistical mapping between the features $\mathbf{x}$ and the target y. However, an AI based on statistical learning, although overall successful in applications, presents two aspects that seem to be worth discussing: (i) limitations, and (ii) potential failures. I start by discussing limitations, to then speak about what circumstances can make statistical learning unable to predict well, thus preventing the possibility to perform successfully intelligent tasks.

2.9.1 ML Limitations

There exists a wide debate about ML limitations, from different angles and perspectives. In this section, I focus on four main concerns raised by some authors about the suitability of using ML for specific purposes. I discuss ethics, causality issues, scope, and finally interpretability related to ML applications.

---

[2] To clarify this point, consider the following illustrative example. Assume that the true value to predict in the population is equal to 10. Suppose there are three predictors, L_1, L_2, and L_3, with zero prediction variances, but with means respectively equal to $\mu_1 = 15$, $\mu_2 = 20$, and $\mu_3 = 25$. The squared bias of L_1 is thus $b_1^2 = (15 - 10)^2 = 25$, of L_2 is $b_2^2 = (20 - 10)^2 = 100$, and of L_3 is $b_3^2 = (25 - 10)^2 = 225$. As the ensemble prediction is an average of the predictions from L_1, L_2, and L_3, its squared bias is equal to $[(15 + 20 + 25)/3 - 10]^2 = 100$. This value, equal to that of L_2, is not the smallest, as L_1 performs better obtaining a mean square error equal to 25.

Ethics. When decisions are taken by a machine, who is responsible for their effect? Automated decision-making poses a series of relevant ethical questions, that human beings have somehow to address today and in the near future. Also, the effectiveness of AI algorithms based on ML requires lots of data (more later on) and thus needs individual information with remarkable privacy issues to handle. The relationship between AI solutions and human moral values is at the top of the regulatory agenda of many countries and will become an increasingly pressing issue in the near future.

Causality. The statistical mapping between y and $\mathbf{x}$ found out by a generic ML algorithm is based on correlation, not on causality. This means that we have to be very careful in drawing causal conclusions from features' partial effects when using ML methods. ML models, by themselves, are not meant as causal models, although they can be used also for carrying out causal inference in specific settings, as we will discuss at length when we will present the causal Lasso.

Scope. Should we use ML methods in whatever analytic context? Even when the aim is pure prediction, ML algorithms can neglect fundamental theory. Take the example of weather forecasts: should we predict future weather conditions without relying on Newton's laws of motion? This is an open question, whose response goes however beyond the ends of this book. But it is somewhat an ontological question, which concerns the fundamental underpinnings of the scientific method, an aspect that may have a relevant impact on the pace and direction of future scientific developments.

Interpretability. We saw that, in ML modeling, there exists a fundamental trade-off between predictability and interpretability. But in some research contexts–especially in the social sciences–interpretability is sometimes more important than predictability. In many settings, even when used for complexity reduction, ML methods can be pure *black boxes*, lending poor value added to the understanding of the phenomenon under analysis. This is partly linked to the previous points, concerning the weak causal meaning and the unsuited scientific scope of some ML methods. Good prediction without correct understanding can also be scientifically dangerous, leading to an incorrect interpretation of the true mechanisms underlying the functioning of the phenomena under study. The history of scientific discoveries presents many examples of wrong but highly predictive theories.

2.9.2 ML Failure

Can ML predictions fail? The response is affirmative, and may depend on the main sources of statistical learning, i.e. algorithm development, and data constrains. As for data constrains, we refer to these dimensions: data availability, data quality, data sparseness, and data ordering.

Data availability and rate of convergence. The availability of data is crucial for ML algorithms to work properly. Large amount of observations and features, high-frequency data and detailed spatial data are of the utmost importance for ML to allow for good predictions. In general, Big Data are essential for ML more than ML is for Big Data. This sentence, apparently weird, rests on the high nonparametric nature of ML algorithms. Consider a generic nonparametric estimate of $E(y|\mathbf{x})$, the conditional expectation of y given $\mathbf{x}$. Stone (1982) showed that if $\mathbf{x}$ is p-dimensional and $E(y|\mathbf{x})$ is d times differentiable, then the fastest possible rate of convergence for a nonparametric estimator of an s-th order derivative of $E(y|\mathbf{x})$ is N^{-r} with:

$$r = \frac{d - s}{2d + p} \tag{2.18}$$

According to this equation, the rate r decreases as the order of the derivative increases, and as the dimension of $\mathbf{x}$ increases. It increases the more differentiable $E(y|\mathbf{x})$ is assumed to be, approaching $N^{-1/2}$ if $E(y|\mathbf{x})$ has derivatives of order approaching infinity. With one single feature, if we want to estimate only $E(y|\mathbf{x})$ (i.e., $s = 0$), by assuming the existence of at least the second derivative of $E(y|\mathbf{x})$ (i.e., $d = 2$), we have:

$$r = \frac{2 - 0}{2 \cdot 2 + 1} = \frac{2}{5} = 0.4$$

showing that the fastest nonparametric convergence rate is $N^{-0.4}$. The nonparametric convergence rate is smaller than the rate of convergence of parametric methods that is $N^{-0.5}$. For a regression using $p = 10$ features, the rate is equal to 0.2, a very small rate. As a conclusion, for ML methods to work properly, we need a rather large sample size to obtain reliable prediction estimates. In this sense, ML requires Big Data because Big Data compensate for the slow rate of convergence of ML methods.

Data quality. Poor data quality may also be a problem, even if there is an abundant amount of it. Both the target variable and the features can be, in fact, measured with error, and this may reduce the prediction ability of ML methods. For example, in the social sciences, scholars sometimes rely on row approximations of specific concepts such as intelligence, propensity to bear risks, or specific social attitudes (such as, benevolence, generosity, selfishness, etc.). Data collection and measurement of these concepts may be problematic, thus producing poor predictions even in the presence of a large amount of information. Also in engineering and medicine some signals can be measured with error. This may reduce the capacity of ML algorithms to find hidden structural patterns in the data, and thus predict properly.

Data sparseness. Data sparseness is probably the most relevant cause of prediction failure. It has to do with what is sometimes called in the literature as *interpolation bias*. Put simply, problems of data sparseness arise when the model is trained over data coming from a specific joint distribution–let's say $P(y, \mathbf{x})$–but the new data come from a different joint distribution–$P'(y, \mathbf{x})$. Figure 2.21 illustrates two prediction exercises in the presence of low and high sparseness. In this figure, we aim at predicting the level of cholesterol for an individual based on her age. In the left-hand panel, data are collected both for old and young people, with young people showing, as expected, lower levels of cholesterol. In this case, we train the model over this

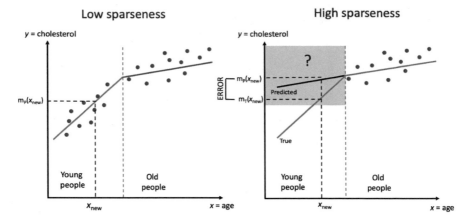

Fig. 2.21 Example of how data sparseness can weaken prediction ability

complete set of data, finding two different slopes for young and old people, with young people having a steeper slope. If we consider a new "young" individual, completely out-of-the-training-sample, characterized by age x_{new}, we see that we are able to generate a correct prediction.

In the right-hand panel, just for illustrative purposes, we eliminated the observations about young people. The training of the model is now made only of old people. If we consider again the prediction for a new "young" instance as before, we see that we commit a severe mistake, as we would be tempted to conclude that the correct prediction is $m_P(x_{new})$, while we know that the true prediction is $m_T(x_{new})$. In this setting, the error comes from the fact that we erroneously projected the line of old people within the area of young people, thus producing a mistake as large as we increase the youth level. This problem has a widespread nature and regards also classification exercises. Whenever the unlabeled data that we want to classify come from a different distribution compared to the training data, we cannot expect ML methods to perform well in classifying correctly brand-new instances.

Problems of sparseness can arise because of *structural breaks* in the phenomena we aim at predicting. For example, consider a time period before and a time period after the Covid-19 pandemic, and suppose one wants to predict labor productivity in the post-pandemic times based on a model trained in the pre-pandemic times. It is clear that the relationship between people's characteristics and their work productivity is expected to be very affected by the pandemic event. For example, people with no good wi-fi connection at home might become less productive during the lockdowns imposed by the government in the most severe phases of the pandemic. Even if the characteristics of the individual remain the same, the pandemic structurally changes the statistical mapping between **x** and *y*. If the training set does not catch these changes on time, through a relatively fast updating process, it is very likely that any attempt to predict correctly people's labor productivity will miserably fail.

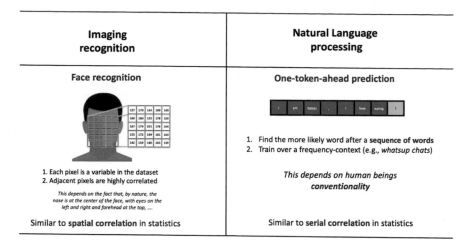

Fig. 2.22 Example of how data ordering can help prediction. Sequential and spatial data

Data ordering. Data ordering is an essential characteristic of data that can be highly relevant for strengthening prediction ability. Data can have some inherent regular structure that characterize them. These regularities can help prediction, as well as their absence can weaken prediction ability. There are various forms of data ordering, but two are particularly important: *spatial* and *sequential* (including *temporal*) orderings. A third type of data ordering, derived from the previous ones, is the *hierarchical structure* of the data. Figure 2.22 shows two examples of datasets characterized by spatial (right-hand panel), and sequential (left-hand panel) data ordering.

In the left-hand panel, we find an example of spatial ordering referring to imaging recognition. Over many observations of faces, there is a high correlation among pixels depending on the natural way specific components of the face are placed within the face framework. For example, the nose is always more or less at the center of the face, the mouth is located below the nose, the eyes above the nose (one on its right, and the other on its left), and so forth. Repeated regular evidence helps prediction by allowing for the ML algorithms to learn very accurately the mapping between the **x** (each pixel becomes a variable) and the target variable y (the face label). This spatial ordering also generates a meaningful hierarchical structure from the initial data by, for example, extracting hierarchically *nested* synthetic features from the initial original features. As will see further on in this book, this hierarchical structure is at the basis of the high ability of *convolutional* neural networks to recognize images.

In the right-hand panel, we have an example of sequential ordering referring to natural language processing (NLP). We consider a simple one-token-ahead prediction, where we train a learner to predict the subsequent word after a certain number of previously spelled words. The conventional way in which we humans speak, and the existence of stereotypical behaviors common to large masses of human beings, make word sequence highly *sequentially* correlated. In this example, we know that, based on the high frequency of times this sequence appears in the real-world speaking, if one is Italian and wants to eat, it is very likely that she will eat pasta or pizza. As will see further on in this book, the existence of these recurrences is at the basis of the high ability of *recurrent* neural networks to forecast time series, or generate new text starting from a specific document (a book, for example).

Mixed orderings–both spatial and sequential, for example–can further help prediction. For example, we might have measures of spatial units observed over time, as in longitudinal datasets. It is clear that, as long as data do not own inherent ordering, prediction accuracy will decrease. This is the situation arising in the social sciences' typical data frames, where regular patterns–including hierarchies–are more rarely met. This explains why prediction ability in the context of the social sciences is weaker than in engineering and in hard sciences more in general.

2.10   Software

In this book, we will make use of three popular software for carrying out ML in practice: Python, R, and Stata.

Python is generally considered the most complete software for ML. It has powerful platforms to carry out both ML and deep learning algorithms (Raschka and Mirjalili 2019). Among them, the most popular are Scikit-learn (Pedregosa et al. 2011) for fitting a large number of ML methods, and `Tensorflow` and `Keras` for fitting neural networks and deep learning techniques more in general. These make Python, which is freeware, probably the most effective and complete software for ML and deep learning available within the community.

R, a freeware software as well, has many packages to run ML, and a specialized all-inclusive package called CARET (short for Classification And REgression Training), which contains a set of functions to streamline the process for creating ML predictive models (Kuhn 2022). The CARET package contains tools for data splitting, preprocessing, feature selection, model optimal tuning using resampling, and variable importance estimation.

Unlike Python and R, Stata (which is a commercial software) does not have dedicated built-in packages for fitting ML algorithms, if one excludes the `lasso` package of Stata 16. Recently, however, the Stata community has developed some popular ML routines that Stata users can suitably exploit. Among them, I mention Schonlau (2005a) implementing a Boosting Stata plugin; Guenther and Schonlau (2016) providing a command fitting Support Vector Machines (SVM); Ahrens et al. (2020b) setting out the `lassopack`, a set of commands for model selection and prediction with regularized regression; Schonlau and Zou (2020a) recently providing a module for the random forests algorithm. More recently, by taking advantage of the new Stata/Python integration interface, Droste (2022) and Cerulli (2022e) have developed `pylearn` (Droste) and `c_ml_stata_cv` and `r_ml_stata_cv` (Cerulli), a set of Stata commands to perform supervised learning in Stata, that are however wrappers of the Python Scikit-learn platform. These commands exhibit a common Stata-like syntax for model estimation and post-estimation.

Other software, like Matlab and SAS, have their own dedicated tools for fitting supervised and unsupervised ML, but we will not use them in this book.

2.11 Conclusions

This chapter has offered an introduction to the statistics of ML, and provided the basis to go through the next chapters of this book. It has discussed several relevant issues, such as: the trade-off between prediction and inference, that between flexibility and interpretability, the meaning of goodness-of-fit and overfitting, the way to optimally tune models' hyper-parameters, and the limitations and failures of ML prediction in real-world applications. All the subjects addressed in this chapter will be reconsidered in the next chapters in light of the specific ML methodologies we will present.

References

Ahrens, A., Hansen, C. B., & Schaffer, M. E. (2020). lassopack: Model selection and prediction with regularized regression in Stata. *The Stata Journal: Promoting Communications on Statistics and Stata, 20*(1), 176–235.

Bottou, L. (1999). On-line learning and stochastic approximations. In *On-line learning in neural networks* (pp. 9–42). Cambridge University Press.

Cerulli, G. (2021). Improving econometric prediction by machine learning. *Applied Economics Letters, 28*(16), 1419–1425.

Cerulli, G. (2022). Machine learning using Stata/Python. *The Stata Journal: Promoting Communications on Statistics and Stata, 22*(4), 1–39.

Droste, M. (2022). Pylearn. Original-date: 2020-04-03T21:21:10Z. https://github.com/mdroste/stata-pylearn

Efron, B. (1979). Bootstrap methods: Another look at the jackknife. *Annals of Statistics 7*, 1–26.

Guenther, N., & Schonlau, M. (2016). Support vector machines. *The Stata Journal: Promoting Communications on Statistics and Stata, 16*(4), 917–937.

Hastie, T., Tibshirani, R., & Friedman, J. (2009). *The Elements of Statistical Learning: Data Mining, Inference, and Prediction* (2nd ed.). Springer Series in Statistics, Springer-Verlag. https://www.springer.com/gp/book/9780387848570

Kuhn, M. (2022). *The caret Package.* https://topepo.github.io/caret/

Kumar, S. (2020). Does ensemble models always improve accuracy?. https://medium.com/analytics-vidhya/does-ensemble-models-always-improve-accuracy-c114cdbdae77

Parisi, G. I., Kemker, R., Part, J. L., Kanan, C., & Wermter, S. (2019). Continual lifelong learning with neural networks: A review. *Neural Networks, 113*, 54–71.

Pedregosa, F., Varoquaux, G., Gramfort, A., Michel, V., Thirion, B., Grisel, O., Blondel, M., Prettenhofer, P., Weiss, R., Dubourg, V., Vanderplas, J., Passos, A., & Cournapeau, D. (2011). Scikit-learn: Machine learning in Python. *Journal of Machine Learning Research, 12*, 2825–2830.

Raschka, S., & Mirjalili, V. (2019). *Python machine learning—3rd edition*. Packt Publishing.

Schonlau, M. (2005). Boosted regression (boosting): An introductory tutorial and a Stata plugin. *The Stata Journal: Promoting communications on statistics and Stata, 5*(3), 330–354.

Schonlau, M., & Zou, R. Y. (2020). The random forest algorithm for statistical learning. *The Stata Journal: Promoting Communications on Statistics and Stata, 20*(1), 3–29.

Silverman, B. W. (1986). *Density Estimation for Statistics and Data Analysis*. Routledge & CRC Press.

Stone, C. J. (1982). Optimal global rates of convergence for nonparametric regression. *The Annals of Statistics, 10* (4), 1040–1053. Publisher: Institute of Mathematical Statistics.

Van der Laan, M. J., & Rose, S. (2011). *Targeted learning: Causal inference for observational and experimental data* (2011th ed.). Springer.

van der Laan, M. J., Polley, E. C., & Hubbard, A. E. (2007). Super learner. *Statistical applications in genetics and molecular biology, 6*, Article 25.

Zhou, Z.-H. (2012). *Ensemble methods: Foundations and algorithms* (1st ed.). Chapman and Hall/CRC.

Chapter 3
Model Selection and Regularization

3.1 Introduction

This chapter presents regularization and selection methods for linear and nonlinear (parametric) models. These are important Machine Learning techniques as they allow for targeting three distinct objectives: (1) prediction improvement; (2) model identification and causal inference in high-dimensional data settings; (3) feature-importance detection. We set off by discussing model selection for improving prediction accuracy and model identification and estimation in high-dimensional data settings, i.e., datasets where p (the number of features) is much larger than N (the number of cases). Then, we present regularized linear models (also known as *shrinkage* methods) focusing on Lasso, Ridge, and Elastic-net regressions. In a related manner, we also address regularized nonlinear models that are extensions of the linear ones to generalized linear models (GLMs). Subsequently, we illustrate optimal Subset selection algorithms that are pure computational approaches to optimal modeling, and feature-importance extraction. In this section, we illustrate three methods: exhaustive (or best) selection, forward, and backward stepwise selection. After delving into the statistical properties of regularized regression, we discuss causal inference in high-dimensional settings, both with an exogenous and endogenous treatment.

The second part of the chapter is fully dedicated to the Stata, R, and Python implementations of the methods presented in the first part. Given the extensive work recently performed by the Stata community to provide users with a easy-to-apply set of commands for carrying out regularized methods, we will devote special attention to the Stata implementation. We will present several applications using the Stata built-in LASSO package, the community-distributed package lassopack, and some Stata wrappers of Python and R functions to carry out techniques not developed in Stata. Stata implementations will also cover inferential Lasso and regularized regression with time-series and panel data for optimal forecasting purposes. The R implementation part presents the glmnet package, an all-encompassing package to fit regularized regression/classification. This package, that is an extension of the popular glm package to fit generalized linear models, covers a large set of distributional

G. Cerulli, *Fundamentals of Supervised Machine Learning*, Statistics and Computing, https://doi.org/10.1007/978-3-031-41337-7_3

assumptions about the target variables. In particular, unlike the Stata LASSO suite, `glmnet` allows also for estimating multinomial regularized classification. Finally, the Python applications will make use of the Scikit-learn general Machine Learning platform to run specialized penalized regression and classification models. Specifically, we will focus on the function (or method) `GridSearchCV()` for easily carrying out K-fold cross-validation in a Python environment.

3.2   Model Selection and Prediction

As emphasized in the introductory chapter, statistical modeling can incur two different types of errors: the sampling error and the specification error (Theil, 1957; Wold & Faxer, 1957). The sampling error occurs as we consider just a portion (the sample) of the entire population, while the specification error depends on misspecifying the model whenever our specification choice is different from the true data generating process (DGP). The sampling error can be generally "cured" by increasing the sample size, thus reducing the uncertainty of parameters' estimates and increasing statistical tests' power. But the pernicious consequences of the specification error are more subtle, as they can imply either inconsistent parameter estimates ("bad" inference), or model-reduced predictive power. Let's focus here on the latter.

Among various types of specification errors, the most common concerns an incorrect number of variables to consider as predictors in a regression (or classification) model. From a model predictive performance (thus abstracting here from inferential considerations), what are the consequences of this type of specification error? To reply to this question, consider the traditional linear regression with p predictors:

$$y = \beta_0 + \beta_1 x_1 + \beta_2 x_2 + \cdots + \beta_p x_p + \epsilon \qquad (3.1)$$

Assuming that $x_1, \ldots, x_p$ are exogenous, i.e., uncorrelated with the error term ϵ, the model's prediction is defined as

$$E(y|x_1, \ldots, x_p) = \beta_0 + \beta_1 x_1 + \beta_2 x_2 + \cdots + \beta_p x_p \qquad (3.2)$$

An increase in the number of predictors (a larger p) increases the model's degrees of freedom (i.e., model complexity) thus reducing prediction bias at the expense of an increase in the variance of the estimated parameters. This latter effect produces an increase in the prediction variance, as the prediction is clearly a function of the estimated parameters. Increasing the number of parameters (number of predictors) thus entails a bias–variance trade-off requiring, as stressed in the previous chapter, an optimal tuning of the model based on an appropriate loss function minimization.

Optimal parameters' tuning may suggest selecting only a few of the many initial candidate predictors by including in the model only those providing a larger predictive power. Methods like Lasso, Elastic-net, and Subset selection follow this principle. Nonetheless, for predictive performance to increase, it may be not necessary to reduce the number of predictors to include in the model to a smaller set. It may be sometimes better to just shrink regression parameters (also called "weights", in the Machine Learning jargon) so as to give larger weight to the features lending larger predictive accuracy. This is the philosophy laying behind the Ridge estimator, as we will discuss in the following sections.

3.3 Prediction in High-Dimensional Settings

Regularized regressions are popular statistical learning methods as they allow to estimate (generalized) linear models also in the case of high-dimensional data, i.e., cases where $p \gg N$ (Hastie et al., 2015). Consider again the linear regression of Eq. (3.1). We may have in this case three different situations:

- $p \ll N$: all the parameters are point-identified, and least-squares estimates have low variance. Prediction is feasible, and test accuracy can be acceptable.
- $p \cong N$: all the parameters are point-identified, and least-squares estimates have high variance. Prediction is feasible, but test accuracy may be poor.
- $p \gg N$: all the parameters are not identified, least squares have no unique solution, and coefficient estimates have infinite variance. Prediction is infeasible.

By constraining regression coefficients, regularized regression restores prediction feasibility also in the case of high-dimensional data. Moreover, as running regularized regression generally shrinks estimates' variance at the expense of a small increase in their bias, it allows for larger prediction accuracy even when $p \approx N$. In the literature, various types of regularized methods have been proposed. The most popular are Lasso and Ridge regressions, with the Elastic-net representing a compromise between the two. The next section provides a short introduction to regularized linear models.

3.4 Regularized Linear Models

As suggested in the previous section, least-squares regression may be unsatisfactory whenever we want to increase model predictive performance, especially if p is close or even larger than N (high-dimensional data), and would like to carry out feature selection. Regularized regression can offer a solution on these issues (Hastie et al., 2009).

To set the stage, consider the least-squares optimization problem. As known, least squares minimize the residual sum of squares (RSS) over the parameter vector β. In this case, the RSS formula is

$$\text{RSS}(\beta) = \sum_{i=1}^{N} \left(y_i - \beta_0 + \sum_{j=1}^{p} \beta_j x_{ij} \right)^2 \qquad (3.3)$$

Regularized (linear) regression minimizes a penalized version of Eq. (3.3), by adding up a penalization function $P(\beta)$:

$$\text{RSS}(\beta) + \lambda P(\beta) \tag{3.4}$$

where λ is a (non-negative) penalization parameter that increases the degree of penalization as long as it gets larger. Depending on the form of the penalization function, we may obtain different regularization approaches. Among all possible forms, the literature has focused on three popular regularized approaches: Ridge, Lasso, and Best subset selection. Specifically, we obtain the Ridge estimator by assuming

$$P(\beta) = \sum_{j=1}^{p} \beta_j^2 \tag{3.5}$$

We obtain the Lasso by assuming

$$P(\beta) = \sum_{j=1}^{p} |\beta_j| \tag{3.6}$$

Finally, we obtain the Best subset selection by putting

$$P(\beta) = \sum_{j=1}^{p} I(\beta_j \neq 0) \tag{3.7}$$

where $I(A)$ is an indicator function taking value one if A is true, and zero otherwise. It is useful to observe that the penalization term depends on the type of parameters' vector norm assumed to penalize the least squares. A general form for the penalization term is $\sum_{j=1}^{q} |\beta_j|^q$, with the Ridge penalization obtained for $q = 2$, and the Lasso for $q = 1$. Other forms of penalization can be obtained by different choices of $q > 0$, as visualized in Fig. 3.12 of Hastie et al. (2009) that plots the contours of the penalization functions at various values of q. In the next sections, we provide a short account of the three regularizing schemes cited above and discuss their different characteristics.

3.4.1 Ridge

In matrix form, the least-squares RSS can be written as $(\mathbf{y} - \mathbf{x}\beta)'(\mathbf{y} - \mathbf{x}\beta)$. By adding a quadratic penalty ($q = 2$), the Ridge regression penalized RSS is equal to $(\mathbf{y} - \mathbf{x}\beta)'(\mathbf{y} - \mathbf{x}\beta) + \lambda\beta'\beta$ (Hoerl et al., 2007; Gorman & Toman, 1966). If we derive the last expression over β, it is straightforward to show that

$$\widehat{\beta}_{ridge}(\lambda) = (\mathbf{x}'\mathbf{x} + \lambda\mathbf{I})^{-1}\mathbf{x}'\mathbf{y} \tag{3.8}$$

The constant $\lambda \geq 0$ is the *penalization* parameter and allows for identifying a family of models, all indexed by different choices of λ. Interestingly, when $\lambda = 0$ the Ridge solution coincides with the ordinary least-squares estimate, and when $\lambda \to \infty$ all the coefficients approach zero, although they are not exactly zero. Compared to least squares, Ridge's parameters are smaller in size, the reason why models such as that are also called *shrinking* regressions, with λ known as the *shrinking* parameter.

As with any other regularizing approach, the selection of λ is central in Ridge regression. This is a flexibility parameter and, as such, it induces a bias–variance trade-off. When λ is small and close to zero, the Ridge estimates approach the least-squares estimates, and the model is highly flexible as containing all the parameters with the largest size. In this case, the prediction variance (which depends also on the size of the coefficients) is high, but the bias is small. As long as λ increases, the model becomes less flexible, as the size of the coefficients shrinks. In this case, the prediction variance decreases but at the expense of a larger bias. As showed by Hoerl & Kennard (1970), there always exists a $\lambda > 0$ such that the MSE performed by the Ridge is less than that of the least-squares estimate. In this sense, the Ridge regression is generally preferable to least squares when the objective is pure prediction.

Ridge regression allows for prediction improvement but has the drawback of not being a feature selector. Even if it shrinks coefficients toward zero, the model still includes all of the features (except when $\lambda = \infty$, in which case all predictors are equal to zero). In comparison, the Lasso approach—by setting some coefficients exactly equal to zero—has the advantage to perform feature selection, by still maintaining a better prediction performance compared to least squares. In a few words, Lasso has larger interpretability when compared to Ridge.

3.4.2 Lasso

Lasso (least absolute shrinkage and selection operator) regression (Tibshirani, 1996) is a regularization method obtained when the penalized RSS assumes the following form:

$$(\mathbf{y} - \mathbf{x}\beta)'(\mathbf{y} - \mathbf{x}\beta) + \lambda \sum_{j=1}^{p} |\beta_j| \tag{3.9}$$

Compared to Ridge regression, Lasso is a feature selector as the minimization of its penalized RSS generates a *sparse* solution, i.e., a solution where some of the estimated parameters are exactly equal to zero. Unlike Ridge, however, computing the Lasso coefficients is more complicated as the minimization of the Lasso objective function entails a nonconvex optimization problem. In general, there is not a close solution to the Lasso problem (Efron et al., 2004). As in the case of the Ridge regression, λ is the flexibility parameter entailing a bias–variance trade-off. Also for Lasso, in fact, when λ increases, the model gets less flexible exhibiting higher bias, but smaller variance. In contrast, when λ gets smaller, the model becomes more flexible, the bias reduces, but at the expense of a larger prediction variance. As for the Ridge regression, when $\lambda = 0$ the Lasso coincides with the least squares, but when $\lambda = \infty$ no variables are retained in the model, thus producing an empty degenerate model.

Although popular, the Lasso regression presents a couple of important drawbacks. First, in high-dimensional data, where $p \gg N$, the Lasso can select at most N features. Second, in the presence of grouped variables, i.e., variables generally highly correlated, the Lasso has a tendency to select one variable from each group neglecting all the others (Efron et al., 2011).

3.4.3   Elastic-Net

Partly motivated to overcome Lasso's limitations, Elastic-net proposed by Zou and Hastie (2005) is a penalized regression entailing a compromise between Ridge and Lasso regressions. More specifically, the Elastic-net penalization term is a weighted average (with weight α) of the Lasso and the Ridge penalizations:

$$P(\beta) = \alpha \sum_{j=1}^{p} \beta_j^2 + (1 - \alpha) \sum_{j=1}^{p} |\beta_j| \qquad (3.10)$$

Elastic-net owns the benefits of both methods, without inheriting their drawbacks. On the one hand, the Elastic-net is—like Lasso—a model selector, as it provides corner solutions, and thus coefficients exactly equal to zero; on the other hand—like Ridge—it shrinks the coefficients of correlated features, without dropping some of them out by default. Finally, it accommodates models that are intermediate between Ridge and Lasso, without relying on generalized shrinking regression with $q \in (0, 1)$, whose computational solution may be problematic.

3.5   The Geometry of Regularized Regression

There is an interesting and popular geometrical representation of the Lasso, Ridge, and Elastic-net solutions when we look at them with the lens of constrained optimization (Hastie et al., 2015). Assuming to have only two regression parameters, β_1 and β_0, the Lasso, Ridge, and Elastic-net solution problems can be suitably reformulated as a constrained optimization problem. As for the Lasso, we have

$$\min_{\beta_1, \beta_2} \left\{ \sum_{i=1}^{N} (y_i - \beta_1 - \beta_2 x_i)^2 \right\} \quad \text{s.t.} \quad \sum_{j=1}^{2} |\beta_j| \leq \tau \qquad (3.11)$$

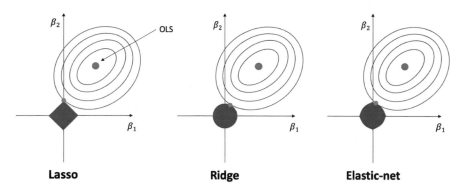

Fig. 3.1 Comparison of Lasso, Ridge, and Elastic-net penalization solutions

The Ridge optimal parameters are obtained by

$$\min_{\beta_1,\beta_2} \left\{ \sum_{i=1}^{N}(y_i - \beta_1 - \beta_2 x_i)^2 \right\} \text{ s.t. } \sum_{j=1}^{2}\beta_j^2 \le \tau \qquad (3.12)$$

Finally, the Elastic-net optimization problem can be represented as

$$\min_{\beta_1,\beta_2} \left\{ \sum_{i=1}^{N}(y_i - \beta_1 - \beta_2 x_i)^2 \right\} \text{ s.t. } \alpha \sum_{j=1}^{2}\beta_j^2 + (1-\alpha)\sum_{j=1}^{2}|\beta_j| \le \tau \qquad (3.13)$$

In these representations, τ is the constraining bound and there exists a one-to-one relationship between this parameter and the penalization parameter λ, with τ increasing when λ decreases. Figure 3.1 shows a graphical representation of the solution of the previous three problems. In every panel, we plot the contour plots of the RSS (ellipses), and of one of the constraining functions when the parameter τ is set to a specific value. The least-squares solution is placed at the center of each ellipse, and the RSS expands outwards by increasing its value (its minimum is at the least-squares solution). The Lasso performs corner (or angle) solutions, with the parameter β_1 set exactly equal to 0. The Ridge parameters are smaller in size compared to the least-squares parameters, but they are never exactly equal to zero. The Elastic-net solution may be, under certain circumstances, an angle or quasi-angle solution (as in this case). Observe that, by enlarging τ, the area of the constrained contour plots magnifies, finally including the least-squares solution. This is when $\lambda = 0$, and no penalization is binding.

3.6 Comparing Ridge, Lasso, and Best Subset Solutions

Unlike Ridge, both Lasso and Best subset selection do not own an explicit solution.
Comparing the coefficients provided by these three methods is thus tricky. There
is however a special case in which comparison is possible, and this is the case of
orthonormal features, i.e., when the matrix of features **x** is unit diagonal (Wieringen,
2020; Hastie et al., 2009). This is the case in which all the features are uncorrelated
with the same unit of measurement (i.e., unit norm). In this special (and unrealistic)
case, it can be shown that the Ridge, Lasso, and Best subset selection coefficients are
transformations of the least-squares coefficients, with their difference depending only
on applying a different type of transformation. Specifically, these transformations are

$$\widehat{\beta}_j^{\text{ridge}} = (1 + N\lambda)^{-1}\widehat{\beta}_j^{\text{ols}} \tag{3.14}$$

$$\widehat{\beta}_j^{\text{lasso}} = \widehat{\beta}_j^{\text{ols}} \max\left(0; 1 - \frac{N\lambda}{|\widehat{\beta}_j^{\text{ols}}|}\right) \tag{3.15}$$

$$\widehat{\beta}_j^{\text{subset}} = \widehat{\beta}_j^{\text{ols}} \text{I}\left(|\widehat{\beta}_j^{\text{ols}}| \geq \sqrt{N\lambda}\right) \tag{3.16}$$

Figure 3.2 sets out the graphical representation of the three solutions. In the
y-axis, we have the regularized solution, while in the x-axis the least-squares coefficient $\widehat{\beta}_j^{\text{ols}}$. Thus, the 45° line colored in red directly represents the least-squares
solution. It is easy to see that Ridge regression shrinks coefficients by a proportional
constant factor $(1 + N\lambda)^{-1}$ and no coefficients are set exactly equal to zero. Lasso
applies to the least-squares coefficients the so-called "soft thresholding" transformation. This transformation pushes the coefficients' value toward zero, by setting them
exactly equal to zero when they are smaller than a certain threshold. Best subset
selection, finally, applies to the least-squares coefficients the so-called "hard thresholding" function, as coefficients are the same as the least squares, but then put to
zero after a certain threshold is reached.

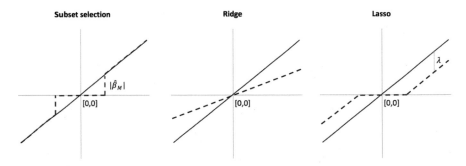

Fig. 3.2 Lasso, Ridge, and Subset selection solutions in the case of orthonormal features

3.7 Choosing the Optimal Tuning Parameters

Previous discussion makes it clear that the choice of the hyper-parameter λ (and also α in the case of the Elastic-net) is a central task. Parameters' estimation may be severely affected by an unwise choice of the tuning hyper-parameters, resulting in poor prediction performance of the regularization approaches so far considered. As widely discussed in the previous chapter, the general criterion followed for achieving optimal hyper-parameters' tuning is that of minimizing the out-of-sample prediction error of the model. In this direction, the K-fold cross-validation (CV) is an all-encompassing approach valid for pretty much all Machine Learning approaches, and regularized regressions are not an exception in this vein. However, as regularized regressions are part of the larger class of generalized linear models (GLMs), a less computationally expensive alternative is available in this case, namely using information criteria (AIC, BIC, EBIC, and Adjusted-R^2). Therefore, CV and information criteria can be suitably used for any Elastic-net regression, including the Ridge and Lasso sub-cases. In the case of Lasso, however, other two model selection options are available in the literature, i.e., adaptive Lasso, and plugin estimation. It is useful to present both.

3.7.1 Adaptive Lasso

Proposed by Zou (2006), the adaptive Lasso is a weighted variant of the standard Lasso, with the penalization term including coefficient-specific penalty weights ω_j:

$$\widehat{\beta}_{\text{adalasso}} = \arg\min_{\beta} \left\{ \sum_{i=1}^{n} (y_i - \mathbf{x}_i \beta)^2 + \lambda \sum_{j=1}^{p} \omega_j |\beta_j| \right\}$$

The weights are chosen so that variables with higher (lower) coefficients have smaller (higher) weights. A possible choice of the weighting scheme is $\omega_j = 1/|\tilde{\beta}_j|^\delta$, with $\tilde{\beta}_j$ obtained from a first-step least squares (or Ridge, or Lasso regression), and δ a power to choose beforehand (typically equal to one). In general, adaptive Lasso estimation is carried out by a pathwise approach. This entails estimating sequentially multiple Lasso regressions. After each Lasso, variables with zero coefficients are dropped out, and a new Lasso is estimated using weights derived from the previous Lasso estimation. The finite number of steps is decided in advance. Typically, this procedure leads to select fewer predictors than CV.

The motivation for adopting adaptive Lasso resides in two related aspects. First, as a feature selector, the Lasso solution highly depends on the size of the coefficients. Lasso can be too sensitive in retaining larger coefficients, and too prone to discard coefficients with a smaller size but that can be relevant to keep in the model. An inverse weighting scheme is a way for attenuating this undesirable property of the

Fig. 3.3 Adaptive Lasso
solution

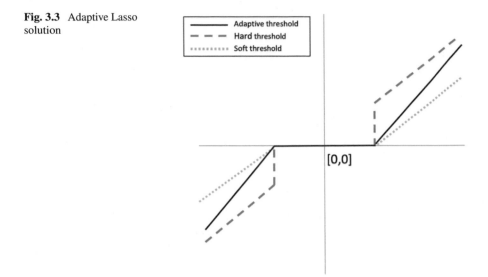

Lasso solution. Second, Lasso coefficients can be sometimes very different from those of the least squares (with the comparison possible only in low-dimensional settings, of course). This may lead to a model that, although more predictive than least squares, can be poorly interpretable. Attenuating the distance between Lasso and least-squares estimates can thus be desirable in many analytical contexts. This property of the adaptive Lasso can be better caught in Fig. 3.3, where the adaptive Lasso solution is compared—in the case of orthonormal features—with standard Lasso (*soft threshold*), and Best subset selection (*hard threshold*). It is easy to see that the adaptive Lasso configures as a compromise between the Lasso and Best subset selection, getting closer to the least squares as long as the coefficient increases in modulus.

3.7.2 Plugin Estimation

The plugin estimation (or rigorous penalization) is a theory-driven approach to find the Lasso optimal penalization parameter and penalization weights. Unlike cross-validation, the plugin does not rely on a pure computational approach, but derives analytic formulas for λ and ω_j based on Lasso's asymptotic theory. This theory provides conditions for optimal rate of convergence for prediction and parameter estimation in a Lasso regression, and assures to select models whose size (number of nonzero coefficients) is of the same order as the true model. The most relevant of these conditions is the so-called *approximate sparsity*. The literature on high-dimensional linear models typically assumes *exact sparsity*, which means that (1) in the true model, only s out of the p variables considered in the regression have coefficients truly different from zero, with $s \ll N$; (2) the linear approximating function of the true (and unknown) conditional mean population has a zero approximation error.

Following Belloni et al. (2014a), indicate the true conditional expectation of y by $f(\mathbf{w}_i)$, where $\mathbf{w}_i$ are elementary variables (single distinct features). A linear sparse approximation of $f(\cdot)$ is $\mathbf{x}_i \beta_0$ where $\mathbf{x}_i$ are transformations of $\mathbf{w}_i$ (as, for example, polynomial functions), and β_0 is the target parameter to estimate with only $s \ll N$ components different from zero. The approximation error is thus

$$r_i = f(\mathbf{w}_i) - \mathbf{x}_i \beta_0 \tag{3.17}$$

Approximate sparsity allows this error to be small, but not exactly zero. It is a less stringent assumption than exact sparsity, which assumes $r_i = 0$. Besides approximate sparsity, Lasso asymptotic properties rely on two further fundamental conditions, namely (i) restricted sparse eigenvalue condition and (ii) regularization event (Belloni et al., 2012). Given these conditions, the Lasso estimation of β_0 minimizes the approximation error when λ and ω_j are chosen properly. By assuming no specific distribution of the error in the Lasso equation, for the case of homoskedastic errors we have

$$\lambda = 2c\sigma\sqrt{N}\Phi^{-1}\left(\frac{1-\gamma}{2p}\right) \quad \omega_j = \sqrt{1/n \sum_i x_{ij}^2} \tag{3.18}$$

while for heteroskedastic errors:

$$\lambda = 2c\sqrt{N}\Phi^{-1}\left(\frac{1-\gamma}{2p}\right) \quad \omega_j = \sqrt{1/n \sum_i x_{ij}^2 \epsilon_i^2} \tag{3.19}$$

The parameter c has a fixed value, and γ requires to converge toward zero when N increases. Typical choices of these parameters are $c = 1.1$, and $\gamma = 0.1/log$ $(\max[p, N])$. Estimating σ (i.e., the error variance) and the linear heteroskedastic penalty loadings follows two specialized iterative algorithms described in Belloni & Chernozhukov (2011) and in Belloni et al. (2014b).

3.8 Optimal Subset Selection

We have seen that Lasso applies to the least-squares coefficients the so-called "soft thresholding" transformation, pushing the coefficients' value toward zero, or exactly to zero when the coefficients are smaller than a certain cutoff. Best subset selection (Hocking & Leslie, 1967) applies to the least-squares coefficients the so-called "hard thresholding" function. In this case, coefficients are the same as the least squares, but some are put to zero after a certain threshold is reached. Carrying out Best subset selection entails estimating a total of 2^p models, to then find among them the best one. Unfortunately, as it is not computationally straightforward to select the

best model from among so many possible alternatives, less computationally intensive strategies known as *stepwise approaches* have been proposed in the literature.
In the next sections, we present three popular procedures for optimal subset selection: (i) Best (or exhaustive) subset selection, (ii) Forward stepwise selection, and
(iii) Backward stepwise selection. The exposition of these subjects draws on Hastie
(2009, Chap. 3).

3.8.1   Best (or Exhaustive) Subset Selection

An exhaustive approach explores all the grid of possible models that we can estimate
with p features available. This means that we have to consider all p models of size one
(i.e., containing only one predictor), all $\binom{p}{2} = p(p-1)/2$ of size two (i.e., containing
only two predictors), and so forth. The total number of models to estimate is 2^p. Just to
give an idea of the computational burden entailed by an exhaustive search, with $p =$
10 we have to estimate 1,024 models, with $p = 20$ the estimation burden increases
dramatically to 1,048,576, and with $p = 30$ it reaches the insurmountable number
of 1,074,000,000 models to run. In a regression with many interactions, reaching a
number of predictors larger than 30 is however common. As a consequence, running
Exhaustive subset selection may be infeasible (Hocking & Leslie, 1967). Table 3.1
sets out the procedure to apply Exhaustive stepwise selection. As clearly visible, we
proceed in two steps: (i) in the first step, we compare models of the same size using
standard in-sample goodness of fit measures, as the R^2 or the RSS. This is a correct
procedure, as we are comparing models of the same complexity (or flexibility),
having in fact the same number of degrees of freedom; (ii) in the second step, we
compare models with different size, entailing that we need to adjust for a different
flexibility when comparing the predictive performance of each model; in this case,
we rely on information criteria, or cross-validated prediction accuracy to pick up the
final model with optimal size.

Table 3.1 Description for the Exhaustive subset selection algorithm

Procedure. *Exhaustive stepwise selection*
1. Let $\mathcal{S}_0$ be the model with no features

2. For $k = 1, ..., p$:
     (a) Consider all $\binom{p}{k}$ with k features
     (b) Among all these $\binom{p}{k}$ models, select the *best* one and call it $\mathcal{S}_k$
         Best model: the one with smallest RSS, or highest R^2

3. Among $\{\mathcal{S}_0, \ldots, \mathcal{S}_p\}$, pick the best model using cross-validated prediction   accuracy, or information criteria

The previous procedure can be easily extended to categorical, countable, or fractional outcomes as long as one implements a generalized linear model (GLM) version of the two-step procedure. In this case, for the first-step models' comparison, instead of using the R^2, one can use $-2\log(L_S)$, i.e., the deviance of model S that is equal to minus twice the negative log-likelihood of the model, where L_S is the maximum log-likelihood reached by the maximum-likelihood fit. Similarly, in the second step, instead of using for example Mallow's C_p information criterion, one can use the Akaike information criterion (AIC) or the Bayesian one (BIC), or a similar test-accuracy cross-validation approach.

3.8.2 Forward Stepwise Selection

Stepwise approaches can reduce sensibly the computational burden implied by running an exhaustive search (Roecker, 1991). This takes place through a guided search that explores only a portion of all the possible options of the grid. Although computationally feasible, Forward stepwise selection (as well as the Backward one) does not assure to ultimately select the optimal model, i.e., the one we would have selected had the exploration run exhaustively. As the leading criterion, stepwise approaches perform a sort of *competition* among models at every model size, where the winning model is retained as the basis of the subsequent step, where a new competition takes place.

The Forward stepwise selection starts with a *null* model containing zero features and adds one feature at a time, retaining the model containing the added feature and that performs better than the others. As in the case of exhaustive search—at the same model size (*first step*)—comparison is made using in-sample prediction criteria, whereas at different sizes (*second step*), we use information criteria or cross-validated prediction accuracy. The number of models fitted by the Forward stepwise selection is $1 + p(p + 1)/2$, which entails a tremendous reduction in a computational burden compared to the exhaustive search, as visible in Table 3.3. Table 3.2 illustrates the Forward stepwise selection procedure. However, to better understand the selection process operated by this algorithm, we focus on the illustrative example of Table 3.4 referring to the case of only $p = 4$ predictors. Let's comment on this example.

Starting from step 1, we observe that for $k = 0$ four models with only one predictor are estimated. Among them, the model containing only the feature x_3 wins the competition (thus marked with "*"). As a consequence, for $k = 1$, we can form three models made of two features and containing x_3. Among these models, the winning one is (x_3, x_1), and at $k = 3$ we consider only the two models made of three features containing (x_3, x_1). Among these models, the winner is (x_3, x_1, x_4). The last model is (x_3, x_1, x_2, x_4), containing all features. In step 2, we compare the four optimal models spotted at each size k. We see that—based on a pre-specified out-of-sample fitting criterion—the final winner is model (x_3, x_1, x_4) resulting in the best model among the estimated eleven ones.

Table 3.2 Description for the Forward stepwise selection algorithm

Procedure. *Forward stepwise selection*
1. Let S_0 be the model with no features
2. For $k = 0, \ldots, p - 1$: 　　　　(a) Consider all $p - k$ models augmenting S_k by one additional predictor 　　　　(b) Among these $p - k$ models, select the *best* one and call it S_{k+1} 　　　　　　　Best model: the one with smallest RSS, or highest R^2
3. Among $\{S_0, \ldots, S_p\}$, pick the best model using cross-validated prediction 　　accuracy, or information criteria

Table 3.3 Comparison of the computational burden of optimal subset selection algorithms

Number of features	Exhaustive	Forward	Backward
5	32	16	16
10	1,024	56	56
15	32,768	121	121
20	1,048,576	211	211
25	33,554,432	326	326
30	1,073,741,824	466	466
35	34,359,738,368	631	631
40	1.09951E + 12	821	821
45	3.51844E + 13	1,036	1,036
50	1.1259E + 15	1,276	1,276

Table 3.4 Example of a Forward stepwise selection procedure with $p = 4$ features

Step 1. Best model for every size k			
$k = 0$	$k = 1$	$k = 2$	$k = 3$
Model: x_1	Model: x_3, x_1 (*)	Model: x_3, x_1, x_2	Model: x_3, x_1, x_2, x_4 (*)
Model: x_2	Model: x_3, x_2	Model: x_3, x_1, x_4 (*)	
Model: x_3 (*)	Model: x_3, x_4		
Model: x_4			
Step 2. Best model among different model sizes			
Model: x_3	Model: x_3, x_1	Model: x_3, x_1, x_4 (*)	Model: x_3, x_1, x_2, x_4

We can adopt Forward stepwise selection also for high-dimensional data, which is also when $p \gg N$, although in this case the largest model that we can fit has size $N - 1$. This depends on the Forward stepwise approach to use least squares (or GLMs for non-Gaussian target variables) that require a number of features larger than the number of observations. When using a GLM, as seen for the exhaustive case, the deviance replaces the R^2 (first step), and the AIC Mallow's C_p (second step).

3.8.3 Backward Stepwise Selection

Backward stepwise selection works similar to its Forward companion, but it starts from considering in step 1 the *full* model, i.e., the model made of all p predictors. Subsequently, it forms all the models of size $p - 1$ selecting the best performing one among them according to a pre-set in-sample prediction criterion. By considering only the features of this model, the procedure goes on by building all the possible models of size $p - 2$, finding the best, and then replicating the competition until features are no longer available. In step 2, Backward stepwise selection puts in competition all models that were selected at every size. As in the Forward case, the best model is selected according to the best out-of-sample prediction accuracy (i.e., via information criteria or cross-validation). Table 3.5 illustrates in detail the Backward stepwise selection procedure.

Table 3.6 sets out an illustrative application with four predictors. It is easier here to understand how the algorithm actually works. At the very beginning of step 1, we estimate the full model (x_1, x_2, x_3, x_4), and then we form all the possible models of size three, based on the previous four features. We select model (x_1, x_3, x_4), and then form all possible models of size two based on (x_1, x_3, x_4). As the winner at this run is model (x_1, x_3), we compare only two models of size one, i.e., (x_1) and (x_3) which is the winner. In step 2, we compare the four models spotted as optimal in step 1, and

Table 3.5 Description for the Backward stepwise selection procedure

Procedure. *Backward stepwise selection*

1. Let $\mathcal{S}_p$ be the full model made of all p features

2. For $k = p, p - 1, \ldots, 1$:
        (a) Consider all k models containing all but one of the features in $\mathcal{S}_k$
        (b) Among these k models, select the *best* one, and call it $\mathcal{S}_{k-1}$
           Best model: the one with smallest RSS, or highest R^2

3. Among $\{\mathcal{S}_0, \ldots, \mathcal{S}_p\}$, pick the best model using cross-validated prediction accuracy, or information criteria

Table 3.6 Example of a Backward stepwise selection procedure with $p = 4$ features

Step 1. Best model for every size k			
$k = 4$	$k = 3$	$k = 2$	$k = 1$
Model: x_1, x_2, x_3, x_4 (*)	Model: x_1, x_2, x_3	Model: x_1, x_3 (*)	Model: x_1
	Model: x_1, x_2, x_4	Model: x_1, x_4	Model: x_3 (*)
	Model: x_1, x_3, x_4 (*)	Model: x_3, x_4	
	Model: x_2, x_3, x_4		
Step 2. Best model among different model sizes			
Model: x_1, x_2, x_3, x_4	Model: x_1, x_3, x_4	Model: x_1, x_3 (*)	Model: x_3

select as the final winner (x_1, x_3) according to its out-of-sample prediction accuracy. Different from the Forward approach, Backward stepwise selection requires that $n > p$, and thus it is not well suited for high-dimensional data. Finally, also this approach can accommodate non-Gaussian outcomes by carrying out a GLM version of the two-step procedure.

3.9   Statistical Properties of Regularized Regression

Although it is not our intention to provide a complete account of the huge literature on the statistical properties of regularized regression models, it is useful to set out some important findings that can help us understand the properties of the methods reviewed in this chapter. This allows us to compare the different methods and is of support with regard to the inferential part that we will cover in Sect. 3.10.

By mainly referring to Lasso, we focus here on two related (asymptotic) properties of regularized regressions, i.e., prediction *rate of convergence* and *support recovery*. Both are key properties. Based on asymptotic theory, theorems on regularized regressions' rate of convergence provide guidance to establish both prediction-consistency and the speed at which such consistency is achieved. In a related manner, support recovery concerns the ability (in particular for the Lasso) to select *only* the predictors belonging to the true data generating process (DGP). We refer, in this case, to the model's *specification-consistency*.

Highly based on the works of Hastie et al. (2015); Bühlmann & van de Geer (2011); Loh & Wainwright (2017); Geer & Bühlmann (2009); Wang et al. (2014), and Fan & Li (2001), the next two subsections present a few important results on the prediction rate of convergence and support recovery for regularized regressions. The third and last subsection presents the *oracle property* and its relation to the Lasso regression.

3.9.1 Rate of Convergence

To begin with, consider the standard linear regression model in (3.1). In matrix form, it can be written as $\mathbf{y} = \mathbf{x}\beta + \epsilon$, where we assume exogeneity of $\mathbf{x}$, and normality of the error term having zero mean and variance σ^2. Under these assumptions, it can be showed that the in-sample prediction error (or simply *risk*) of the least-squares estimator is

$$\frac{1}{N}\mathrm{E}\|\mathbf{x}\widehat{\beta}_{\mathrm{ols}} - \mathbf{x}\beta\|_2^2 = \sigma^2 \frac{p}{N} \tag{3.20}$$

Consider now a high-dimensional setting, where however the true underlying coefficients have support $S = \mathrm{supp}(\beta)$, with $s_0 = |S| < p$ (i.e., we assume model *sparsity*). Then, the estimator provided by the least squares on S is

$$\widehat{\beta}_{\mathrm{oracle}}^S = (\mathbf{x}_S'\mathbf{x}_S)^{-1}\mathbf{x}_S'\mathbf{y}$$
$$\widehat{\beta}_{\mathrm{oracle}}^{-S} = 0$$

that is called *oracle estimator*. Based on least-squares results, the in-sample risk of this estimator is

$$\frac{1}{N}\mathrm{E}\|\mathbf{x}\widehat{\beta}_{\mathrm{oracle}} - \mathbf{x}\beta\|_2^2 = \sigma^2 \frac{s_0}{N} \tag{3.21}$$

In this setting, Foster & George (1994) showed that for the Best subset selection we have

$$\frac{1}{N}\mathrm{E}\|\mathbf{x}\widehat{\beta}_{\mathrm{subset}} - \mathbf{x}\beta\|_2^2 \leqslant s_0 \frac{\log(p)}{N} \tag{3.22}$$

as N and p go to infinity. This is an important finding, as it shows that, for a fixed p, Best subset selection has a rate of convergence of the same order of the oracle estimator. We say that the Best subset selection exhibits an *optimal* rate of convergence. Unfortunately, when it comes to the Lasso, we have that

$$\mathrm{E}\|\mathbf{x}\widehat{\beta}_{\mathrm{lasso}} - \mathbf{x}\beta\|_2^2 \leqslant s_0 \sqrt{\frac{\log(p)}{N}} \tag{3.23}$$

which signals a much lower rate of convergence compared to the optimal one. Faster rate of convergence for Lasso prediction can be achieved, but at the cost of further assumptions on the DGP, and in particular about the degree of correlation of the predictors. There are two versions of the Lasso faster rate property, one under the assumption that $\mathbf{x}$ satisfies the *compatibility condition*, and the other under the so-called *restricted eigenvalue condition*. Both provide a rate of convergence for Lasso of the same order of the Best subset selection. Unfortunately, both conditions entail that the truly active predictors must not be too correlated, a condition that is not common to find in usual applications, especially when the size of p is large, and predictors measure similar or related phenomena.

3.9.2 Support Recovery

We wonder whether the Lasso, as a model selector, is able to recover the features belonging to the true model DGP. More specifically, we would like to know under what conditions a Lasso regression selects the true support with the correct sign of the coefficients (i.e., support recovery). Wainwright (2009) proposed a proof for support recovery called the *primal-dual witness method*. Without entering into technical details, it is sufficient here to state that Lasso may exhibit support recovery under three assumptions: (i) *minimum eigenvalue*, (ii) *mutual incoherence*, and (iii) *minimum signal*. All these three assumptions restrict strongly the correlation structure of the model predictors. By and large, the minimum eigenvalue condition imposes some limitations on the correlations of the features, with particular emphasis on those belonging to the underlying support S. The mutual incoherence states, approximately, that the inactive variables must not be well-explained by the true active variables. The minimum signal condition, lastly, requires that the coefficients in the true DGP have to own a sufficiently high size to be detected and thus selected. It is thus clear that the restrictions needed for the Lasso to achieve support recovery are much tighter than those required for faster rate of convergence, leading to conclude that it is unlikely that Lasso is able to provide specification-consistency when both N and p go to infinity.

3.9.3 Oracle Property

As a model selector, the Lasso is not in general specification-consistent. An interesting exception is however represented by the Adaptive Lasso proposed by Zou (2006). By assuming a fixed p, the author shows that, if the initial estimator used to compute the adaptive weights is $\sqrt{n}$-consistent for the regression coefficients (so that, a least squares is sufficient), then the Adaptive Lasso satisfies the so-called *oracle property* for feature selection and coefficients' estimate. More specifically, the oracle property entails that (i) the model is able to select the exact coefficients' support S; (ii) the estimated parameters are $\sqrt{n}$-consistent and converge in distribution to the same (normal) distribution of the oracle estimator (which is the least-squares estimator on the correct support). Zou (2006) shows that, while Lasso does not own the oracle property, the Adaptive Lasso does. A limitation of this model is that it applies only when N goes to infinity with p held fixed. Therefore, this finding cannot be extended to high-dimensional asymptotic settings where both N and p go to infinity. Heuristically, however, it lends support to the higher robustness of the Adaptive Lasso over the standard Lasso when specification-consistency may be an issue.

3.10   Causal Inference

In the previous section, we stated that Lasso does not in general enjoys the oracle property. It means that it is unsuitable to select the true model with probability one. The model selected by Lasso may or may not contain variables belonging to the true model. Likewise, Lasso may include variables that are not part of the true model. Why does it occur? The intuitive reason is that Lasso is sensitive to variables whose contribution to the reduction of the prediction error is large, independent of whether these variables belong to the true model. These variables are those highly correlated to the outcome. Figure 3.4 sets out an example whose results are simulated using a sample of 1000 observations. We consider a small structural model, where the target variable y is determined by the features x_2 and x_3 through the (true) equation $y = ax_2 + ax_3 + \epsilon_y$, where all the RHS variables are standard normal. The variable x_1 is the only exogenous variable of the model (as not receiving any head of at least one arrow), directly determining x_2 and x_3, while not having a direct effect on y. Hence, x_1 is not part of the true model of y. We also assume that $x_2 = bx_1 + \epsilon_2$ and $x_3 = bx_1 + \epsilon_3$. All the error terms are assumed standard normal, mutually uncorrelated, and uncorrelated to any other observed variable of the model. By substituting the equations of x_2 and x_3 into the equation of y, we obtain that $y = 2abx_1 + \epsilon$, with $\epsilon = ae_2 + ae_3 + \epsilon_y$. The covariance between y and x_1 is thus equal to $Cov(x_1; y) = Cov(x_1; 2abx_1 + \epsilon) = Cov(x_1; 2abx_1) + Cov(x_1; \epsilon) = 2abVar(x_1)$. As x_1 has unit variance, we have that $Cov(x_1; y) = 2ab$. In Fig. 3.4, we set $a = 10$, with $b = 1$ in the left-hand panel, and $b = 2$ in the right-hand panel. These parameters correspond to a correlation between y and x_1 respectively of 20 and 40. When the correlation is relatively low (i.e., 20) as in the left-hand panel, the Lasso correctly excludes feature x_1; but when the correlation is relatively high (i.e., 40) as in the right-hand panel, feature x_1 is erroneously selected. A two-time larger correlation thus induces a specification error on the part of Lasso.

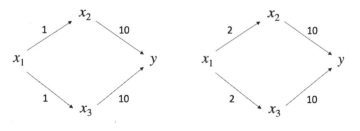

[$a = 10, b = 1$], selected: x_2, x_3 [$a = 10, b = 2$], selected: x_1, x_2, x_3

Fig. 3.4 Lasso feature selection in a simple structural model. Left-hand panel: given its low indirect effect on y (due to set $b = 1$), x_1 is correctly not selected. Right-hand panel: given its higher indirect effect on y (due to set $b = 2$), x_1 is erroneously selected. All errors and variables are standard normal. Errors are mutually uncorrelated, and uncorrelated with all the observed variables

Although Lasso is not consistent for support recovery, this does not exclude to use it for causal inference, as long as we consider some adjustment of the standard Lasso implementation. It is clear that carrying out exhaustive causal inference, that is, one for every Lasso selected feature, is too demanding and also incorrect (Leeb & Pötscher, 2008). However, focusing on a few pre-fixed features, one could wonder whether it is possible for these features to provide correct inference, that is, correct parameter size, sign, and statistical significance. Fortunately, the response is positive.

In the literature, various methods have been proposed to deal with inference for Lasso (Belloni et al., 2014a; Belloni & Chernozhukov, 2011; Belloni et al., 2014b; Chernozhukov et al., 2015). Three recent popular approaches (implemented in Stata) are (i) partialing-out, (ii) double-selection, and (iii) cross-fit partialing-out. In what follows, we present these three approaches in the case of a (conditionally) exogenous treatment. Section 3.10.4 will extend them to the case of an endogenous treatment, presenting a Lasso version of the popular instrumental-variables (IV) estimator.

3.10.1 Partialing-Out

Consider an inferential setting where we are interested in singling out the causal effect of a treatment variable d on a target variable y, conditional on a large set of (observable) control variables $\mathbf{x} = [x_1, x_2, \ldots, x_p]$, where p may be much larger than N:

$$y = \alpha d + \mathbf{x}\beta + \epsilon \tag{3.24}$$

Assuming conditional independence, i.e., the exogeneity of d conditional on $\mathbf{x}$ entails also that $E(\epsilon|\mathbf{x}) = 0$. The parameter of interest is α, and we would like to provide correct inference on it. A naïve approach would be that of running a Lasso regression of (3.24) without exposing d to selection, i.e. operating the selection process only on $\mathbf{x}$. This approach would be highly biased. The reason is that we might miss controls that have a strong predictive power on d, but only small impact on y. What solution can we adopt in such circumstances? A possible solution is the *partialing-out* approach, a method employed for consistent estimation of the linear component of partially linear models (Robinson, 1988). To understand why this method works, take the expectation of (3.24) conditional on $\mathbf{x}$:

$$E(y|\mathbf{x}) = \alpha E(d|\mathbf{x}) + \mathbf{x}\beta \tag{3.25}$$

that follows from the fact that $E(\epsilon|\mathbf{x}) = 0$. By subtracting now Eqs. (3.24) and (3.25), we obtain

$$y - E(y|\mathbf{x}) = \alpha(d - E(d|\mathbf{x})) + \epsilon \tag{3.26}$$

Suppose that consistent estimates of the two predictions $E(y|\mathbf{x})$ and $E(d|\mathbf{x})$ are available, and call them $\widehat{y}_{\mathbf{x}}$ and $\widehat{d}_{\mathbf{x}}$. By plugging in these estimates into Eq. (3.26), we obtain

Table 3.7 Description of the partialing-out procedure

Procedure. *Partialing-out*
1. Run a Lasso of d on $\mathbf{x}$. Let $\tilde{\mathbf{x}}_d$ be the features selected
2. Regress d on $\tilde{\mathbf{x}}_d$, and let $\tilde{d}$ be the residuals from this regression
3. Run a Lasso of y on $\mathbf{x}$. Let $\tilde{\mathbf{x}}_y$ be the features selected
4. Regress y on $\tilde{\mathbf{x}}_d$, and let $\tilde{y}$ be the residuals from this regression
5. Regress $\tilde{y}$ on $\tilde{d}$, thus obtaining correct inference for α

Table 3.8 Description of the double-selection procedure

Procedure. *Double-selection*
1. Run a Lasso of d on $\mathbf{x}$
2. Run a Lasso of y on $\mathbf{x}$
3. Let $\tilde{\mathbf{x}}$ be the union of the features selected in steps 1 and 2
4. Regress y on d and $\tilde{\mathbf{x}}$, thus obtaining correct inference for α

$$\tilde{y}_{\mathbf{x}} = \alpha \tilde{d}_{\mathbf{x}} + \epsilon \tag{3.27}$$

where $\tilde{y}_{\mathbf{x}} = y - \widehat{y}_{\mathbf{x}}$ and $\tilde{d}_{\mathbf{x}} = d - \widehat{y}_{\mathbf{x}}$. The parameter α can be consistently recovered by a least-squares estimation of Eq. (3.27), and classical inference can thus be applied. As stressed in Sect. 3.9.1, Lasso is consistent for prediction and thus a Lasso estimate of both $\widehat{y}_{\mathbf{x}}$ and $\widehat{d}_{\mathbf{x}}$ is a valid strategy. This suggests to implement the procedure illustrated in Table 3.7. Observe, however, that this procedure considers the prediction from the post least squares instead of the prediction directly obtained from the Lasso. Both alternatives are valid.

3.10.2 Double-Selection

The double-selection procedure is similar to partialing-out, but follows an even simpler approach, as set out in Table 3.8 (Chernozhukov et al., 2015). This procedure clearly stresses the variable omission problem that we can incur if erroneously using the naïve approach.

3.10.3 Cross-Fit Partialing-Out

The cross-fit partialing-out is an extension of the partialing-out using a split-sample procedure. In the literature, it is also known as Double Machine Learning (DML) (Chernozhukov et al., 2018). The procedure is illustrated in Table 3.9.

Table 3.9 Description of the Cross-fit partialing-out procedure

Procedure. *Cross-fit partialing-out*
1. Split randomly the original dataset into two equal-sized subsamples 1 and 2
2. Partialing-out 1
a. on sample 1, carry out the partialing-out obtaining $[\tilde{\mathbf{x}}_{d1}, \tilde{\boldsymbol{\beta}}_1]$ and $[\tilde{\mathbf{x}}_{y1}, \tilde{\boldsymbol{\gamma}}_1]$
b. on sample 2, impute $\tilde{d} = d - \tilde{\mathbf{x}}_{d1}\tilde{\boldsymbol{\beta}}_1$ and $\tilde{y} = y - \tilde{\mathbf{x}}_{y1}\tilde{\boldsymbol{\gamma}}_1$
3. Partialing-out 2
a. on sample 2, carry out the partialing-out obtaining $[\tilde{\mathbf{x}}_{d2}, \tilde{\boldsymbol{\beta}}_2]$ and $[\tilde{\mathbf{x}}_{y2}, \tilde{\boldsymbol{\gamma}}_2]$
b. on sample 1, impute $\tilde{d} = d - \tilde{\mathbf{x}}_{d2}\tilde{\boldsymbol{\beta}}_2$ and $\tilde{y} = y - \tilde{\mathbf{x}}_{y2}\tilde{\boldsymbol{\gamma}}_2$
4. Regress $\tilde{y}$ on $\tilde{d}$ in the full sample, thus obtaining correct inference for α

The cross-fit partialing-out is a more robust approach to inference than partialing-out and double-selection. This depends on the fact that the estimated coefficients are retrieved from one subsample and used in another independent sample. Its robustness can also be increased using more than one split. Despite the procedure of Table 3.9 considering only a twofold split, a k-fold approach is in general preferable, with k typically chosen as equal to 5 or 10.

3.10.4 Lasso with Endogenous Treatment

Consider again Eq. (3.24):

$$y = \alpha d + \mathbf{x}\beta + \epsilon$$

but assume this time that there exists correlation between d and ϵ, even if we condition on the vector of observable features $\mathbf{x}$. Assume for the moment that $\mathbf{x}$ is low-dimensional.

Under these assumptions, d is endogenous and both least squares and regularized least squares would lead to a biased estimation of α. In this case, a possible way to recover a consistent estimate of α is to apply an instrumental-variables (IV) approach, provided that at least as many instruments as the number of endogenous variables are available. For Eq. (3.24), at least one instrument for d would suffice. Suppose, however, that such instruments—all collected into the vector $\mathbf{z}$—are available in a high-dimensional way, so that the number of instruments may be even larger than the sample size. This case may arise, for example, when we have categorical instruments and lots of interactions are used as generated instruments. How can we retrieve a consistent estimation of α in such a setting?

The first thing to notice is that the IV estimates are severely biased when the number of instruments is large and/or the available (or generated) instruments are weak, that is, poorly correlated with the endogenous predictors. Selecting only a few instruments, those having the highest correlation with the endogenous d, would

Table 3.10 Lasso IV procedure with **z** high-dimensional and **x** low-dimensional

Procedure. *Lasso IV* 1
1. Run a Lasso of d on **z** and **x**. Let $\tilde{z}_d$ and $\tilde{x}_d$ be the instruments and features selected
2. Regress d on $\tilde{z}_d$ and $\tilde{x}_d$, and let $\widehat{d}$ be the linear prediction of this regression
3. Run an IV regression of y of d and **x**, using $[\widehat{d},\mathbf{x}]$ as (optimal) instruments for $[d,\mathbf{x}]$
4. The IV estimate of step 3 provides correct inference for α

Table 3.11 Lasso IV procedure with both **z** and **x** high-dimensional

Procedure. *Lasso IV* 2
1. Run a linear Lasso of y on **x**. Let $\tilde{x}_y$ be the features selected
2. Regress y on $\tilde{x}_y$, and let $\tilde{y}$ be the residuals from this regression
3. Run a Lasso of d on **z** and **x**. Let $\tilde{z}_d$ and $\tilde{x}_d$ be the instruments and features selected
4. Regress d on $\tilde{z}_d$ and $\tilde{x}_d$, and let $\widehat{d}$ be the linear prediction of this regression
5. Run a Lasso of $\widehat{d}$ on the controls **x**, and denote the selected controls by $\check{x}_d$
6. Regress $\widehat{d}$ on the controls $\check{x}_d$. Let $\check{\beta}$ be the estimated coefficients, and $\check{d}$ the residuals
7. Compute the residual of the level, i.e., $\tilde{d} = d - \check{x}_d\check{\beta}$
8. Run an IV regression of $\tilde{y}$ on $\tilde{d}$, using as instrument $\check{d}$

therefore be a wise strategy to pursue. A Lasso regression of d over the entire bundle of the available exogenous variables, i.e., **z** and **x**, would allow to select the most correlated instruments, by disregarding those with too low prediction power over d. As the first-stage IV regression, the regression of the endogenous variable(s) on all the exogenous ones (including the original predictors) involves a prediction task; using a Lasso estimate is in this case fully appropriate. The idea is to slightly modify the usual optimal-instrument IV procedure to accommodate the presence of high-dimensional instruments. Table 3.10 illustrates the Lasso IV procedure when **z** is high-dimensional and **x** is low-dimensional. When both **z** and **x** are high-dimensional, the Lasso IV procedure is a slight modification of the partialing-out procedure set out in Table 3.7. In this case, we have to consider that also y requires a Lasso regression, by accounting for the high-dimensional nature of both **z** and **x**. The procedure is a little more complicated, but the steps are straightforward to understand if one recognizes that the optimal (not the initial) instruments are used in the final IV regression. Table 3.11 sets out the procedure for this specific case.

These IV procedures can be easily generalized to more than one endogenous predictor, and extended to the case in which one or more additional variables are not subject to selection. In this case the model takes on this form:

$$y = \mathbf{d}\alpha + \mathbf{x}_1\beta_1 + \mathbf{x}_2\beta_2 + \epsilon \tag{3.28}$$

where **d** is the vector of endogenous treatments, and $\mathbf{x}_1$ and $\mathbf{x}_2$ the predictors respectively not subject and subject to selection, with $\mathbf{x} = [\mathbf{x}_1, \mathbf{x}_2]$. Finally, it is worth stressing that a second type of Lasso IV procedure is available, that is, the IV version of the cross-fit partialing-out illustrated in Table 3.9. This procedure is only a little more articulated than the one presented in Table 3.11 that is in fact based only on partialing-out. Similar to the exogenous treatment case, the IV cross-fit partialing-out provides further robustness, as based on a less demanding sparsity condition. For a detailed exposition, see Chernozhukov et al. (2018) and the section on methods and formulas of the Stata Lasso intro documentation.

3.11   Regularized Nonlinear Models

Regularized regression can be extended to nonlinear models by GLM penalized estimation encompassing target outcomes that can be binary, categorical (i.e., multinomial), and countable. The extension is straightforward as long as we consider the minimization of a generalized penalized objective function that in the case of Lasso (with Ridge, and Elastic-net following a similar approach) takes on this form:

$$Q_L = \frac{1}{N} \sum_{i=1}^{N} \theta_i h(y_i; \beta_0 + \mathbf{x}_i \boldsymbol{\beta}) + \lambda \sum_{j=1}^{p} \omega_j |\beta_j| \tag{3.29}$$

where θ_i is an observation-level weight, and $h(\cdot)$ is the likelihood contribution for the specific GLM considered.

For the Logit case, we have

$$h(y_i; \beta_0 + \mathbf{x}_i \boldsymbol{\beta}) = -y_i(\beta_0 + \mathbf{x}_i \boldsymbol{\beta}) + \ln\left[1 + \exp(\beta_0 + \mathbf{x}_i \boldsymbol{\beta})\right] \tag{3.30}$$

For the Poisson case, we have

$$h(y_i; \beta_0 + \mathbf{x}_i \boldsymbol{\beta}) = -y_i(\beta_0 + \mathbf{x}_i \boldsymbol{\beta}) + \exp\left[\exp(\beta_0 + \mathbf{x}_i \boldsymbol{\beta})\right] \tag{3.31}$$

For the Probit case, we have

$$h(y_i; \beta_0 + \mathbf{x}_i \boldsymbol{\beta}) = -y_i \ln[\Phi(\beta_0 + \mathbf{x}_i \boldsymbol{\beta})] - (1 - y_i)\ln[1 - \Phi(\beta_0 + \mathbf{x}_i \boldsymbol{\beta})] \tag{3.32}$$

For the Multinomial logit case, the objective function is slightly different. Specifically, assume to have K classes with $y_i \in [1, \ldots, K]$. With no loss of generality, set $\beta_K = \mathbf{0}_p$ as reference class, where $\mathbf{0}_p$ is the zero vector of length p, which makes the model identifiable. The log-likelihood of this model is

$$Q_{M_{Lasso}} = -\sum_{i=1}^{N} \sum_{k=1}^{K} y_{ik} \ln(p_{ik}) + \lambda \sum_{k=1}^{K-1} \sum_{j=1}^{p} |\beta_{kj}| \tag{3.33}$$

where

$$p_{ik} = \frac{\exp(\beta_0 + \mathbf{x}_i \boldsymbol{\beta})}{\sum_{k=1}^{K} \exp(\beta_0 + \mathbf{x}_i \boldsymbol{\beta})} \tag{3.34}$$

Equation (3.34) is called *softmax* function, and it is the generalization of the logistic function to multiple classes. It measures the probability of a unit i to belong to class k conditional on its characteristics, that is, $p_{ik} = P(y_i = k|X_i)$.

The linear and nonlinear Lasso (and Elastic-net) objective function minimization problem is not trivial as it entails a nonconvex optimization problem requiring specialized solving algorithms. A popular approach is the so-called *coordinate descent* algorithm. For a technical exposition of the coordinate descent algorithm and its computational implementation, we refer to Fu (1998), Daubechies et al. (2004), Friedman et al. (2010), and Friedman et al. (2007).

Finally, inference for nonlinear regularized regression is also possible, and follows the lines outlined for the linear case, with both patialing-out and cross-fit partialing-out implementable for a set of GLMs.

3.12 Stata Implementation

Stata provides a dedicated package, LASSO, to regularized regression and, more specifically, Lasso regression (see [LASSO] Lasso intro). All the regularized methods reviewed in the previous sections, with the exclusion of the optimal subset selection approaches, are implemented in Stata including Ridge, Lasso, and Elastic-net, all available in their adaptive form. Lasso inference (under both treatment exogeneity and endogeneity) is also implemented within this package. Table 3.12 sets out the Stata commands that implement standard Lasso, Ridge, and Elastic-net as well as the Lasso causal inference models based on the partialing-out (PO), double-selection (DS), and cross-fit partialing-out (XPO) estimators.

3.12.1 The lasso *command*

The lasso command is the workhorse of the LASSO package, as the other commands follow a similar structure. Its syntax takes this form:

lasso *model depvar* $\big[$*(alwaysvars)*$\big]$*othervars* $\big[$ *if* $\big]$ $\big[$ *in* $\big]$ $\big[$ *weight* $\big]$ $\big[$, *options* $\big]$

where *model* is a type to choose among linear, logit, probit, or poisson; *alwaysvars* are variables that are always kept within the model; *othervars* are variables that will be included in or excluded from the model.

The lasso command has several options. The most important is the selection(*sel_method*) option, used to select a value of the Lasso penalty parameter λ within a grid of possible values. The selection method options are cv (the

Table 3.12 Lasso and related estimators available in Stata 17

Basic regularized regression commands

Model	Lasso	Elastic-net	Square-root Lasso
Linear	`lasso linear`	`elasticnet linear`	`sqrtlasso`
Probit	`lasso probit`	`elasticnet probit`	
Logit	`lasso logit`	`elasticnet logit`	
Poisson	`lasso poisson`	`elasticnet poisson`	

Lasso commands for causal inference

Model	Partialing-out	Double-selection	Cross-fit partialing-out
Linear	`poregress`	`dsregress`	`xporegress`
Logit	`pologit`	`dslogit`	`xpologit`
Poisson	`popoisson`	`dspoisson`	`xpopoisson`
Linear IV	`poivregress`		`xpoivregress`

default, selecting λ using K-fold cross-validation), `adaptive` (selecting λ using an adaptive Lasso), `plugin` (selecting the penalty parameter using the plugin formula), and `none` (not selecting λ). Observe that the `adaptive` option uses an iterated cross-validation, thus it is not a stand-alone approach for tuning, but only one allowing for estimating adaptive Lasso. These options have then several suboptions that can be used for tuning the model according to the specific tuning method selected.

By default, the `lasso` command does not provide all the needed statistics and graphics to evaluate the selected model, including the obtained coefficients. For this task, `lasso` provides dedicated post-estimation commands. The most relevant are

- `cvplot` for graphing the cross-validation function;
- `coefpath` for graphing the Lasso coefficient path-diagram;
- `lassogof` reporting fit statistics to evaluate the predictive performance of a model;
- `lassocoef` listing the selected variables in the model.

Less relevant post-estimation commands, but useful however for a deeper inspection of the results, are

- `lassoknots` displaying a knot table for the features entering or leaving the model, and related measures of fit;
- `lassoselect` selecting a different model from the one chosen by the estimation command;
- `lassoinfo` displaying Lasso information such as the dependent variable, selection method, and number of nonzero coefficients for one or more models.

The `elasticnet` command follows a similar syntax, with the only difference of providing a tuning not only for the penalty parameter λ but also for the elastic parameter α.

3.12.1.1   The `splitsample` Command

The `LASSO` package provides also some utility-commands to make model validation easier. Among them, the `splitsample` command comes in handy to generate random splits of the initial dataset by setting a given number of samples, and specified proportions for each sample. The syntax of this command is

`splitsample` [*varlist*][*if*] [*in*] , `generate`(*newvar* [, `replace`]) [*options*]

where the `generate()` option creates a new sample identifier variable as *newvar*, optionally replacing an existing one. Two options are often used:

- `nsplit`(*#*), splitting the initial dataset into # random samples of equal size;
- `split`(*numlist*), specifying a *numlist* of proportions or ratios for the split.

3.12.2   The `lassopack` Suite

The `lassopack` suite, developed by Ahrens et al. (2020a), is a community-contributed set of Stata programs for fitting penalized regression methods. By and large, it overlaps with the Stata built-in suite `LASSO`, but comparatively it presents some additional features, as well as some limitations. One important additional feature is the possibility to run Elastic-net regressions (including Lasso and Ridge, of course) choosing the best penalization and elastic parameters (respectively, λ and α) by minimizing one of the following information criteria: AIC, BIC, AICc, or EBIC. This can be carried out using the `lasso2` command.

`lassopack` allows also for fitting K-fold cross-validation via the `cvlasso` command. Interestingly, this latter command supports also *rolling* cross-validation for time-series and longitudinal (panel) data, as well as fixed-effect estimation.

Finally, `lassopack` includes the `rlasso` command, implementing the theory-driven penalization for the Lasso (and square-root Lasso), both for cross-section and panel data.

Besides previous commands, the `lassopack` suite incorporates the commands `lassologit`, `cvlassologit`, and `rlassologit` for logistic Lasso regression. A limitation is that `lassopack` provides no further regularized regression methods for nonlinear models.

In this section, we focus on the `lasso2` and `cvlasso` commands. The `lasso2` syntax is quite standard:

`lasso2` *depvar varlist* [*if*] [*in*] [, *options*]

with many options available (see the related documentation). Among them, the most relevant are `alpha`(*real*), the elastic-net parameter; `adaptive`, calling for the adaptive Lasso estimator; `lambda`(*numlist*), a scalar lambda value or list of descending lambda values; ic(*string*), setting which information criterion is used for model validation and then shown in the output, with *string* replaced by aic, bic, aicc, or ebic (the default); partial(*varlist*) calling for variables in *varlist* to be partialed out prior to estimation; fe, applying within-transformation prior to estimation; and plotpath(*method*) plotting the coefficients' path as a function of the L1-norm (norm), lambda (lambda), or the log of lambda (lnlambda).

The syntax of cvlasso is similar:

cvlasso *depvar varlist* [*if*] [*in*] [, *options*]

with some specialized options. Among them, the most relevant are nfolds(*integer*), specifying the number of folds used for K-fold cross-validation (default is 10); rolling for carrying out rolling h-step ahead cross-validation in a time-series setting; h(*integer*) changing the time-series forecasting horizon (default is 1); origin(*integer*), controlling the number of observations in the first training dataset for time-series cross-validation; fixedwindow, ensuring that the size of the training dataset is always the same for time-series cross-validation; seed(*real*), setting the seed for the generation of a random fold variable.

3.12.3   The subset *Command*

Neither official Stata commands nor Python programs are available for fitting optimal subset selection. We describe here the community-contributed Stata command subset (Cerulli, 2020), a wrapper for the R software function regsubsets(), carrying out "exhaustive" (or "best"), "backward", and "forward" stepwise subset selection in a linear model with many predictors. As said in Sect. 3.8.2, the "forward" algorithm can be also used when the number of predictors is larger than the sample size, i.e., in high-dimensional settings. This command provides both the optimal subset of covariates for each specific size of the model, and the overall optimal size. The latter one is found using three criteria as validation approaches: adjusted-R^2, Mallow's CP, and BIC.

The subset syntax is

subset *depvar* [*varlist*], model(*modeltype*) rversion(*R_version*)

[index_values(*filename*) matrix_results(*filename*)

optimal_vars(*filename*) nvmax(*number*)]

The option model(*modeltype*) specifies the model to be estimated. It is always required to specify one model, with *modeltype* taking on three suboptions: best_subset (fitting best subset selection), backward (fitting backward stepwise selection), and forward (fitting forward stepwise selection). The option rversion(*R_version*) specifies the R version installed in the user's operating system, with typical value as, for example, "3.6.0". The option index_values (*filename*) specifies the name of the Stata data file containing the values of the adjusted-R^2, Mallow's CP, and BIC used for finding the final optimal number of covariates. The option `matrix_results`(*filename*) specifies the name of the Stata data file containing the matrix of results, i.e., the optimal number of covariates for each model number of covariates. The option `optimal_vars`(*filename*) specifies the name of the Stata data file containing the name of the optimal covariates at each model number of covariates. Finally, the option `nvmax`(*number*) specifies the maximum order of the best set of variables to show as the result.

For correctly running the `subset` command in Stata, the user must install in her machine both the freeware software R and `RStudio`.

3.12.4 Application S1: Linear Regularized Regression

To familiarize with regularized regression, I consider the `boston_house` dataset, collecting information from the U.S. Census Service on housing in the area of Boston (Massachusetts). This dataset, made of 506 observations and 15 variables, has been extensively used in the literature to compare different ML algorithms. Data come from Harrison & Rubinfeld (1978). Table 3.13 reports variables' names and descriptions.

As emphasized in the previous chapter, the implementation steps typically adopted in Machine Learning predictive studies are as follows:

1. split the initial sample into a *training* and a *testing* (or validation) dataset;
2. using the training dataset, fit the model using one or more (competing) learners;

Table 3.13 Description of the `boston_house` dataset

Variable	Meaning
medv	Median value of owner-occupied homes in $1000s
crim	Per capita crime rate by town
zn	Proportion of residential land zoned for lots over 25,000 sq.ft
indus	Proportion of non-retail business acres per town
chas	Charles river dummy variable (= 1 if tract bounds river; 0 otherwise)
nox	Nitric oxides concentration (parts per 10 million)
rm	Average number of rooms per dwelling
age	Proportion of owner-occupied units built prior to 1940
dis	Weighted distances to five boston employment centers
rad	Index of accessibility to radial highways
tax	Full-value property-tax rate per $10,000
ptratio	Pupil–teacher ratio by town
b	1000(bk–0.63)^2 where bk is the proportion of blacks by town
lstat	% lower status of the population

3. using the testing dataset, estimate the out-of-sample predictive performance
   (i.e., the mean squared error—MSE—for regression, and the mean classification
   error—MCE—for classification) produced by each learner;
4. select the learner producing the smallest out-of-sample error.

We begin this application by generating the testing and the training dataset using the
`splitsample` command:

```
───────────────────────────────── Stata code and output ─────────────────
. * Load the dataset
. sysuse boston_house.dta

. * Split the sample into testing and training datasets
. splitsample , generate(sample) split(.75 .25) rseed(12345)

. * Define a label
. label define slabel 1 "Training" 2 "Validation"

. * Attach the label
. label values sample slabel

. * Tabulate the sample
. tabulate sample

      sample │      Freq.     Percent        Cum.
─────────────┼───────────────────────────────────
    Training │        380       75.10       75.10
  Validation │        126       24.90      100.00
─────────────┼───────────────────────────────────
       Total │        506      100.00
```

The above one-way table shows that the initial dataset has been split, as requested,
in two portions of 75 and 25%, respectively.

In this application, we consider a regression of the variable `medv` (median value
of owner-occupied homes in \$1000s) on all the other fourteen variables of the
`boston_house` dataset. For the sake of comparison, we set off by computing ordinary least-squares (OLS) estimates using the training data, by then storing the results
as `ols`. Subsequently, we use the `lassogof` postestimation command with option
`over(sample)` to compute the in-sample (i.e. training) and out-of-sample (i.e., testing or validation) estimates of the MSE. The code to carry out these steps is

```
───────────────────────────────── Stata code and output ─────────────────
. * Define the target
. global y "medv"

. * Define the features
. global X "crim zn indus chas nox rm age dis rad tax ptratio b lstat"

. * Run OLS on the training dataset
. quietly regress $y $X if sample==1

. * Save estimates results as "ols"
. estimates store ols

. * Compute OLS training and testing MSE
. lassogof ols, over(sample)
Penalized coefficients

─────────────────────────────────────────────────────────────────
Name                  sample │      MSE    R-squared       Obs
────────────────────────────┼────────────────────────────────────
ols                         │
               Training      │  21.45167       0.7319       380
             Validation      │  23.73229       0.7565       126
─────────────────────────────────────────────────────────────────
```

We can immediately observe that, as expected, the training MSE (21.45) is smaller than the validation MSE (23.73), whereas the opposite happens to the R-squared, being it in fact an accuracy (not an error) measure. We then run a linear Lasso on the training dataset, store the estimates obtained by cross-validation as `lasso_cv`, and then plot the cross-validation (CV) function:

```
──────────────────────────────── Stata code and output ────────────────────────────────
. * Run a linear Lasso on the training dataset
. lasso linear $y $X if sample==1, nolog rseed(12345)

Lasso linear model                            No. of obs          =        380
                                              No. of covariates   =         13
Selection: Cross-validation                   No. of CV folds     =         10

                                      No. of    Out-of-        CV mean
                                      nonzero     sample     prediction
         ID    Description    lambda    coef.   R-squared         error

          1   first lambda  6.630447        0      0.0049       79.62319
         64  lambda before  .0188838       13      0.6960       24.32759
       * 65 selected lambda  .0172062       13      0.6960        24.3274
         66   lambda after  .0156776       13      0.6960       24.32756
         76    last lambda  .0061836       13      0.6959       24.33049

* lambda selected by cross-validation.

. * Store estimates obtained by cross-validation as "lasso_cv"
. estimates store lasso_cv

. * Plot the CV function
. cvplot, minmax
```

Figure 3.5 shows the Lasso cross-validation plot, with the CV optimal λ identified by the dashed vertical line. The `lasso` command uses by default a tenfold cross-

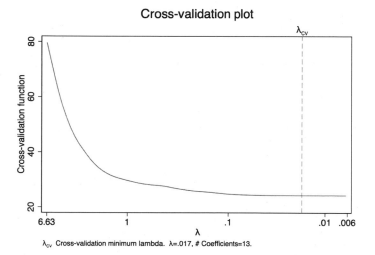

Fig. 3.5 Lasso cross-validation plot. Linear Lasso of the median value of the houses as a function of various housing characteristics

validation, which selects 13 out of 14 initial variables. The optimal λ is equal to .017 and the cross-validated MSE is equal to 24.33.

We can find the optimal λ also using the plug-in approach. This is computationally faster, as the plug-in approach requires to compute only a formula, in place of a computational procedure as in the standard and adaptive Lassos. As maintained by Belloni et al. (2016, 2012), the variables retained by plug-in Lasso are expected to be closer to those of the true data generating process. However, the plug-in Lasso may retain also variables characterized by very small coefficients.

By specifying either the `selection(adaptive)` or `selection(plugin)` options in the `lasso` command, we can estimate both the adaptive and plug-in Lassos. We store the obtained results as `lasso_adapt` and `lasso_plugin`, respectively. For the adaptive Lasso, we type

```
──────────────────────────────── Stata code and output ────────────────────────────────
. * Run adaptive Lasso on the training dataset
. lasso linear $y $X if sample==1, nolog rseed(12345) selection(adaptive)

Lasso linear model                    No. of obs          =     380
                                      No. of covariates   =      13
Selection: Adaptive                   No. of lasso steps  =       2

Final adaptive step results
```

ID	Description	lambda	No. of nonzero coef.	Out-of-sample R-squared	CV mean prediction error
77	first lambda	139.9178	0	0.0049	79.62319
159	lambda before	.0680365	11	0.6991	24.07801
* 160	selected lambda	.0619923	11	0.6991	24.07506

```
* lambda selected by cross-validation in final adaptive step.
. * Store estimates as "lasso_adapt"
. estimates store lasso_adapt
```

while for the plug-in Lasso, we type

```
──────────────────────────────── Stata code and output ────────────────────────────────
. * Run plug-in Lasso on the training dataset
. lasso linear $y $X if sample==1, nolog rseed(12345) selection(plugin)

Lasso linear model                    No. of obs          =     380
                                      No. of covariates   =      13
Selection: Plugin heteroskedastic
```

ID	Description	lambda	No. of nonzero coef.	In-sample R-squared	BIC
* 1	selected lambda	.1815368	5	0.6386	2392.544

```
* lambda selected by plugin formula assuming heteroskedastic errors.
. * Store estimates as "lasso_plugin"
. estimates store lasso_plugin
```

The number of selected predictors is now 11 for adaptive, and 5 for the plug-in Lasso, which appears as a more parsimonious approach. We can display the penalized coefficients of the non-standardized variables (sorted by absolute value of the penalized coefficients of the standardized variables) using the `lassocoef` posterimation command:

Stata code and output
```
. * Compare the coefficients
. lassocoef lasso_cv lasso_adapt lasso_plugin, display(coef, standardized) ///
> sort(coef, standardized) nolegend
```

	lasso_cv	lasso_adapt	lasso_plugin
lstat	-3.894611	-3.897588	-3.665931
dis	-2.921093	-3.02119	
rad	2.461494	2.620345	
rm	2.4281	2.412034	1.748285
ptratio	-1.948724	-1.961869	-1.411607
nox	-1.897616	-1.921439	-.0761126
tax	-1.888991	-2.022748	
zn	1.085225	1.116376	
b	.8228499	.8254961	.196253
chas	.7592363	.7538754	
crim	-.6940454	-.7104728	
indus	.0334029		
age	.0306566		
_cons	0	0	0

The variable `lstat` (percentage of lower status of the population) presents the largest coefficient with a negative impact on the median value of the houses. Following `lstat`, the variables `dis` (distances to employment centers) and `rad` (accessibility to highways) have in turn high magnitude with expected opposite sign, although they are not selected by the plug-in Lasso. Finally, houses value is positively related to the average number of rooms per dwelling (`rm`), a feature selected by all estimated Lassos.

We can also run an Elastic-net regression, where now we cross-validate the model on both the parameters λ and α. As done above, we store the estimates thus obtained as `elasticnet`. Finally, we also (quietly) estimate a Ridge regression (obtained by setting $\alpha = 0$) and store estimates as `ridge`:

Stata code and output
```
. * Run an Elasti-net regression on the training dataset
. elasticnet linear $y $X if sample==1, alpha(.25 .5 .75) nolog rseed(12345)
```

Elastic net linear model   No. of obs   =   380
                              No. of covariates =   13
Selection: Cross-validation   No. of CV folds   =   10

alpha	ID	Description	lambda	No. of nonzero coef.	Out-of-sample R-squared	CV mean prediction error
0.750						
	1	first lambda	26.52179	0	0.0031	80.26784
	113	last lambda	.0008841	13	0.6959	24.33078
0.500						
	114	first lambda	26.52179	0	0.0031	80.26784
	226	last lambda	.0008841	13	0.6959	24.33004

```
0.250
          227 │    first lambda    26.52179      0    0.0027    80.23639
          310 │   lambda before     .013128     13    0.6961     24.3186
      *   311 │ selected lambda    .0119617     13    0.6961    24.31817
          312 │    lambda after    .0108991     13    0.6961    24.31823
          339 │     last lambda    .0008841     13    0.6960    24.32931

  * alpha and lambda selected by cross-validation.
  . * Store estimates as "elasticnet"
  . estimates store elasticnet
  . * Run a Ridge on the training dataset
  . qui elasticnet linear $y $X if sample==1, alpha(0) nolog rseed(12345)
  . * Store estimates as "elasticnet"
  . estimates store ridge
```

Using the validation sample, we can compare the out-of-sample predictive performance of all the previous estimators just by typing

```
                              ──────── Stata code and output ────────
  . * Compare learners predictive accuracy on the validation sample
  . lassogof lasso_cv lasso_adapt lasso_plugin elasticnet ridge if sample==2
  Penalized coefficients

         Name │       MSE   R-squared       Obs
  ────────────┼──────────────────────────────────
      lasso_cv │  23.77881      0.7560       126
   lasso_adapt │  23.73091      0.7565       126
  lasso_plugin │  36.34681      0.6270       126
    elasticnet │  23.80884      0.7557       126
         ridge │  33.83159      0.6528       126
```

Results suggest that the adaptive and CV Lassos are the most predictive learners, with the worst performance reached by Ridge.

We finally show how to obtain the Lasso coefficients path-diagram. We consider the adaptive Lasso, and we use the postestimation command coefpath:

```
                              ──────── Stata code and output ────────
  . * Run quietly adaptive Lasso on the training dataset
  . qui lasso linear $y $X if sample==1, nolog rseed(12345) selection(adaptive)
  . * Put optimal lmabda in a global
  . global opt_lamda=e(lambda_sel)
  . * Plot the Lasso coefficients path-diagram
  . coefpath, lineopts(lwidth(medthick)) ///
  > legend(on position(6) rows(3) cols(1)) xsize(5) ///
  > xunits(rlnlambda) xline($opt_lamda) scheme(s2mono)
```

Figure 3.6 sets out the path-diagram. We see that two variables stand out with opposite signs, both in terms of size and in terms of persistence in being retained over increasing values of λ; they are lstat (the percentage of lower status of the population) and rm (the average number of rooms per dwelling).

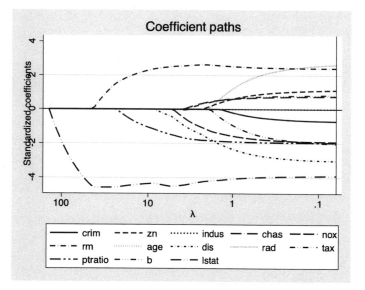

Fig. 3.6 Lasso coefficients path-diagram. Linear adaptive Lasso of the median value of the houses as a function of various housing characteristics

3.12.5   Application S2: Nonlinear Lasso

In this application, we compare the prediction performance of four popular nonlinear learners typically used when the target variable is binary: Probit, Logit, Lasso Logit, and Lasso Probit. We consider the `titanic` dataset containing information about the famous Titanic disaster. This dataset reports passengers' survival status, along with many other information regarding passengers, such as sex, age, and socio-economic status. Table 3.14 lists the variables contained in the dataset with their name and meaning. The total number of passengers is 891 (with 342 survived, and 549 not survived to the disaster), and the number of variables is 11 (excluding passengers' identifier). Notice that the dataset does not contain information on the Titanic crew. Data source is from the *Encyclopedia Titanica*, and one of the original sources is Eaton and Haas (1995).

We start by splitting the dataset into a training and a testing sample, and specifying the survivorship variable as dependent on `age`, `sibsp`, `parch`, `fare`, `pclass`, and `sex`. Then, we run a Logit and (quietly) a Probit regression on the training dataset, and save the estimates for later comparison:

Table 3.14 Description of the `titanic` dataset. Number of passengers 891; number of variables 11 (excluding passengers' id)

Variable	Meaning
`passengerid`	Identifier of the passenger
`name`	Name of the passenger
`survived`	1 = survived, 0 = non-survived
`pclass`	Passenger socio-economic status (SES): 1 = upper, 2 = middle, 3 = lower
`age`	Age of the passenger
`sibsp`	Number of siblings/spouses
`parch`	Number of parents/children
`ticket`	Identifier of the passenger ticket
`fare`	Fare of the passenger
`cabin`	Identifier of the passenger cabin
`embarked`	Port of Embarkation: C = Cherbourg, Q = Queenstown, S = Southampton
`sex`	Sex of the passenger: 1 = female, 2 = male

```
───────────────────────── Stata code and output ─────────────────────────
. * Load the "titanic" dataset
. sysuse titanic , clear

. drop if survived==.
(0 observations deleted)

. * Split the data into a "training" and a "testing" dataset
. set seed 1234

. splitsample, generate(sample) split(70 30)

. label define svalues 1 "training" 2 "testing"

. label values sample svalues

. * Form the y and the X
. global y "survived"

. global X "age sibsp parch fare i.pclass i.sex"

. * Fit a Logit model to the training dataset
. logit $y $X if sample==1 , nolog or

Logistic regression                             Number of obs    =        500
                                                LR chi2(7)       =     239.34
                                                Prob > chi2      =     0.0000
Log likelihood = -219.11724                     Pseudo R2        =     0.3532

─────────────────────────────────────────────────────────────────────────────
    survived │ Odds Ratio   Std. Err.      z    P>|z|     [95% Conf. Interval]
─────────────┼───────────────────────────────────────────────────────────────
         age │  .9504841    .0093881    -5.14   0.000     .9322607    .9690638
       sibsp │  .6162954    .0959732    -3.11   0.002      .454187    .8362635
       parch │  .9358086     .131128    -0.47   0.636     .7110735    1.231572
        fare │  .9989561    .0042875    -0.24   0.808     .9905881    1.007395
             │
      pclass │
      middle │   .207774    .0877699    -3.72   0.000     .0907859    .4755148
       lower │  .0648478    .0288633    -6.15   0.000     .0271039    .1551524
             │
         sex │
        male │  .0666854    .0183474    -9.84   0.000     .0388898    .1143472
       _cons │  127.0779    84.20588     7.31   0.000     34.67662    465.6969
─────────────────────────────────────────────────────────────────────────────
Note: _cons estimates baseline odds.
```

```
. estimates store logit
. * Fit a Probit model to the training dataset
. qui probit $y $X if sample==1 , nolog
. estimates store probit
```

From the Logit regression results listed above, we observe that all the variables considered have a negative impact on survivorship (odds rations are smaller than one). The variables having the largest negative effect on survivorship are being male (male), and belonging to a lower ticket class (lower).

Then, we run a Lasso Logit on the training dataset, compute the optimal λ, and set out the CV plot and the coefficients path-diagram. Similarly, we also run a Lasso Probit:

──────────────────────────────── Stata code and output ────────────────────────────────

```
. * Fit Lasso Logit (defaul cv with k-fold=10)
. lasso logit $y $X if sample==1 , rseed(1234) nolog
Lasso logit model                        No. of obs        =        500
                                         No. of covariates =          9
Selection: Cross-validation              No. of CV folds   =         10
```

ID	Description	lambda	No. of nonzero coef.	Out-of- sample dev. ratio	CV mean deviance
1	first lambda	.2660966	0	0.0032	1.359466
39	lambda before	.0077569	6	0.3276	.9112184
* 40	selected lambda	.0070678	6	0.3277	.911027
41	lambda after	.0064399	6	0.3277	.9110913
49	last lambda	.0030595	6	0.3264	.9128157

```
* lambda selected by cross-validation.
. * Put the optimal lambda in a global
. global opt_lamda=e(lambda_sel)
. * Plot the CV plot
. cvplot , saving("cvplot2",replace) name("cvplot2", replace)
(file cvplot2.gph saved)
. * Plot the Lasso coefficients path-diagram
. coefpath, lineopts(lwidth(medthick)) ///
> legend(on position(6) rows(4) cols(2) size(small)) xsize(5.5) ///
> xunits(rlnlambda) xline($opt_lamda) scheme(s2mono) ///
> saving("coefplot2", replace) name("coefplot2", replace)
(file coefplot2.gph saved)
```

Figure 3.7 sets out the Lasso Logit cross-validation plot that, similar to the linear case, singles out the optimal penalization parameter (equal to 0.0071). In a related manner, Fig. 3.8 displays the Lasso Logit coefficients' path-diagram. It is easy to spot a larger relative importance of two variables, namely sex and pclass. As expected, both have a strong impact on survivorship, with the variable sex emerging as the most relevant both in terms of size and persistence over the grid of λs.

For the sake of comparison, we list the Lasso Logit and Lasso Probit coefficients:

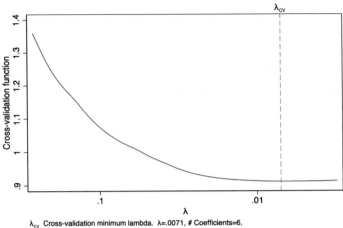

Fig. 3.7 Lasso Logit cross-validation plot. Logit Lasso of the probability to survive the Titanic disaster as a function of various passengers' characteristics

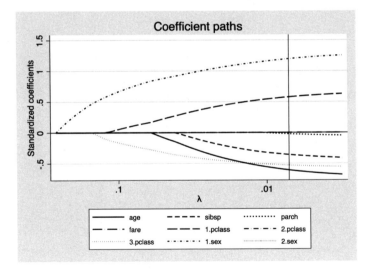

Fig. 3.8 Lasso Logit coefficients path-diagram. Logit Lasso of the probability to survive the Titanic disaster as a function of various passengers' characteristics

```
──────────────────────────────── Stata code and output ────────────────
. * Fit (quietly) Lasso Logit
. qui xi: lasso logit  $y $X if sample==1 , rseed(1234) nolog
. estimates store lasso_logit
. * Fit (quietly) Lasso Probit
. qui xi: lasso probit $y $X if sample==1 , rseed(1234) nolog
. estimates store lasso_probit
. * Show standardized coefficients for Lasso Probit and Lasso Logit
. lassocoef lasso_logit lasso_probit, ///
> sort(coef, standardized) display(coef, standardized) nolegend
```

	lasso_logit	lasso_probit
_Isex_2	-1.225872	-.7317573
_Ipclass_3	-1.196456	-.677981
age	-.6375805	-.3551229
_Ipclass_2	-.5318217	-.3020808
_cons	-.4429623	-.2536148
sibsp	-.3849556	-.2280996
parch	-.0387236	-.0338211

Although different in absolute size, both sets of coefficients present similar proportions among their relative size, thus providing a similar importance ranking of the features. Finally, we compare the predictive performance of all four learners as follows:

```
──────────────────────────────── Stata code and output ────────────────
. * Compare the out-of-sample performance of the four learners
. lassogof logit probit lasso_logit lasso_probit, over(sample)
Penalized coefficients
```

Name	sample	Deviance	Deviance ratio	Obs
logit				
	training	.7022988	0.4769	624
	testing	.7544903	0.4205	267
probit				
	training	.7047837	0.4750	624
	testing	.7549647	0.4201	267
lasso_logit				
	training	.8788649	0.3515	500
	testing	.9294478	0.3062	214
lasso_probit				
	training	.8818497	0.3493	500
	testing	.9305773	0.3054	214

The out-of-sample (or test) deviance ratios are notably worse than the in-sample values. Observe that the deviance ratio for nonlinear models is analogous to R-squared for linear models. The test deviance ratio is smaller for the Lasso estimates than for the unregularized estimates. For example, the test deviance ratio of the Lasso Logit is 0.306, whereas that of the standard Logit is 0.420. This is a sizable difference suggesting that the Lasso is not explaining a larger amount of the total survivorship variance. However, to better appreciate the different performance accuracy between the two learners, it is more convenient to compute the test error rate, as it is a more informative predictive measure than the deviance ratio:

```
────────────────────────────────── Stata code and output ──────────────
. * Estimate the test error rate for Logit
. qui xi: logit  $y $X if sample==1 , nolog
. predict pr_logit if sample==2 , pr
(677 missing values generated)
. gen class_logit = floor(0.5+pr_logit) if sample==2
(677 missing values generated)
. gen match_logit = (class_logit == survived) if sample==2 & class_logit!=.
(677 missing values generated)
. summarize match_logit
```

Variable	Obs	Mean	Std. Dev.	Min	Max
match_logit	214	.7570093	.4298951	0	1

```
. * Estimate the test error rate for Lasso Logit
. qui xi: lasso logit  $y $X if sample==1 , rseed(1234) nolog
. predict pr_lasso_logit if sample==2 , pr
(option penalized assumed; Pr(survived) with penalized coefficients)
. gen class_lasso_logit = floor(0.5+pr_lasso_logit) if sample==2
(677 missing values generated)
. gen match_l_logit = (class_lasso_logit == survived) if sample==2 & class_lasso_logit!=.
(677 missing values generated)
. summarize match_l_logit
```

Variable	Obs	Mean	Std. Dev.	Min	Max
match_l_lo~t	214	.7803738	.4149641	0	1

We see that the Lasso Logit outperforms the standard Logit by allowing for a 2.3 probability points larger accuracy. This result suggests that, on average, the probability to predict correctly the Titanic disaster survivorship using the selected features and the Lasso approach is of 78%.

3.12.6 Application S3: Multinomial Lasso

Regularized regression can be extended to models with more than two categorical outcomes. In many contexts, in fact, the dependent variable is qualitative and takes on more than two classes defining, for example, different types of preferences or satisfaction ratings, different market strategies, various levels of salary, or types of diseases. Of course, the examples are as large as one can imagine.

Multinomial Logit (or Probit) regressions are extensions of Logit (Probit) models to a multi-class target variable. Regularized Multinomial Logit (or Probit) are the corresponding penalized versions. Unfortunately, the LASSO package does not provide us with an implementation of Regularized Multinomial models. We can however exploit the Stata integration with Python to accomplish this task using the Scikit-learn platform, a popular Python API for supervised and unsupervised Machine Learning. We develop here an application using the Scikit-learn function `LogisticRegression()` that implements both standard and regularized multinomial Logit.

In this application, we consider a Logit Multinomial Lasso applied to a high-dimensional medical dataset concerning DNA micro-array-based tumor gene expression profiles. Data source comes from Ramaswamy et al. (2001), and collects information on 14-cancer gene expression for 16,063 genes, 144 training observations, and 54 test observations. In this application, however, we consider only the training dataset and 4,180 genes, and call it `cancer_genes.dta`. This dataset is clearly high-dimensional, with $N = 144$ and $p = 4,180$.

With a dataset of this kind at hand, we can answer many research questions regarding the relationship between gene expression and type of tumor. For example, we could predict whether a new cancer tissue is of a specific type based on its genetic micro-array; or we could understand what genes are possibly more related to certain types of cancer. In this application, we try to reply to both research questions. First of all, however, it is important to stress that gene expression can take both positive and negative values, with negative values signaling an under-expressed gene, and positive values an over-expressed gene in a specific cancer. For this reason, we first transform all the 4,180 features (genes' expression variables) in binary (0/1) variables, with one indicating over-expression and zero indicating under-expression in the specific cancer considered (located by row). In what follows, we run a piece of code carrying out this transformation, tabulating our cancer-type target variable, listing a portion of the dataset, and finally forming our training and testing datasets (for later use, we also save the original target variable in a dataset called `y_data_test`):

```
───────────────────────────────── Stata code and output ─────────────
. * Open the "cancer_genes" dataset
. use cancer_genes , clear
. * Replace all the genetic variables so that: under-expression=0, over-expression=1
. qui{
. * Tabulate the cancer-type target variable ("y")
. tab y

          y |      Freq.     Percent        Cum.

     breast |          8        5.56        5.56
   prostate |          8        5.56       11.11
       lung |          8        5.56       16.67
collerectal |          8        5.56       22.22
   lymphoma |         16       11.11       33.33
    bladder |          8        5.56       38.89
   melanoma |          8        5.56       44.44
     uterus |          8        5.56       50.00
   leukemia |         24       16.67       66.67
      renal |          8        5.56       72.22
   pancreas |          8        5.56       77.78
      ovary |          8        5.56       83.33
       meso |          8        5.56       88.89
        cns |         16       11.11      100.00

      Total |        144      100.00

. * List the first 10 features and first 20 observations
. list y v1-v10 in 1/20
```

		y	v1	v2	v3	v4	v5	v6	v7	v8	v9	v10
1.	breast	0	0	0	1	0	0	0	0	0	1	
2.	breast	0	0	0	0	0	0	0	0	0	1	
3.	breast	1	0	0	0	0	0	1	0	1	1	
4.	breast	0	0	0	0	0	0	0	0	0	1	
5.	breast	0	0	0	1	0	0	1	0	1	1	
6.	breast	0	0	0	0	0	0	1	0	1	1	
7.	breast	0	1	1	1	0	0	1	0	0	1	
8.	breast	0	0	1	1	0	0	0	0	0	1	
9.	prostate	0	0	0	1	0	0	1	0	1	0	
10.	prostate	1	0	0	1	0	0	0	0	1	1	
11.	prostate	0	0	0	0	0	0	0	0	0	1	
12.	prostate	0	0	0	0	0	0	0	0	0	1	
13.	prostate	1	0	0	0	0	0	1	0	0	1	
14.	prostate	1	0	0	0	0	0	1	0	1	1	
15.	prostate	0	0	0	1	0	0	1	0	0	1	
16.	prostate	0	0	0	0	0	0	1	0	1	1	
17.	lung	0	0	0	0	0	0	1	0	1	1	
18.	lung	1	0	0	0	0	0	0	0	0	1	
19.	lung	0	0	0	0	0	0	0	0	0	0	
20.	lung	1	0	0	0	0	0	1	0	0	1	

```
. * Split the sample into training and testing datasets
. splitsample , generate(sample) split(.75 .25) rseed(12345)
. label define slabel 1 "train" 2 "test"
. label values sample slabel
. * Form the training dataset and save it as "data_train"
. preserve
. keep if sample==1
(36 observations deleted)
. drop sample
. lab drop _all
. save data_train , replace
file data_train.dta saved
. restore
. * Form the testing dataset and save it as "data_test"
. preserve
. keep if sample==2
(108 observations deleted)
. drop sample
. drop y
. lab drop _all
. save data_test , replace
file data_test.dta saved
. restore
. * Save the original "y" of the testing dataset
. preserve
. keep if sample==2
(108 observations deleted)
. gen index=_n-1
. order index
. keep index y sample
. lab drop _all
. save y_data_test , replace
file y_data_test.dta saved
. restore
```

We can now run our Python script `lasso_mln.py` to estimate a Logit Multinomial Lasso. This script takes as arguments the training and the testing datasets (namely, `data_train` and `data_test`, respectively), and operates a tuning of the penalty parameter λ over a pre-fixed grid of values by selecting the best λ via cross-validation:

```
——————————————————— Stata code and output ———————————————————
. * Run the Python script "C_RM_book_ML.py" for multinomial Lasso
. *python script "C_RM_book_ML.py"
```

This script returns the main outputs as Stata datasets; in particular, we obtain

- `lasso_mln_cv.dta`: containing the cross-validation results;
- `lasso_mln_coefs.dta`: containing the regression coefficients;
- `lasso_mln_train_pred.dta`: containing the in-sample (or training) predicted labels and probabilities;
- `lasso_mln_test_pred.dta`: containing the out-of-sample (or testing) predicted labels and probabilities.

We start by focusing on the Lasso coefficients considering the `lasso_mln_coefs` dataset containing the high-dimensional βs of our estimated regularized multinomial. An important index that we can calculate using this dataset is the so-called *sparsity index*, i.e., the percentage of coefficients having exactly zero value. The larger this index, the larger the penalization implied by the CV optimal tuning of λ. The `lasso_mln_coefs` dataset has 14 rows (number of cancer-types) and 4,180 coefficients (number of genes), so we reshape it to obtain all the coefficients stacked vertically in a single variable v. As in the logit regression, coefficients do not have an immediate interpretation, thus we consider the odds ratio (OR). For cancer type k and gene v, we have

$$\mathrm{OR}(k, v) = \frac{\mathrm{odds}(k, v = 1)}{\mathrm{odds}(k, v = 0)} = \frac{\left(\frac{Prob(cancer=k|v=1)}{1 - Prob(cancer=k|v=1)} \right)}{\left(\frac{Prob(cancer=k|v=0)}{1 - Prob(cancer=k|v=0)} \right)} = \frac{e^{\beta_0 + \beta_v \cdot 1}}{e^{\beta_0 + \beta_v \cdot 0}} = e^{\beta_v} \quad (3.35)$$

For example, a value of e^{β_v} of 1.1 means that the odds at $v = 1$ are 10% larger than those at $v = 0$. It signals that the presence of gene v increases the odds ratio by a 10%. Thus, the relative likelihood of observing cancer k instead of another cancer type in the subgroup having gene v is 1.1 times larger than the relative likelihood of observing cancer k instead of another cancer type in the subgroup not having gene v.

To obtain the odds ratios by cancer and by gene, we thus apply the exponential transformation of the coefficients v, thus obtaining the variable OR. The percentage of ones in this variable is the sparsity index. We can finally save this dataset as `lasso_mln_coefs_long.dta`. The Stata code to implement these steps is

```
──────────────────────── Stata code and output ────────────────────
. * Load the obtained multinomial Lasso coefficients, reshape them, and form the odds ratios
. use lasso_coefs , clear
. qui{
. replace v = exp(v) // odds ratios
(58,520 real changes made)
. rename v OR
. * Compute sparsity
. count
  58,520
. global N=r(N)
. count if OR==1
  53,612
. global N1=r(N)
. global S=$N1/$N*100
. display "The coefficients sparsity is equal to " $S
The coefficients sparsity is equal to 91.613124
. tostring index , gen(cancer) force
cancer generated as str2
. save lasso_mln_coefs_long , replace
file lasso_mln_coefs_long.dta saved
```

We see that the sparsity index is sizable and equal to 91.6%, which means that the largest majority of genes are not related to the probability to observe a given type of cancer. Catching coefficients' relative importance in high-dimensional settings may be complicated, given their very large number. An appropriate graphical representation comes therefore in handy. For example, we can graphically eyeball all odds ratios different from one by means of a heat map giving a more intense color to more relevant odds ratios by cancer-type. We can obtain this map in Stata by typing:

```
──────────────────────── Stata code and output ────────────────────
. * Graph the heatplot of the coefficients
. use lasso_mln_coefs_long , clear
. #delimit ;
delimiter now ;
. heatplot OR j cancer if OR!=1, color(PiYG , reverse)
> plotregion(style(none)) scheme(s1mono) xlabel(1(1)14 , labsize(vsmall)
> )
> levels(25) ylabel(0(500)4200 , labsize(vsmall))
> xtitle("Cancer-type" , size(small)) ytitle("Genes")
> note(" " "Legend: 1 breast; 2 prostate; 3 lung; 4 collerectal;
> 5 lymphoma; 6 bladder; 7 melanoma; 8 uterus;"
> "9 leukemia; 10 renal; 11 pancreas; 12 ovary; 13 meso; 14 cns");
. #delimit cr
delimiter now cr
```

Figure 3.9 shows the obtained cancer/genes coefficients heat map. We can observe that odds ratios are in general rather small. Cancer 4 (collerectal), 7 (melanoma), and 8 (uterus) are characterized by darker colors, thus signaling an overall greater impact of genes on these types of cancers. To go more in depth in this analysis, we can list those cancer–gene pairs having an odd ratio larger than 1.1. In Stata, we type

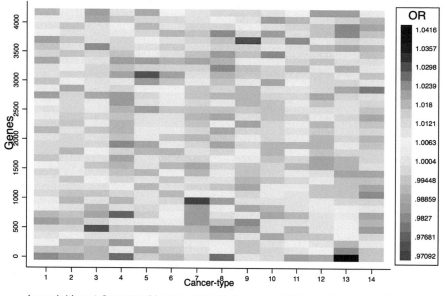

Legend: 1 breast; 2 prostate; 3 lung; 4 collerectal; 5 lymphoma; 6 bladder; 7 melanoma; 8 uterus; 9 leukemia; 10 renal; 11 pancreas; 12 ovary; 13 meso; 14 cns

Fig. 3.9   The cancer/genes coefficients heatmap. OR: odds ratios

```
                                            Stata code and output
. * Display the most relevant genes by cancer
. list index j if OR>1.1

            index       j

4135.           0    4135
9231.           2     871
14207.          3    1667
19055.          4    2335
19217.          4    2497

19821.          4    3101
20429.          4    3709
27139.          6    2059
34398.          8     958
52782.         12    2622

56933.         13    2593
```

We can see that only a few pairs are listed. They are those in which gene j has a substantial positive impact on increasing the relative likelihood to be associated with a specific cancer type. For example, the presence of gene 958 has odds to be associated to uterus cancer larger than 10% than the odds of its absence, signaling that this gene may have an impact in manifesting this cancer type. It is however important to stress that Lasso coefficients have to be interpreted as associations and not as causal effects. Lasso causal effect analysis requires a different estimation setting that we addressed in detail in Sect. 3.10 and subsections.

As mentioned at the beginning of this section, we can also check the ability of DNA micro-arrays to predict cancer types. To this end, we estimate the test error of this multinomial regression, by running this simple code:

```
———————————————————————— Stata code and output ————————————
. use lasso_mln_test_pred , clear

. rename _0 y_test_pred

. merge 1:1 index using "y_data_test"
(note: variable index was long, now double to accommodate using data´s values)

    Result                          # of obs.

    not matched                          0
    matched                             36    (_merge==3)

. keep index y y_test_pred

. gen test_err=(y != y_test_pred)

. sum test_err
    Variable |       Obs        Mean    Std. Dev.       Min        Max

    test_err |        36    .3888889    .4944132         0          1
. display "The % test prediction error is " r(mean)*100
The % test prediction error is 38.888889
```

We obtain a test error of around 39% or, equivalently, a test accuracy of around 61%. This means that, on average, there is a 61% probability to predict the correct cancer-type given the set of genetic information considered in this dataset. This accuracy may be unsatisfactory, thus suggesting trying other learners in the hope of increasing its magnitude.

3.12.7 Application S4: Inferential Lasso Under Exogeneity

As listed in Table 3.12, Stata has dedicated commands for carrying out inferential Lasso under the assumption of treatment (conditional) exogeneity. For linear models (i.e., models with a continuous outcome), these commands are `poregress` for fitting the partialing-out model, `dsregress` for fitting the double-selection, and `xporegress` for fitting the cross-fit partialing-out. For a binary outcome, corresponding commands are `pologit`, `dslogit`, and `xpologit`. Similarly, for a countable outcome, the three commands are `popoisson`, `dspoisson`, and `xpopoisson`. As the syntax for these commands is fairly similar, I set out here only the syntax for `poregress`:

`poregress` *depvar varsofinterest* [*if*] [*in*], `controls`([(*alwaysvars*)]*othervars*) [*options*]

where *varsofinterest* are the *causal* variables of interest, those for which coefficients and their standard errors are estimated; *alwaysvars* are variables that are always kept within the model, but no standard errors are provided for them; finally, *othervars* are

Table 3.15 Dataset `nlsw88`: National Longitudinal Survey of Mature and Young Women. Year 1988

Variable	Meaning
Idcode	NLS id
Age	Age in current year
Race	Race
Married	Married
Never_married	Never married
Grade	Current grade completed
Collgrad	College graduate
South	Lives in south
Smsa	Lives in SMSA
c_city	Lives in central city
Industry	Type of industry
Occupation	Type of occupation
Union	Union membership
Wage	Hourly wage
Hours	Usual hours worked
ttl_exp	Total work experience
Tenure	Job tenure (years)

variables that will be included in or excluded from the model (again, no standard errors are provided for them). The `xporegress` command (as well as `dsregress` and `xporegress`) has several options. As in the case of the `lasso` command, the most important is the `selection(sel_method)` option, used to select a value of the Lasso penalty parameter λ. The selection method options are `plugin` (the default, selecting the penalty parameter using the plugin formula); `cv` (selecting λ using K-fold cross-validation); and `adaptive` (selecting λ using an adaptive Lasso). A wider description of all the options can be found in the manual.

In this section, we consider a Lasso inferential application under treatment exogeneity, often referred to as "selection-on-observables". We employ the `nlsw88` dataset, an extract from the U.S. National Longitudinal Survey of Mature and Young Women for the year 1988. This dataset collects information on 16 variables for 2,246 American women. Table 3.15 sets out the variables we are going to use, with names and meaning. Although not high-dimensional, this dataset is however suitable to explore the use and potential of an inferential Lasso analysis using Stata.

In this application, we are interested in measuring the impact of having a union membership on the log of women hourly salary. The treatment is thus the (binary) variable `union`, and the outcome the log of the variable `wage`. Various control variables, both numerical and categorical, are used to make the treatment exogenous under the assumption of selection-on-observables.

To enrich our analysis, however, we focus on the wage differential effect induced by having a union membership in different occupational types. This entails to introduce an interaction term between our treatment (union) and the categorical variable occupation. To better clarify our treatment setting, however, we first provide a simple presentation of the treatment model we use in this application, to then fit it in Stata.

Suppose that the variable occupation takes on only three categories $\{0, 1, 2\}$ (the extension to more than three categories is straightforward). By defining as d_0, d_1, and d_2 the three corresponding categorical dummies, and choosing category 0 as our baseline, we can write down our treatment model as follows:

$$y = \mu + \alpha w + \beta_1 d_1 + \beta_2 d_2 + \gamma_1 w d_1 + \gamma_1 w d_2 + g(\mathbf{x}) + \epsilon \qquad (3.36)$$

where w is the union membership 0/1 treatment (1 = having a union affiliation), and $g(\cdot)$ a parametric function of the controls. Observe that, for avoiding the dummy trap, the baseline categorical dummy d_0 is excluded from the equation. Under selection-on-observables, the average treatment effect (ATE) conditional on d_1, d_2, and $\mathbf{x}$ is

$$\text{ATE}(d_1, d_2, \mathbf{x}) = \text{E}(y|w = 1; d_1, d_2, \mathbf{x}) - \text{E}(y|w = 0; d_1, d_2, \mathbf{x}) = \alpha + \gamma_1 d_1 + \gamma_2 d_2 \quad (3.37)$$

and the ATE is thus equal to

$$\text{ATE} = \text{E}_{(d_1, d_2)}[\text{ATE}(d_1, d_2)] = \alpha + \gamma_1 \text{prob}(d_1 = 1) + \gamma_2 \text{prob}(d_2 = 1) \qquad (3.38)$$

As a consequence, the effect in each occupational type is defined as

$$\text{ATE}(d_0 = 1) = \alpha$$
$$\text{ATE}(d_1 = 1) = \alpha + \gamma_1$$
$$\text{ATE}(d_2 = 1) = \alpha + \gamma_2$$

with

$$\gamma_1 = \text{ATE}(d_1 = 1) - \text{ATE}(d_0 = 1)$$
$$\gamma_2 = \text{ATE}(d_2 = 1) - \text{ATE}(d_0 = 1)$$

It follows that the occupational j's interaction parameter γ_j measures the difference of the effect of having a union affiliation between occupation j and the occupation chosen as baseline. In this application, we consider "managers and administratives" as our occupational baseline category. This choice depends on this category to present a low difference-in-mean (DIM) between treated (unionized) and untreated (non-unionized) women, a value that we can approximately consider equal to zero. In this case, whenever α in Eq. (3.36) is zero, we can interpret γ_j directly as the average treatment effect for occupational type j.

In Stata, we ran three distinct linear Lasso inferential regressions, i.e., partialing-out, cross-fit partialing-out, and double-selection, with the default plugin option. Our

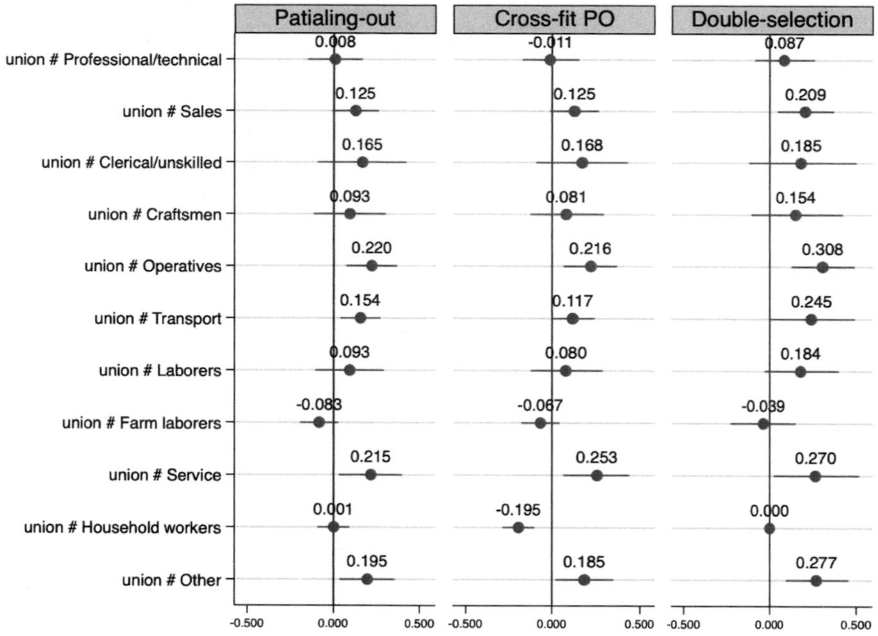

Fig. 3.10 Lasso treatment effect of union membership on the log-salary by occupation type. Occupation baseline category: "managers and administratives"

list *varsofinterest* comprises all the union/occupation interaction dummies chosen as our *causal* variables of interest. After running the regressions, and obtaining the tables of results, we also plot the coefficients and their significance intervals as set out in Fig. 3.10. The whole Stata code to carry out this analysis, including producing the figure, is as follows:

```
                                  Stata code and output
. * Open the dataset
. sysuse nlsw88 , clear
(NLSW, 1988 extract)

. * Generate the log of wage
. gen ln_wage=ln(wage)

. * Define the main macros
. global y "ln_wage" // outcome

. global T "1bn.union 1bn.union#ib(2).occupation" // treatments

. global X "age hours ttl_exp tenure" // controls (continuous)

. global F "race married never_married grade collgrad south smsa c_city industry occupation" //
controls (categorical)

. * Partialing-out with plugin lambda
. poregress  $y $T , controls($X i.($F)) nolog
note: 1.union#9.occupation identifies no observations in the sample

Partialing-out linear model        Number of obs              =        1,848
                                   Number of controls         =           60
                                   Number of selected controls =          25
                                   Wald chi2(10)              =       346.58
                                   Prob > chi2                =       0.0000
```

ln_wage	Coef.	Robust Std. Err.	z	P>|z|	[95% Conf. Interval]	
union						
union	-.0051961	.0636927	-0.08	0.935	-.1300314	.1196392
union#occupation						
union#Professional/technical	.0084059	.0802134	0.10	0.917	-.1488094	.1656212
union#Sales	.1253328	.0689783	1.82	0.069	-.0098622	.2605279
union#Clerical/unskilled	.1650043	.1314113	1.26	0.209	-.0925572	.4225658
union#Craftsmen	.0926305	.1059854	0.87	0.382	-.1150971	.3003581
union#Operatives	.2196868	.0754372	2.91	0.004	.0718327	.3675409
union#Transport	.1538317	.0589935	2.61	0.009	.0382065	.2694568
union#Laborers	.0929673	.1008438	0.92	0.357	-.1046828	.2906175
union#Farmers	0	(empty)				
union#Farm laborers	-.0826418	.0565164	-1.46	0.144	-.1934119	.0281283
union#Service	.2147789	.0939313	2.29	0.022	.0306768	.3988809
union#Household workers	.0012028	.0467578	0.03	0.979	-.0904408	.0928465
union#Other	.1952772	.0812596	2.40	0.016	.0360113	.354543

```
Note: Chi-squared test is a Wald test of the coefficients of the variables
      of interest jointly equal to zero. Lassos select controls for model
      estimation. Type lassoinfo to see number of selected variables in each
      lasso.

. estimates store POREG

. * Double selection with plugin lambda
. dsregress $y $T , controls($X i.($F)) nolog
note: 1.union#9.occupation identifies no observations in the sample

Double-selection linear model      Number of obs              =      1,848
                                   Number of controls         =         60
                                   Number of selected controls =        25
                                   Wald chi2(11)              =      62.81
                                   Prob > chi2                =     0.0000
```

ln_wage	Coef.	Robust Std. Err.	z	P>|z|	[95% Conf. Interval]	
union						
union	-.0827125	.07487	-1.10	0.269	-.229455	.06403
union#occupation						
union#Professional/technical	.0866792	.0900195	0.96	0.336	-.0897557	.2631141
union#Sales	.2091363	.0820398	2.55	0.011	.0483413	.3699313
union#Clerical/unskilled	.1846555	.15847	1.17	0.244	-.1259399	.495251
union#Craftsmen	.1537593	.1361918	1.13	0.259	-.1131717	.4206902
union#Operatives	.3083499	.0906984	3.40	0.001	.1305843	.4861154
union#Transport	.2448303	.1244627	1.97	0.049	.0008879	.4887726
union#Laborers	.184073	.1095816	1.68	0.093	-.0307031	.3988491
union#Farmers	0	(empty)				
union#Farm laborers	-.0386696	.0997395	-0.39	0.698	-.2341555	.1568163
union#Service	.2702514	.1250341	2.16	0.031	.0251891	.5153137
union#Household workers	0	(omitted)				
union#Other	.2768218	.0915392	3.02	0.002	.0974083	.4562353

```
Note: Chi-squared test is a Wald test of the coefficients of the variables
      of interest jointly equal to zero. Lassos select controls for model
      estimation. Type lassoinfo to see number of selected variables in each
      lasso.

. estimates store DSREG

. * Cross-fit partialing-out with plugin lambda
. xporegress $y $T , controls($X i.($F)) nolog
note: 1.union#9.occupation identifies no observations in the sample

Cross-fit partialing-out           Number of obs              =      1,848
linear model                       Number of controls         =         60
```

```
                    Number of selected controls   =        35
                    Number of folds in cross-fit  =        10
                    Number of resamples           =         1
                    Wald chi2(12)                 =    4215.65
                    Prob > chi2                   =     0.0000
```

	Coef.	Robust Std. Err.	z	P>|z|	[95% Conf. Interval]	
ln_wage						
union						
union	.0035919	.0644984	0.06	0.956	-.1228226	.1300065
union#occupation						
union#Professional/technical	-.0113451	.0828052	-0.14	0.891	-.1736403	.15095
union#Sales	.1247172	.0694343	1.80	0.072	-.0113715	.2608058
union#Clerical/unskilled	.1678263	.1330311	1.26	0.207	-.0929098	.4285625
union#Craftsmen	.0808227	.1061156	0.76	0.446	-.1271601	.2888056
union#Operatives	.2163074	.0763972	2.83	0.005	.0665717	.3660432
union#Transport	.116845	.0617148	1.89	0.058	-.0041137	.2378037
union#Laborers	.079801	.103953	0.77	0.443	-.1239432	.2835451
union#Farmers	0	(empty)				
union#Farm laborers	-.0671534	.0572588	-1.17	0.241	-.1793787	.0450719
union#Service	.2530173	.0950136	2.66	0.008	.0667941	.4392404
union#Household workers	-.1951381	.0476768	-4.09	0.000	-.2885829	-.1016933
union#Other	.1845231	.0830097	2.22	0.026	.0218271	.3472192

```
Note: Chi-squared test is a Wald test of the coefficients of the variables
      of interest jointly equal to zero. Lassos select controls for model
      estimation. Type lassoinfo to see number of selected variables in each
      lasso.
. estimates store XPOREG

. * Plot the graph of the treatment coefficients
. coefplot POREG  , bylabel(Patialing-out)  ///
>    || XPOREG , bylabel(Cross-fit PO) ///
>    || DSREG  , bylabel(Double-selection) || ///
>    , drop(1.union _cons) xline(0) mlabel format(%9.3f) mlabposition(12) mlabgap(*2)  ///
>    plotregion(style(none)) scheme(s1mono) byopts(row(1)) ///
>    xlabel(,labsize(vsmall)) ylabel(,labsize(small))
```

As our reference, we choose to consider the results from the cross-fit partialing-out. Looking at the obtained coefficients, we can observe that α, i.e., the average treatment effect of having a union membership for the baseline occupational type "managers and administratives" is as expected close to zero. This allows us to interpret the interaction coefficients (the γ_js) directly as the effect of unionization in each remaining occupational type. For example, the occupational type "operatives" obtains a sizable significant value of 0.216, meaning that on average and when adjusting for observable confounders, having a union affiliation yields around a 22% larger salary than not having it.

Figure 3.10 shows the results in a more compact way for all the three Lasso inferential regressions. Over all the three approaches, results are fairly similar, and we observe positive and significant effects of union affiliation also for women working in service activities, in sales, and in the all-encompassing residual occupation denoted by "other".

3.12.8 Application S5: Inferential Lasso Under Endogeneity

Stata fits linear Lasso instrumental-variables regression through two commands: `poivregress` for partialing-out and `xpoivregress` for cross-fit partialing-out. The syntax of both commands is similar. Here, we set out the syntax of `poivregress`:

`poivregress` *depvar* [*exovars*] (*endovars* = *instrumvars*) [*if*] [*in*],

    `controls`([(*alwaysvars*)]*othervars*) [*options*]

where coefficients and standard errors are estimated for the exogenous variables, *exovars*, and the endogenous variables, *endovars*. The set of instrumental variables, *instrumvars*, may be high-dimensional.

In this section, we run our application using a dataset from Angrist & Krueger (1991) concerning a paper published in *The Quarterly Journal of Economics* titled "Does Compulsory School Attendance Affect Schooling and Earnings?". In this paper, the authors investigate the traditional returns on education relying on a wage equation with individual educational attainment in the RHS along with a series of confounders. As usual in this literature, education is assumed to be endogenous mainly because of an omitted variables' occurrence, potentially producing biased least-squares estimates. Relying on an instrumental-variables (IV) estimation, by properly instrumenting years of education, seems a wise strategy in this context. Among various IV candidates, Angrist and Krueger select the individual "season of birth" as an instrument for education. This variable is naturally exogenous, as being born in one specific season of the year is a random event; at the same time, it is also related to educational attainment because of compulsory education rules. Indeed, this relationship draws on the fact that those born in the beginning of the year start school at an older age, and can therefore drop out after completing less schooling than individuals born closer to the end of the year. They find that around 25 percent of potential dropouts continue to attend school because of compulsory schooling laws. As a consequence, the "quarter of birth" appears as a good candidate for instrumenting education.

The baseline regression estimated by Angrist and Krueger is

$$\log(wage_i) = \alpha \cdot education_i + g(controls_i) + error_i \qquad (3.39)$$

where α is the parameter of interest, and $g(\cdot)$ a parametric function of the exogenous control variables. The error term might be potentially correlated with educational attainment as it may contain, for example, an unobserved ability component having contemporaneous effect on both earnings and education.

Different U.S. states have different laws concerning compulsory school age. Therefore, individual years of education vary depending on the state, on when one was born, and on when the cutoff binds. The authors estimate Eq. (3.39) using various controls (year dummies, place-of-birth state dummies) and exploit as instruments the "quarter-of-birth" (QOB), plus interactions of QOB with YOB ("year-of-birth") and POB ("place-of-birth") dummies. We are therefore in a setting with many available instruments, thus running the risk to obtain inefficient estimates of the main treatment effect, or even biased in finite samples. We can thus regard this one as a setting with high-dimensional instruments. Reducing the number of instruments to those that really matter in predicting the endogenous variable seems appropriate, and fitting thus an IV Lasso a suitable choice.

In this Stata application, we run three IV models, that is, standard two-stage least squares (2SLS), Lasso partialing-out, and Lasso cross-fit partialing-out. As the sample size of Angrist and Krueger dataset is rather huge (around 330,000 observations), for illustrative purposes, we consider a random extract of the original dataset. We then compare the results of the three methods. The Stata code is as follows:

```
──────────────────────────────────── Stata code and output ────────────────────────────
. * Open the dataset
. use wageduc , clear
. * Extract randomly a subsample
. set seed 1010
. gen u=runiform()
. sort u
. keep in 1/10000
(319,509 observations deleted)
. drop u
. * Save the dataset thus obtained
. save wageduc10000 , replace
file wageduc10000.dta saved
. * Form model´s specification
. global X "i.yob"
. global w "edu"
. global y "lnwage"
. global Z "i.qob i.yob#i.qob i.pob#i.qob"
. * Fit a standard 2sls regression and store the estimates
. ivregress 2sls $y $X ($w = $Z) , nolog
may only specify nolog and log with iterative GMM estimator
r(198);

end of do-file

r(198);
. do "/var/folders/n5/5h3q21nx41gbwprhx27_jbpm0000gp/T//SD12085.000000"
. xpoivregress $y ($w = $Z) , controls($X) rseed(1010) nolog
2o.pob#1o.qob omitted
2o.pob#3o.qob omitted
2o.pob#1o.qob omitted
2o.pob#3o.qob omitted
56o.pob#2o.qob omitted
2o.pob#4o.qob omitted
2o.pob#2o.qob omitted
15o.pob#2o.qob omitted
──Break──
r(1);

end of do-file
──Break──
r(1);
. do "/var/folders/n5/5h3q21nx41gbwprhx27_jbpm0000gp/T//SD12085.000000"
. * Open the dataset
. use wageduc , clear
. * Extract randomly a subsample
. set seed 1010
. gen u=runiform()
```

```
. sort u
. keep in 1/10000
(319,509 observations deleted)
. drop u
. * Save the dataset thus obtained
. save wageduc10000 , replace
file wageduc10000.dta saved
. * Form model's specification
. global X "i.yob"
. global w "edu"
. global y "lnwage"
. global Z "i.qob i.yob#i.qob i.pob#i.qob"
. * Fit a standard 2sls regression and store the estimates
. ivregress 2sls $y $X ($w = $Z)
note: 2.pob#1.qob identifies no observations in the sample
note: 2.pob#3.qob identifies no observations in the sample
```

Instrumental variables (2SLS) regression	Number of obs	=	10,000
	Wald chi2(10)	=	279.28
	Prob > chi2	=	0.0000
	R-squared	=	0.0565
	Root MSE	=	.65681

lnwage	Coef.	Std. Err.	z	P>|z|	[95% Conf. Interval]	
edu	.1211208	.0073629	16.45	0.000	.1066899	.1355518
yob						
31	.000584	.0299203	0.02	0.984	-.0580586	.0592266
32	-.0300556	.0296682	-1.01	0.311	-.0882042	.0280929
33	-.0001469	.029744	-0.00	0.996	-.058444	.0581502
34	-.0282784	.0299901	-0.94	0.346	-.0870579	.030501
35	.0192223	.0295842	0.65	0.516	-.0387617	.0772063
36	-.0229111	.0294317	-0.78	0.436	-.0805962	.034774
37	-.0674544	.0288046	-2.34	0.019	-.1239104	-.0109984
38	-.0517775	.0287423	-1.80	0.072	-.1081114	.0045564
39	-.0939179	.0289603	-3.24	0.001	-.150679	-.0371569
_cons	4.383456	.0948762	46.20	0.000	4.197502	4.56941

```
Instrumented:  edu
Instruments:   31.yob 32.yob 33.yob 34.yob 35.yob 36.yob 37.yob 38.yob
               39.yob 2.qob 3.qob 4.qob 31.yob#2.qob 31.yob#3.qob
               31.yob#4.qob 32.yob#2.qob 32.yob#3.qob 32.yob#4.qob
               33.yob#2.qob 33.yob#3.qob 33.yob#4.qob 34.yob#2.qob
               34.yob#3.qob 34.yob#4.qob 35.yob#2.qob 35.yob#3.qob
               35.yob#4.qob 36.yob#2.qob 36.yob#3.qob 36.yob#4.qob
               37.yob#2.qob 37.yob#3.qob 37.yob#4.qob 38.yob#2.qob
               38.yob#3.qob 38.yob#4.qob 39.yob#2.qob 39.yob#3.qob
               39.yob#4.qob 2.pob#2.qob 2.pob#4.qob 4.pob#1b.qob 4.pob#2.qob
               4.pob#3.qob 4.pob#4.qob 5.pob#1b.qob 5.pob#2.qob 5.pob#3.qob
               5.pob#4.qob 6.pob#1b.qob 6.pob#2.qob 6.pob#3.qob 6.pob#4.qob
               8.pob#1b.qob 8.pob#2.qob 8.pob#3.qob 8.pob#4.qob 9.pob#1b.qob
               9.pob#2.qob 9.pob#3.qob 9.pob#4.qob 10.pob#1b.qob
               10.pob#2.qob 10.pob#3.qob 10.pob#4.qob 11.pob#1b.qob
               11.pob#2.qob 11.pob#3.qob 11.pob#4.qob 12.pob#1b.qob
               12.pob#2.qob 12.pob#3.qob 12.pob#4.qob 13.pob#1b.qob
               13.pob#2.qob 13.pob#3.qob 13.pob#4.qob 15.pob#1b.qob
               15.pob#2.qob 15.pob#3.qob 15.pob#4.qob 16.pob#1b.qob
               16.pob#2.qob 16.pob#3.qob 16.pob#4.qob 17.pob#1b.qob
               17.pob#2.qob 17.pob#3.qob 17.pob#4.qob 18.pob#1b.qob
               18.pob#2.qob 18.pob#3.qob 18.pob#4.qob 19.pob#1b.qob
               19.pob#2.qob 19.pob#3.qob 19.pob#4.qob 20.pob#1b.qob
               20.pob#2.qob 20.pob#3.qob 20.pob#4.qob 21.pob#1b.qob
               21.pob#2.qob 21.pob#3.qob 21.pob#4.qob 22.pob#1b.qob
```

```
                    22.pob#2.qob 22.pob#3.qob 22.pob#4.qob 23.pob#1b.qob
                    23.pob#2.qob 23.pob#3.qob 23.pob#4.qob 24.pob#1b.qob
                    24.pob#2.qob 24.pob#3.qob 24.pob#4.qob 25.pob#1b.qob
                    25.pob#2.qob 25.pob#3.qob 25.pob#4.qob 26.pob#1b.qob
                    26.pob#2.qob 26.pob#3.qob 26.pob#4.qob 27.pob#1b.qob
                    27.pob#2.qob 27.pob#3.qob 27.pob#4.qob 28.pob#1b.qob
                    28.pob#2.qob 28.pob#3.qob 28.pob#4.qob 29.pob#1b.qob
                    29.pob#2.qob 29.pob#3.qob 29.pob#4.qob 30.pob#1b.qob
                    30.pob#2.qob 30.pob#3.qob 30.pob#4.qob 31.pob#1b.qob
                    31.pob#2.qob 31.pob#3.qob 31.pob#4.qob 32.pob#1b.qob
                    32.pob#2.qob 32.pob#3.qob 32.pob#4.qob 33.pob#1b.qob
                    33.pob#2.qob 33.pob#3.qob 33.pob#4.qob 34.pob#1b.qob
                    34.pob#2.qob 34.pob#3.qob 34.pob#4.qob 35.pob#1b.qob
                    35.pob#2.qob 35.pob#3.qob 35.pob#4.qob 36.pob#1b.qob
                    36.pob#2.qob 36.pob#3.qob 36.pob#4.qob 37.pob#1b.qob
                    37.pob#2.qob 37.pob#3.qob 37.pob#4.qob 38.pob#1b.qob
                    38.pob#2.qob 38.pob#3.qob 38.pob#4.qob 39.pob#1b.qob
                    39.pob#2.qob 39.pob#3.qob 39.pob#4.qob 40.pob#1b.qob
                    40.pob#2.qob 40.pob#3.qob 40.pob#4.qob 41.pob#1b.qob
                    41.pob#2.qob 41.pob#3.qob 41.pob#4.qob 42.pob#1b.qob
                    42.pob#2.qob 42.pob#3.qob 42.pob#4.qob 44.pob#1b.qob
                    44.pob#2.qob 44.pob#3.qob 44.pob#4.qob 45.pob#1b.qob
                    45.pob#2.qob 45.pob#3.qob 45.pob#4.qob 46.pob#1b.qob
                    46.pob#2.qob 46.pob#3.qob 46.pob#4.qob 47.pob#1b.qob
                    47.pob#2.qob 47.pob#3.qob 47.pob#4.qob 48.pob#1b.qob
                    48.pob#2.qob 48.pob#3.qob 48.pob#4.qob 49.pob#1b.qob
                    49.pob#2.qob 49.pob#3.qob 49.pob#4.qob 50.pob#1b.qob
                    50.pob#2.qob 50.pob#3.qob 50.pob#4.qob 51.pob#1b.qob
                    51.pob#2.qob 51.pob#3.qob 51.pob#4.qob 53.pob#1b.qob
                    53.pob#2.qob 53.pob#3.qob 53.pob#4.qob 54.pob#1b.qob
                    54.pob#2.qob 54.pob#3.qob 54.pob#4.qob 55.pob#1b.qob
                    55.pob#2.qob 55.pob#3.qob 55.pob#4.qob 56.pob#1b.qob
                    56.pob#2.qob 56.pob#3.qob 56.pob#4.qob

. estimates store sls

. * Fit a Lasso partialing-out regression and store the estimates
. poivregress $y ($w = $Z) , controls($X) rseed(1010) nolog
2o.pob#1o.qob omitted
2o.pob#3o.qob omitted
2o.pob#1o.qob omitted
2o.pob#3o.qob omitted

Partialing-out IV linear model        Number of obs              =      10,000
                                      Number of controls         =          10
                                      Number of instruments      =         246
                                      Number of selected controls =          0
                                      Number of selected instruments =       3
                                      Wald chi2(1)               =       32.33
                                      Prob > chi2                =      0.0000
```

		Robust			
lnwage	Coef.	Std. Err.	z	P>|z|	[95% Conf. Interval]
edu	.1390114	.0244494	5.69	0.000	.0910916 .1869313

```
Endogenous:    edu
Note: Chi-squared test is a Wald test of the coefficients of the variables
      of interest jointly equal to zero. Lassos select controls for model
      estimation. Type lassoinfo to see number of selected variables in each
      lasso.

. estimates store poiv

. * Fit a Lasso cross-fit partialing-out regression and store the estimates
. xpoivregress $y ($w = $Z) , controls($X) rseed(1010) nolog
2o.pob#1o.qob omitted
2o.pob#3o.qob omitted
2o.pob#1o.qob omitted
2o.pob#3o.qob omitted
```

```
56o.pob#2o.qob omitted
2o.pob#4o.qob omitted
2o.pob#2o.qob omitted
15o.pob#2o.qob omitted
```

Cross-fit partialing-out	Number of obs	=	10,000
IV linear model	Number of controls	=	10
	Number of instruments	=	246
	Number of selected controls	=	0
	Number of selected instruments	=	5
	Number of folds in cross-fit	=	10
	Number of resamples	=	1
	Wald chi2(1)	=	19.94
	Prob > chi2	=	0.0000

lnwage	Coef.	Robust Std. Err.	z	P>|z|	[95% Conf. Interval]	
edu	.1556948	.034864	4.47	0.000	.0873625	.224027

```
Endogenous:    edu
Note: Chi-squared test is a Wald test of the coefficients of the variables
      of interest jointly equal to zero. Lassos select controls for model
      estimation. Type lassoinfo to see number of selected variables in each
      lasso.
. estimates store xpoiv

. * Compare the three models
. estimates table sls poiv xpoiv ,  ///
> b(%9.5f) star keep($w) title("Comparison between 2sls, poivregress, xpoivregress")

Comparison between 2sls, poivregress, xpoivregress
```

Variable	sls	poiv	xpoiv
edu	0.12112***	0.13901***	0.15569***

```
                 legend: * p<0.05; ** p<0.01; *** p<0.001
```

The size and significance of the estimates of the parameter α, as shown by the comparative table at the end, are fairly in line across all three methods, with the cross-fit partialing-out obtaining the largest magnitude (around 0.15). As wages are in logarithmic scale, a value of the education parameter of 0.15 means that each year of education increases wages by 15%. Interestingly, the cross-fit approach selected 5 out of the initial 246 instruments, while the partialing-out only 3. Both use all of the 10 considered controls. Given the large number of instruments, we consider the Lasso estimated effects as more precise than the one obtained by 2SLS, as it is well known that—in finite sample—the use of a large set of instruments increases the bias of the 2SLS estimation. Observe finally that the two Lasso commands return the selected instruments and controls in the e-return macros e(inst_sel) and e(controls_select), respectively.

3.12.9 Application S6: Optimal Subset Selection

In this section, we illustrate a Stata application of the optimal subset selection using the `subset` command described in Sect. 3.12.3. We consider the `hitters` dataset, a collection of information on major league baseball players observed from the 1986 and 1987 seasons. This dataset comes from the StatLib library which is maintained at Carnegie Mellon University. Our aim is to predict baseball players' salary conditional on a series of information about players' performance in the previous year. The dataset's description, with variables' name and meaning, is reported in Table 3.16. The dataset contains 21 features measured on 322 baseball players.

To begin with, we fit exhaustive (or best) subset selection for the log of baseball players' salary as a function of all the performance features contained in the dataset. Therefore, we consider a model with 19 features, with the subset algorithm fitting $2^{19} = 524,288$ linear regressions. The below code performs all the needed steps to fit this model:

Table 3.16 Description of the Hitters dataset. Number of observations: 322. Number of features: 21

Variable	Meaning
AtBat	Number of times at bat in 1986
Hits	Number of hits in 1986
HmRun	Number of home runs in 1986
Runs	Number of runs in 1986
RBI	Number of runs batted in 1986
Walks	Number of walks in 1986
Years	Number of years in the major leagues
CAtBat	Number of times at bat during his career
CHits	Number of hits during his career
CHmRun	Number of home runs during his career
CRuns	Number of runs during his career
CRBI	Number of runs batted in during his career
CWalks	Number of walks during his career
League	A factor with levels A and N indicating player's league at the end of 1986
Division	A factor with levels E and W indicating player's division at the end of 1986
PutOuts	Number of put outs in 1986
Assists	Number of assists in 1986
Errors	Number of errors in 1986
Salary	1987 annual salary on opening day in thousands of dollars
NewLeague	A factor with levels A and N indicating player's league at the beginning of 1987

```
──────────────────────── Stata code and output ────────────────────────
. * Load the dataset
. sysuse hitters , clear

. * Generate the log of salary
. gen ln_Salary=ln(Salary)
(59 missing values generated)

. * Tranform factor-variables into strings for R to read them
. global VARS "League Division NewLeague"

. foreach V of global VARS{
  2. decode `V´ , gen(`V´2)
  3. drop `V´
  4. rename `V´2 `V´
  5. }

. * Set outcome and features
. global y "ln_Salary"

. global X AtBat Hits HmRun Runs RBI Walks Years CAtBat ///
> CHits CHmRun CRuns CRBI CWalks League Division ///
> PutOuts Assists Errors NewLeague

. * Estimate by "best subset selection"
. quietly{

. * Look at the optimally selected features by fit-index
. use my_opt_vars , clear
(Written by R.               )

. list
```

	AdjR2	CP	BIC
1.	AtBat	AtBat	Hits
2.	Hits	Hits	Walks
3.	HmRun	Walks	Years
4.	Walks	Years	.
5.	Years	CRuns	.
6.	CRuns	CWalks	.
7.	CWalks	LeagueN	.
8.	LeagueN	DivisionW	.
9.	DivisionW	PutOuts	.
10.	PutOuts	.	.
11.	Assists	.	.
12.	Errors	.	.
13.	NewLeague	.	.

```
. * Plot each test fit-index as a function of the number of features
. use my_ind_values , clear
(Written by R.               )

. * Graph BIC
. qui sum BIC

. qui sum ID if BIC==r(min)

. global minIDBIC=r(mean)

. line BIC ID , lc(red) xlabel(1(1)19) ytitle("BIC") ///
> xtitle("Number of features") ///
> plotregion(style(none)) xline($minIDBIC, lp(dash)) ///
> name(g1, replace) scheme(s1mono)

. * Graph adjR2
. qui sum adjR2

. qui sum ID if adjR2==r(max)

. global maxIDadjR2=r(mean)

. line adjR2 ID , lc(red) xlabel(1(1)19) ytitle("adjR2") ///
> xtitle("Number of features") ///
> plotregion(style(none)) xline($maxIDadjR2, lp(dash)) ///
```

```
> name(g2, replace) scheme(s1mono)
. * Graph CP
. qui sum CP
. qui sum ID if CP==r(min)
. global minIDCP=r(mean)
. line CP ID , lc(red) xlabel(1(1)19) ytitle("CP") ///
> xtitle("Number of features") ///
> plotregion(style(none)) xline($minIDCP, lp(dash)) ///
> name(g3, replace) scheme(s1mono)
. * Graph RSS
. line RSS ID , lc(red) xlabel(1(1)19) ytitle("RSS") ///
> xtitle("Number of features") ///
> plotregion(style(none)) ///
> name(g4, replace) scheme(s1mono)
. * Combine all graphs
. graph combine g1 g2 g3 g4
```

We comment on the previous code. After opening the `hitters` dataset, we first transform all factor variables into strings for R to correctly read them as factor and not as numerical variables. Then, we run `subset` for the specified model with the model option set to `best_subset`. After this, we list the optimally selected features by type of fit-index opening the output dataset `my_opt_vars`. This dataset contains the list of optimally selected features by information criterion, more specifically for BIC, adjusted-R^2, and Mallow's CP (or AIC). We observe that, as expected, the most parsimonious information criterion is the BIC that retains only three variables: `Hits`, `Walks`, and `Years`. The other two criteria provide a larger set of selected features. There is, however, a good overlap among the features picked up by each criterion. The code goes on by opening the output dataset `my_ind_values` containing the optimal value of each information criterion by model size. Using this dataset, we plot each fit-index optimal value as a function of the number of features (i.e., model size). Results are visible in Fig. 3.11, where we also plot the residual sum of squares (RSS), which describes the behavior of the training error.

As expected, the information criteria plots exhibit a U-shaped form, signaling that they are approximating well the behavior of the test error as a function of model complexity (or flexibility) measured by the number of the included features (i.e., model degrees of freedom). Observe that, in the case of the adjusted-R^2, we are plotting test accuracy and not test error, thus obtaining in this case a maximum rather than a minimum. As expected, the pattern of the RSS is decreasing as long as we increase model size. This reflects the training error behavior plagued by the over-fitting phenomenon.

Also, we can estimate the optimal least-squares regression using the features selected, for example, by the BIC. The following code performs this task:

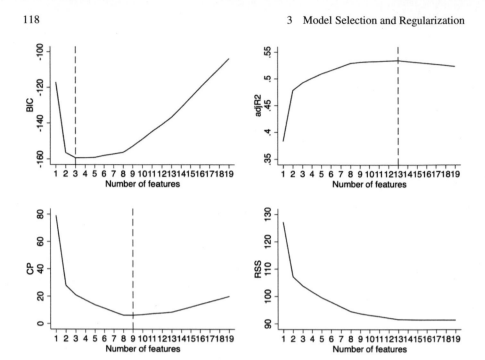

Fig. 3.11 Optimal number of features using exhaustive (or best) subset selection according to different information criteria. The residual sum of squares (RSS) graph shows the training error behavior

```
─────────────────────────────── Stata code and output ───────────────────────────────
. * Load the generated dataset "my_opt_vars"
. use my_opt_vars , clear
(Written by R.                    )

. * Decode the variable BIC -> BICs
. decode BIC , gen(BICs)

. * Store the levels into BICs ///
> * (i.e., the name of the selected predictors) into ///
> * a local called "mysvars" using the option "clean" ///
> * to eliminate undesired quotes
. levelsof BICs , local(mysvars) clean
Hits Walks Years

. global mysvars `mysvars´

. * Re-load the original dataset
. sysuse hitters , clear

. cap gen ln_Salary=ln(Salary)

. * Tranform categorical variables into strings again
. global VARS "League Division NewLeague"

. foreach V of global VARS{
  2. decode `V´ , gen(`V´2)
  3. drop `V´
  4. rename `V´2 `V´
  5. }

. * Run the regression with the "optimal" selected regressors
. reg $y $mysvars

      Source |       SS           df       MS      Number of obs   =       263
─────────────┼──────────────────────────────      F(3, 259)       =     85.85
```

Model	103.288416	3	34.4294719	Prob > F	=	0.0000
Residual	103.865323	259	.401024412	R-squared	=	0.4986
				Adj R-squared	=	0.4928
Total	207.153738	262	.790663123	Root MSE	=	.63326

ln_Salary	Coefficient	Std. err.	t	P>|t|	[95% conf. interval]	
Hits	.0068119	.0010743	6.34	0.000	.0046965	.0089273
Walks	.006582	.0022523	2.92	0.004	.0021469	.0110171
Years	.0944676	.0082603	11.44	0.000	.0782017	.1107334
_cons	4.231361	.1176719	35.96	0.000	3.999645	4.463076

We see that any additional year played in the major league produces a salary increase of around 9.4%, a sizable effect. Although we cannot interpret this result causally, we can conclude that the variable Years has a strong predictive impact on players' salary. We can now use this regression fit to predict the salary to pay to a brand-new player, as this model results in the most predictive one out of the 524,288 models we have compared.

We can run a similar procedure using as model type options either forward or backward with no substantial differences in the subset syntax. Results (not reported here) are very close to those obtained by best subset selection, although they may be different in other datasets.

3.12.10 Application S7: Regularized Regression with Time-Series and Longitudinal Data

Regularized regression can be naturally extended to time-series and longitudinal (panel) data with minor adjustments. With time-series and longitudinal data, however, cross-validation splitting cannot be carried out randomly, as data have an inner sequential ordering. Instead, time-series settings require the so-called rolling h-step ahead cross-validation (Bergmeir et al., 2018). Table 3.17 shows the logic of the rolling cross-validation, in the 1-step ahead (panel A), 2-step ahead (panel B), and 1-step ahead with fixed window (panel C).

Carrying out rolling h-step ahead cross-validation requires to set two parameters, i.e., the order of the steps ahead h, and the origin of the validation. For example, in Table 3.17, panel A shows a 1-step cross-validation ($h=1$), with origin at time $t=3$. This implies that the validation procedure starts with the first training set made of the first three observations of the dataset, then progressing by one observation at time until the last observation is used as a test unit. With a time-series made of 8 observations, a 1-step ahead, and origin at 3, we can perform 5 pairs of training (T) and testing (V) sets. Panel B is similar to panel A, but this time the first observation ahead is skipped, as a 2-step ahead was considered here. Panel C, finally, performs a 1-step ahead cross-validation, but guarantees a fixed window that allows for the training datasets to preserve the same size over the entire procedure.

Table 3.17 Rolling h-step ahead cross-validation

	A 1-step ahead						B 2-step ahead						C 1-step fixed window				
	1	2	3	4	5		1	2	3	4	5		1	2	3	4	5
1	T	T	T	T	T	1	T	T	T	T	T	1	T	.	.	.	.
2	T	T	T	T	T	2	T	T	T	T	T	2	T	T	.	.	.
3	T	T	T	T	T	3	T	T	T	T	T	3	T	T	T	.	.
4	V	T	T	T	T	4	.	T	T	T	T	4	V	T	T	T	.
5	.	V	T	T	T	5	V	.	T	T	T	5	.	V	T	T	T
6	.	.	V	T	T	6	.	V	.	T	T	6	.	.	V	T	T
7	.	.	.	V	T	7	.	.	V	.	T	7	.	.	.	V	T
8	.	.	.	.	V	8	.	.	.	V	.	8	.	.	.	.	V
						9	.	.	.	.	V						

The `cvlasso` command supports cross-validation using the rolling h-step ahead cross-validation for forecasting purposes via the option `rolling`, once data are time-series initialized using `tsset` or `xtset`. The options h(*integer*) and `origin`(*integer*) control the forecasting horizon, and the starting point of the rolling forecast, respectively. The option `fixedwindow` allows for the size of the training dataset to remain the same.

In this application, we consider a subsample of the dataset `stock`, containing 315 daily data on the stock returns of three car manufacturers—Toyota, Nissan, and Honda—from October 2, 2003, to December 31, 2010. The stock return variables are called `toyota`, `nissan`, and `honda`.

We run a Lasso auto-regressive model of Toyota returns as a function of contemporaneous and lagged returns of Toyota itself, and Nissan and Honda returns. As data are daily, we consider seven lags (i.e., one week). We run the `cvlasso` command with 1-step ahead cross-validation ($h=1$), origin at $t=208$ (corresponding to July 20, 2010), and with a training set fixed window. Then, we plot the cross-validation mean squared error prediction error (MSPE) as a function of a grid of the (log of the) penalization parameter λ. The full code is

```
───────────────────────────── Stata code and output ─────────────────────────────
. use stock , clear
(Data from Yahoo! Finance)

. tsset t
        time variable:  t, 1 to 315
                delta:  1 unit

. qui cvlasso toyota L(1/7).toyota nissan L(1/7).nissan honda L(1/7).honda , ///
> rolling origin(208) fixedwindow h(1)

. cvlasso, plotcv plotregion(style(none))

. cvlasso, lopt
Estimate lasso with lambda=1.024 (lopt).

            Selected │       Lasso    Post-est OLS
            ─────────┼──────────────────────────────
              toyota │
                 L5. │     0.0046629      0.0838325
```

nissan	0.1161236	0.1682295
honda		
--.	0.3117006	0.3672821
L1.	-0.0238446	-0.1055994
L2.	0.0507098	0.1479368
L5.	0.0174663	0.0879722
Partialled-out*		
_cons	-0.0003852	-0.0005655

```
. di "The optimal lambda is equal to: " e(lopt)
The optimal lambda is equal to: 1.0242345

. di "The optimal MSPE is equal to: " e(mspemin)
The optimal MSPE is equal to: .00008033
```

The optimal λ is equal to 1.024, and the corresponding (minimized) MSPE is equal to 0.000081. At this level, only a few parameters are retained, the lag number 5 of Toyota, the contemporaneous Nissan returns, the lags 1, 2, and 5, and the contemporaneous returns of Honda. Only the first lag of Honda has a negative effect on Toyota current returns.

Figure 3.12 shows the minimization of the MSPE over the log of λ. We can observe a well-identified minimum with a not too large one standard error confidence interval for the log of λ.

We can perform many different calibrations of our forecasting model. An interesting one may be that of discovering at which horizon our model is able to produce the best forecast. For this purpose, we can form a grid of horizons (from 1 to 7), run

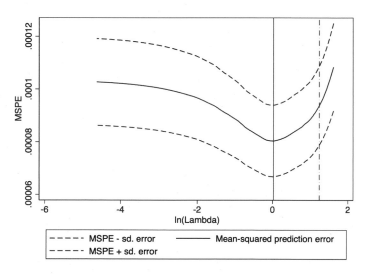

Fig. 3.12 Rolling 1-step ahead cross-validation

the model at every grid point, save the minimum of the MSPE, store all the MPSEs obtain in a matrix, and then plot the result. The following code carries out this task:

```
─────────────────────────────── Stata code and output ───────────────────────────────
. mat M=J(7,2,.)

. local i=1

. forvalues h=1/7{
  2. qui cvlasso toyota L(1/7).toyota nissan L(1/7).nissan honda L(1/7).honda , ///
> rolling origin(208) fixedwindow h(`h')
  3. local mse=e(mspemin)
  4. mat M[`i',1]=`h'
  5. mat M[`i',2]=`mse'
  6. local i=`i'+1
  7. }

. svmat M

. replace M2=M2*10000
(7 real changes made)

. format M2 %9.3g

. twoway dropline  M2 M1 , ylabel(.66(0.05).81) xlabel(0(1)8) ///
> plotregion(style(none)) scheme(s1mono) mlabel(M2) ///
> ytitle(MSPE x 10000) xtitle(Step of forecasting horizon)
```

Figure 3.13 shows that the best forecast is reached at $h = 5$, that is, in the fifth day ahead, with a similar performance also at the sixth and seventh days (for readability, we multiplied the MSPE by a factor of 10,000).

Finally, we provide an illustrative forecasting exercise using this machinery. We split the sample into a training and a test dataset. Specifically, we generate the variable train taking value one for the training dataset (first 300 observations) and zero for the testing dataset (last 15 observations). We fit the previous model on the training dataset using cvlasso with $h = 1$, saving the CV best λ in a global macro; then, we run a standard Lasso (using the lasso2 command) with same specification at the CV optimal λ, and take predictions on the testing dataset. As test error, we compute the mean absolute percentage error (MAPE) on the test dataset. This index is a relative one, ranging from zero and one. Its formula is

$$\text{MAPE} = \frac{|y - \widehat{y}|}{|y|} \cdot 100 \tag{3.40}$$

As test forecasting accuracy we consider $100 - \text{MAPE}$. This is the code performing these steps:

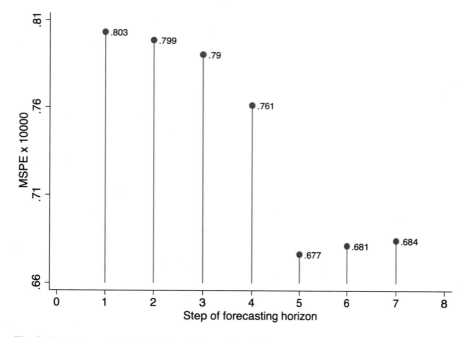

Fig. 3.13 Minimum MSPE at different forecasting horizons

────────────────── Stata code and output ──────────────────

```
. use stock , clear
(Data from Yahoo! Finance)

. gen train=(t<=300)

. tsset t
        time variable:  t, 1 to 315
                delta:  1 unit

. qui cvlasso toyota L(1/7).toyota nissan L(1/7).nissan honda L(1/7).honda in 1/300, ///
> rolling origin(208) fixedwindow h(1)

. qui cvlasso, lopt

. global opt_cv_lambda=e(lopt)

. qui lasso2 toyota L(1/7).toyota nissan L(1/7).nissan honda L(1/7).honda in 1/300 , ///
> lambda($opt_cv_lambda)

. predict double toyota_hat_test in 301/L, xb
Use e(b) from previous lasso2 estimation (lambda=.9085350841123).
(300 missing values generated)

. * Mean Absolute Percentage Error (MAPE) on the test dataset
. gen mape=(abs(toyota-toyota_hat_test))/toyota*100
(300 missing values generated)

. qui sum mape

. global MAPE=round(r(mean), 0.01)

. global Accuracy = 100-$MAPE

. di "The Mean Absolute Percentage Error (MAPE) = " $MAPE
```

```
The Mean Absolute Percentage Error (MAPE) = 19.89

. di "Test accuracy = 1-MAPE = " $Accuracy
Test accuracy = 1-MAPE = 80.11
```

Results show that the test accuracy is equal to 80%. This accuracy could be increased, for example, also by tuning over the Elastic-net parameter α.

The following chunk of code allows us for generating the plot of the out-of-sample Lasso forecast, whose output is displayed in Fig. 3.14.

```
──────────────────────────── Stata code and output ────────────────────
. global lambda_opt=round($opt_cv_lambda,0.01)

. preserve

. keep if t>=301
(300 observations deleted)

. tw (tsline toyota_hat_test , lc(orange) lp(dash) lw(medthick)) ///
>     (tsline toyota , lc(green) lp(dash) lw(medthick)) , ///
> legend(order(1 "Out-of-sample forecast" 2 "Actual data")) plotregion(style(none)) scheme(s1mono) ///
> xtitle("Days") ytitle("Toyota stock returns")   ///
> ylabel(,labsize(small)) ///
> tlabel(, angle(360) labsize(small)) ///
> note( ///
> "- Model: Lasso"  ///
> "- Optimal `lambda´: $lambda_opt" ///
> "- Test-MAPE in % = $Accuracy" ///
> )

. restore
```

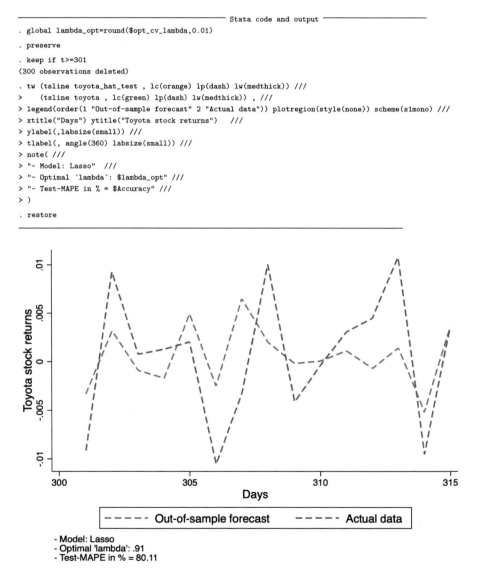

- Model: Lasso
- Optimal 'lambda': .91
- Test-MAPE in % = 80.11

Fig. 3.14 Out-of-sample Lasso forecasting

As expected, the figure shows a better fit in the first forecasting week than in the second one, with the Lasso forecasting providing a quite smoothing pattern.

For the sake of completeness, we finally run a Lasso with 1-step ahead cross-validation in a panel data setting. For this purpose, we use the grunfeld dataset (Grunfeld and Griliches, 1960) for regressing a company's current-year gross investment (invest) as function of the company's 10 prior years market value (mvalue) and the 10 prior years value of the company's plant and equipment (kstock). The dataset includes 10 companies over 20 years, from 1935 through 1954. With a rolling fixed window, the code goes as follows:

──────────────────────── Stata code and output ────────────────────────

```
. * Load the dataset
. sysuse grunfeld, clear

. * xtset the data
. xtset company year
      panel variable:  company (strongly balanced)
       time variable:  year, 1935 to 1954
               delta:  1 year
. * Fit a 1-step ahead CV Lasso with fixed-effects
. cvlasso invest L(1/10).invest mvalue L(1/10).mvalue kstock L(1/10).kstock , ///
> rolling fixedwindow origin(1950) fe
Rolling forecasting cross-validation with 1-step ahead forecasts. Elastic net with alpha=1.
Training from-to (validation point): 1945-1950 (1951), 1946-1951 (1952), 1947-1952 (1953),
1948-1953 (1954).
```

	Lambda	MSPE	st. dev.	
1	18872.211	20990.918	11743.299	
2	17195.656	18395.691	9900.9713	
3	15668.041	16214.459	8439.7665	
4	14276.136	14242.936	7223.1218	
5	13007.883	12603.181	6237.5988	
6	11852.299	11239.131	5440.489	
7	10799.373	10001.358	4742.9243	
8	9839.9865	8945.4939	4186.9475	
9	8965.8292	8047.0393	3744.8772	
10	8169.3296	7298.8384	3380.5422	
11	7443.5889	6669.9393	3082.5275	
12	6782.321	6136.3169	2839.9403	
13	6179.7983	5676.6116	2629.3069	
14	5630.802	5276.7408	2452.7728	
15	5130.5771	4942.1022	2301.1586	^
16	4674.7908	4659.8086	2171.6329	
17	4259.4952	4420.4444	2061.2511	
18	3881.0934	4217.1907	1967.09	
19	3536.3078	4067.013	1900.2491	
20	3222.1519	3949.4015	1846.3927	
21	2935.9048	3873.9801	1806.0486	
22	2675.0872	3824.8529	1783.5335	
23	2437.4398	3744.3448	1746.3502	
24	2220.9044	3662.4328	1708.9739	
25	2023.6053	3601.8414	1687.715	
26	1843.8338	3536.5282	1661.0493	
27	1680.0327	3476.8122	1635.378	
28	1530.7832	3425.6764	1612.7539	
29	1394.7927	3384.2938	1594.1014	
30	1270.8831	3375.7213	1574.6166	*
31	1157.9814	3388.9423	1547.5721	
32	1055.1095	3423.2594	1532.1488	
33	961.37648	3464.751	1520.4705	
34	875.97044	3511.5703	1508.316	
35	798.15164	3597.2843	1517.8686	
36	727.24605	3693.3554	1535.4842	

37	662.63951	3743.5951	1526.782
38	603.77245	3781.5135	1514.5093
39	550.13497	3809.1631	1497.4772
40	501.26249	3832.7227	1485.3611
41	456.73171	3858.4708	1466.7154
42	416.15693	3874.8579	1444.8665
43	379.18669	3886.9812	1418.9082
44	345.50079	3907.8926	1393.1963
45	314.80745	3938.6234	1368.0596
46	286.84083	3951.8554	1338.8332
47	261.35868	3972.3068	1319.7574
48	238.14029	4003.5605	1309.0408
49	216.98457	4027.2992	1307.4483
50	197.70826	4057.1381	1307.7981
51	180.1444	4106.4719	1316.1685
52	164.14087	4149.245	1322.8771
53	149.55904	4242.1446	1359.1322
54	136.27263	4381.0975	1414.4228
55	124.16655	4531.5302	1455.3108
56	113.13593	4674.0809	1493.6194
57	103.08525	4893.6516	1554.7808
58	93.927439	5151.6231	1632.6187
59	85.583184	5347.7117	1670.7953
60	77.98021	5637.9764	1756.2755
61	71.052663	5933.0385	1831.433
62	64.74054	6240.2329	1925.0405
63	58.989169	6532.5331	2046.7886
64	53.748734	6795.7398	2175.6726
65	48.973845	7056.1109	2294.6711
66	44.623143	7279.2067	2388.5182
67	40.658946	7468.0193	2470.1663
68	37.046917	7645.3017	2548.5055
69	33.755771	7854.0271	2610.046
70	30.757002	8076.0008	2680.2903
71	28.024634	8301.2116	2741.7018
72	25.535003	8522.208	2795.7834
73	23.266544	8736.6831	2843.4343
74	21.199608	8942.9901	2885.4649
75	19.316293	9129.4243	2928.3082
76	17.600287	9297.3713	2973.1591
77	16.036726	9449.2532	3013.4836
78	14.612068	9600.8341	3042.8252
79	13.313972	9745.4338	3068.248
80	12.131195	9880.5873	3090.8802
81	11.053493	10006.716	3111.1046
82	10.071531	10129.859	3131.2036
83	9.1768043	10257.957	3158.5711
84	8.3615623	10376.567	3184.4511
85	7.6187442	10492.621	3210.3371
86	6.9419159	10605.893	3235.1068
87	6.3252153	10716.086	3259.244
88	5.7633006	10818.239	3280.9122
89	5.2513048	10917.885	3301.2603
90	4.7847934	11017.065	3321.6522
91	4.3597255	11107.83	3340.2294
92	3.9724195	11191.277	3357.2998
93	3.6195207	11267.929	3372.9756
94	3.2979725	11338.284	3387.3621
95	3.0049897	11402.816	3400.5578
96	2.7380347	11461.968	3412.6548
97	2.4947952	11516.159	3423.7392

* lopt = the lambda that minimizes MSPE.
  Run model: cvlasso, lopt
^ lse = largest lambda for which MSPE is within one standard error of the minimal MSPE.
  Run model: cvlasso, lse

. * Run the optimal selected model by cross-validation

```
. cvlasso, lopt
Estimate lasso with lambda=1270.883 (lopt).
```

Selected	Lasso	Post-est OLS
invest		
L1.	0.6332589	0.6547664
L9.	0.1927027	0.2273480
L10.	0.3754310	0.5440221
mvalue		
--.	0.0834764	0.0838639
L6.	-0.0110594	-0.0478487
L7.	0.0183324	0.0595377
L9.	0.0179779	0.0449245
kstock	0.0594227	0.0423883

```
. * Display the CV plot
. cvlasso, plotcv plotregion(style(none))
```

Figure 3.15 shows the CV plot of this regression, with the optimal level of λ equal to 1270.88. From the regression output table, we see that only a few lags of the predictors are actually retained, thus showing a relevant prediction inefficiency if we were considering instead the entire lag structure.

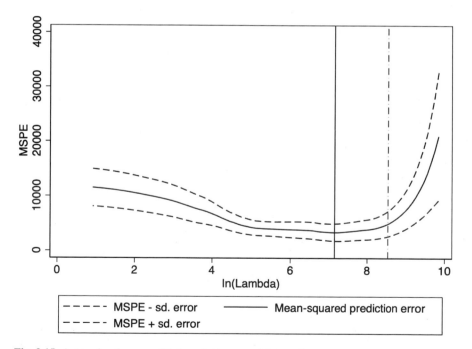

Fig. 3.15 1-step ahead cross-validation plot in a panel data setting

3.13   R Implementation

The software R has a powerful package to fit regularized linear models called `glmnet`. The syntax of the related function `glmnet()` is straightforward and takes on this form:

```
glmnet( x, y, family = familytype, alpha = alpha_value , lambda
    = grid_of_values, standardize = TRUE or FALSE, ... )
```

where

- *familytype*: can be one of `"gaussian"`, `"binomial"`, `"poisson"`, `"multinomial"`, `"cox"`, or `"mgaussian"`;
- *alpha_value*: a number between 0 (Ridge case) and 1 (Lasso case);
- *grid_of_values*: a user supplied lambda sequence;
- `standardize`: `TRUE` if variable have to be standardized, `FALSE` if not;
- ... : other options.

The syntax and use of `glmnet` is similar to its companion function `glm()` for unregularized regression/classification. As in that case, the choice of the family function depends on the nature of the dependent variable y. If y is numerical and continuous, one uses the `"gaussian"` option; if y is binary, the `"binomial"` option, etc. Other specialized functions within the `glmnet` package carry out cross-validation (`cv.glmnet()`), and graphical representation of results (`plot.glmnet()`). Their use will be shown in the applications that follow.

3.13.1   Application R1: Fitting a Gaussian Penalized Regression

In this R application, we consider again the `Hitters` dataset, where we aim at predicting again the log of baseball players' salary as a function of a series of players' performance variables.

As salary is a numerical and continuous variable, we use the `glmnet()` function with the `"gaussian"` family option. After clearing the R environment and setting the working directory, we load the needed packages to fit this model, specifically, the packages ISLR, glmnet, and `foreign`. Also, we load and attach the dataset `Hitters`, eliminate missing values, and describe the dataset.

Then, we form the y and the x matrices from the dataset by noticing that `glmnet()` expects a matrix of predictors as argument, not a dataframe. Specifically, we employ the `model.matrix()` function to create a matrix from a formula by expanding factor variables to a set of dummy variables.

Subsequently, we form a grid for the "lambdas" (100 values), and generate randomly the train and test indexes to form the training and the testing datasets using a 1/2 split of the initial dataset. Also, we identify the "test rows of y" as `y.test`.

Once we have all the ingredients ready, we can fit the Gaussian Lasso to the training dataset via `glmnet()`, obtaining immediately the number of nonzero coefficients for each value of the lambdas considered. We can also plot the Lasso variables as a function of these lambdas.

Then, using the training dataset, we estimate the *optimal* lambda via a tenfold cross-validation (CV) using the function `cv.glmnet()` to then plot the CV results. In particular, we show the optimal lambda and predict y, the log of salary, in the testing dataset using it. At the optional lambda, we also estimate the test prediction error.

Finally, we re-run `glmnet()` over the entire original dataset, estimate the optimal coefficients using the optimal lambda previously found, showing only coefficients different from zero (i.e., those actually retained by Lasso). To conclude, we fit least squares using the (optimal) retained features.

The following R code and output accomplishes all the steps listed above:

───────────────────────────── R code and output ─────────────────────────────

```
# Fitting a lasso gaussian regression using the "glmnet" package

# Clear the R environment
> rm(list = ls())

# Set the directory
> setwd("~/Dropbox/Giovanni/Corsi_Stata_by_G.Cerulli/Course_R_ML_Ircres_WS_2021")

# Load the needed packages
> library(ISLR)
> library(glmnet)
> library(foreign)

# Load and attach the data, eliminate missing values, describe the dataset
> attach(Hitters)
> Hitters=na.omit(Hitters)
> dim(Hitters)
[1] 263   20

# Form y and x matrices
# IMPORTANT: glmnet() expects a matrix of predictors, not a dataframe.
# "model.matrix()" creates matrix from a formula by expanding "factors"
# to a set of dummy variables.
> x=model.matrix(Salary~.,Hitters)[,-1] # -1 -> eliminate the intercept
> y=Hitters$Salary

# Form a grid for the "lambdas" (100 values)
> grid=10^seq(10,-2,length=100)

# Generate randomly 1/2 of rows of x (i.e, the vector "train")
> set.seed(1)
> train=sample(1:nrow(x), nrow(x)/2)

# Generate the "test" rows of x from "train"
> test=(-train)

# Consider only the "test rows of y" and call it "y.test"
> y.test=y[test]

# Estimate LASSO via "glmnet" in the training dataset
> lasso.mod=glmnet(x[train,],y[train],alpha=1,lambda=grid,family="gaussian")
```

```
# Obtain the the number of nonzero coefficients for each value of lambda
> lasso.mod$df
  [1]  0  0  0  0  0  0  0  0  0  0  0  0  0  0  0  0  0  0  0  0  0  0  0  0  0  0  0  0  0  0  0  0  0  0  0  0  0
 [38]  0  0  0  0  0  0  0  0  0  0  0  0  0  0  0  0  0  0  0  0  0  0  0  0  0  0  0  0  0  0  0  0  0  2  3  4  5  5  5  5  7  7  8
 [75] 10  9 13 15 16 17 17 17 18 18 18 17 17 17 17 17 17 17 18 18 19 19 19 19 19 19

# Plot the lasso variables plot as function of lambda
> plot(lasso.mod , xvar="lambda" , label=TRUE)

# Use the training dataset, and estimate the optimal lambda
# via cross-validation (CV) using "cv.glmnet()"
> set.seed(1)
> cv.out=cv.glmnet(x[train,],y[train],alpha=1,nfolds = 10)

# Plot CV results
> plot(cv.out)

# Show the optimal lambda
> bestlam=cv.out$lambda.min
> print(bestlam)
[1] 16.78016

# Predict y in the test dataset using the 'optimal' lambda
> lasso.pred=predict(lasso.mod,s=bestlam,newx=x[test,])

# Estimate the "test prediction error"
> mean((lasso.pred-y.test)^2)
[1] 100743.4

# Compute relative error = 1-|y - y_hat|/y
> mean(1-(abs(lasso.pred-y.test)/y.test))*100
[1] 8.842186

# Re-run "glmnet" over the entire dataset
> out=glmnet(x,y,alpha=1,lambda=grid)

# Estimate the optimal coefficients using the optimal lambda
> lasso.coef=predict(out,type="coefficients",s=bestlam)[1:20,]

# Show the coefficients
> lasso.coef
  (Intercept)         AtBat          Hits         HmRun          Runs           RBI         Walks         Years        CAtBat
   18.5394844     0.0000000     1.8735390     0.0000000     0.0000000     0.0000000     2.2178444     0.0000000     0.0000000
        CHits        CHmRun         CRuns          CRBI        CWalks       LeagueN     DivisionW       PutOuts       Assists
    0.0000000     0.0000000     0.2071252     0.4130132     0.0000000     3.2666677  -103.4845458     0.2204284     0.0000000
       Errors     NewLeagueN
    0.0000000     0.0000000

# Show only coefficients different from zero
> A <- lasso.coef[lasso.coef!=0]
> print(A)
  (Intercept)          Hits         Walks         CRuns          CRBI       LeagueN     DivisionW       PutOuts
   18.5394844     1.8735390     2.2178444     0.2071252     0.4130132     3.2666677  -103.4845458     0.2204284

# Estimate the OLS with the optimal retained features
> B <- names(A)
> B[B=="DivisionW"]<-"Division"
> B[B=="LeagueN"]<-"League"

# Generate the "formula" from the names
> fopt <- paste('y','~',paste(B[-1], collapse=' + ' )) # -1 -> eliminate the intercept
> fit_opt <- lm(fopt, data=Hitters) # optimal linear regression
> summary(fit_opt)

Call:
lm(formula = fopt, data = Hitters)


Residuals:
    Min      1Q  Median      3Q     Max
-874.26 -162.14  -19.01  126.79 2100.32


Coefficients:
            Estimate Std. Error t value Pr(>|t|)
```

```
(Intercept)  -39.0662    65.7792  -0.594 0.553107
Hits           2.0932     0.5706   3.669 0.000297 ###
Walks          2.3991     1.1980   2.003 0.046282 #
CRuns          0.2183     0.1915   1.140 0.255367
CRBI           0.4424     0.1944   2.276 0.023658 #
LeagueN       39.4528    41.0537   0.961 0.337461
DivisionW   -133.4456    40.5836  -3.288 0.001150 ##
PutOuts        0.2580     0.0774   3.333 0.000986 ###
---
Signif. codes:  0 '###' 0.001 '##' 0.01 '#' 0.05 '.' 0.1 ' ' 1

Residual standard error: 326.8 on 255 degrees of freedom
Multiple R-squared:  0.4891,      Adjusted R-squared:  0.4751
F-statistic: 34.87 on 7 and 255 DF,  p-value: < 2.2e-16
```

The best lambda is equal to 16.7, with a relative test error of around 8.84%, and only 7 out of the initial 19 features retained in the model. Figure 3.17 plots the optimal (log of) lambda obtained by the tenfold cross-validation, while the Lasso coefficients' plot is displayed in Fig. 3.16. The interpretation of these results are the same as the ones discussed in the Stata implementation section.

3.13.2 Application R2: Fitting a Multinomial Ridge Classification

Still using the glmnet package and the dataset Hitters, in this application we show how to fit a multinomial Ridge classification. The steps to carry out this fit are similar to those listed in the previous section, but with some differences in the arguments of the glmnet() function.

We start, as usual, by loading the needed dependencies and the dataset. We then generate a categorical variable dSalary built from Salary having three categories (1, 2, 3) corresponding to "low", "medium", and "high" salary. Subsequently, we

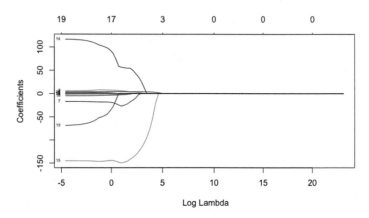

Fig. 3.16 Lasso coefficients' path using glmnet()

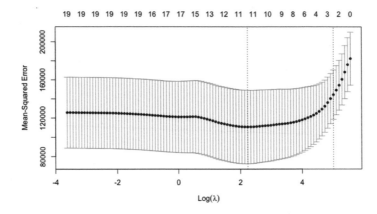

Fig. 3.17 Lasso tenfold cross-validation results using `glmnet()`

generate the matrix of features (x) and the target (y) and perform a default Ridge regression in the case of a multinomial outcome. The options of the `glmnet()` function required to be in this case: `family ="multinomial"` and `alpha = 0`.

After plotting the Elastic-net coefficients graph, we estimate the Ridge multinomial at a pre-fixed lambda equal to 0.3, predict the probabilities and the classes, and estimate the overall "training error rate". By splitting the initial dataset, we also estimate the test error rate.

As we did for the Gaussian Lasso, also for Ridge cross-validation we form a grid of lambdas for finding the optimal one. We then plot CV results, show the optimal lambda, and estimate the test error rate.

By and large, the following R code and output set out how to perform the previous steps:

───────────────────────────────── R code and output ─────────────────────────────────

```
# Fit a multinomial Ridge regression

# Clear all the R environment
> rm(list = ls())

# Change the directory
> setwd("~/Dropbox/Giovanni/Corsi_Stata_by_G.Cerulli/Course_R_ML_Ircres_WS_2021")

# Load the packages to make them active
> library(ISLR)
> library(glmnet)

# Attach the "Hitters" dataset and eliminate the missing values
> attach(Hitters)
> sum(is.na(Hitters$Salary))
[1] 59
> Hitters=na.omit(Hitters)

# Generate a categorical variable 'dSalary' built from 'Salary'
# having three categories (1,2,3) corresponding to
# 'low', 'medium', 'high' salary
> Hitters$dSalary <- 1
> Hitters$dSalary[Hitters$Salary>500] <- 2
> Hitters$dSalary[Hitters$Salary>=800] <- 3
```

```
# Eliminate 'Salary'
> Hitters$Salary<-NULL

# Generate the matrix of features ("x") and the target ("y")
> x=model.matrix(dSalary~.,Hitters)[,-1]
> y <- Hitters$dSalary

# Perform a default Ridge regression in the case of multinomial outcome
> ridge2 <- glmnet(x, y,
+                   ## Multinomial regression
+                   family = "multinomial",
+                   ## alpha = 1 is the lasso penalty, and alpha = 0 the ridge penalty.
+                   alpha = 0)

# Plot the elastic net graph
> plot(ridge2, xvar = "lambda")

# Estimate Ridge at a pre-fixed lambda = 0.3
> ridge2p <- glmnet(x, y,
+                   family = "multinomial",
+                   alpha = 0,
+                   lambda = 0.3)

# Predict 'probabilities' at the pre-fixed lambda
> ridge2p.pred <- predict(ridge2p,type="response",newx=x)

# Predict 'classes' at the pre-fixed lambda
> ridge2p.class <- predict(ridge2p,type="class",newx=x)

# Estimate the overall 'training error rate'
> mean(ridge2p.class!=y)
[1] 0.3117871

# Estimate the test-error

# Generate randomly 1/2 of rows of x (i.e, the vector "train")
# This yields the "row-index"
> set.seed(1)
> train=sample(1:nrow(x), nrow(x)/2) # row-index

# Generate the "test" row-index of x from "train" row-index
> test=(-train)

# Consider only the "test rows of y" and call it "y.test"
> y.test=y[test]

# Form a grid for the "lambdas"
> grid=10^seq(10,-2,length=100)

# Estimate Ridge via "glmnet()"
> ridge.mod=glmnet(x[train,],y[train],
+                   family = "multinomial",
+                   alpha=0,
+                   lambda=grid)

# Obtain the The number of nonzero coefficients for each value of lambda
> ridge.mod$df
  [1] 19 19 19 19 19 19 19 19 19 19 19 19 19 19 19 19 19 19 19 19 19 19 19 19 19 19
 [27] 19 19 19 19 19 19 19 19 19 19 19 19 19 19 19 19 19 19 19 19 19 19 19 19 19 19
 [53] 19 19 19 19 19 19 19 19 19 19 19 19 19 19 19 19 19 19 19 19 19 19 19 19 19 19
 [79] 19 19 19 19 19 19 19 19 19 19 19 19 19 19 19 19 19 19 19 19

# Plot the Ridge variables plot as function of lambda
> plot(ridge.mod)

# Using the training dataset estimate the optimal lambda via CV
> set.seed(1)
> cv.out=cv.glmnet(x[train,],y[train],alpha=0)

# Plot CV results
> plot(cv.out)

# Show the optimal lambda
```

```
> bestlam=cv.out$lambda.min
> bestlam
[1] 0.2387688

# Predict probabilities in the 'test-dataset' using optimal lambda
> ridge.pred.prob=predict(ridge.mod,s=bestlam,type="response",newx=x[test,])

# Predict probabilities in the 'test-dataset' using optimal lambda
> ridge.pred.class=predict(ridge.mod,s=bestlam,type="class",newx=x[test,])

# Estimate the 'test error rate'
> mean(ridge.pred.class!=y.test)
[1] 0.3636364

# Re-run "glmnet" over the entire dataset
> out=glmnet(x,y,alpha=0,lambda=grid)

# Estimate the optimal coefficients using the optimal lambda
> ridge.coef=predict(out,type="coefficients",s=bestlam)

# View the coefficients
> A<-ridge.coef[,1]
> print(A)

  (Intercept)          AtBat           Hits          HmRun           Runs            RBI
 6.660202e-01   2.945127e-04   2.136783e-03   2.590820e-03   9.699307e-04   5.370976e-04
        Walks          Years          CAtBat          CHits         CHmRun          CRuns
 3.178597e-03  -2.133991e-03   2.364863e-05   1.310763e-04   5.443815e-04   2.641133e-04
         CRBI         CWalks         LeagueN       DivisionW        PutOuts         Assists
 1.805291e-04   1.930625e-05   6.873152e-03  -1.628324e-01   3.054495e-04   4.257808e-05
       Errors       NewLeagueN
-2.393060e-03   9.677264e-02
```

We can see that the training error rate is equal to 0.31 while the test error rate is larger and equal to 0.36. For convenience, we show neither the Ridge coefficients plot nor the cross-validation plot in this case. As for the predictions, obtained using the `predict()` function after fitting `glmnet()`, it is useful to observe that the option `type="class"` predicts classes, whereas the option `type="response"` predicts the probabilities of each class.

3.14   Python Implementation

The Python Machine Learning platform Scikit-learn has a series of commands/ functions (called *methods* in Python) to fit lasso and provide graphical representations of the Lasso solution, including cross-validation results. In what follows, I consider two applications: one based on the `lassoCV()` method, and one based on a more general approach to regularized regression/classification using the `ElasticNet()` method. Indeed, although we used `ElasticNet()` in Sect. 3.12.6 from within Stata to fit a multinomial lasso, we did not visualize and explain the underlying Python code. To fill this gap, we provide here a deeper focus on the syntax and semantic of fitting a Machine Learning algorithm in Python using Scikit-learn, with the `ElasticNet()` case representing a proper illustrative example.

Table 3.18 Description of the `diabetes` dataset. Number of variables: 11. Sample size: 442

Name	Description
`diabprog`	Measure of diabetes progression one year after baseline
`age`	Age in years
`sex`	Sex of the patient
`bmi`	Body mass index
`bp`	Average blood pressure
`tc`	Total serum cholesterol
`ldl`	Low-density lipoproteins
`hdl`	High-density lipoproteins
`tch`	Total cholesterol/HDL
`ltg`	Log of serum triglycerides
`glu`	Blood sugar level

3.14.1 Application P1: Fitting Lasso Using the `LassoCV()` Method

In this application, using the `diabetes` dataset, we aim at fitting a Lasso regression. We both provide the Lasso coefficient path and optimal regularization using cross-validation.

The `diabetes` dataset collects ten baseline features (age, sex, body mass index, average blood pressure, and six blood serum measurements) measured over 442 diabetes patients. The response variable of interest (`diabprog`) is a quantitative measure of disease progression one year after baseline. Table 3.18 shows a description of the `diabetes` dataset.

The following Python code fits a Lasso over a pre-defined grid of lambdas and displays the Lasso coefficients' path-diagram using the function `lasso_path()` taking as arguments the target variable y and the array of features X.

```
──────────────────────────────────────── Python code and output ────────────────────
>>> # Import dependencies
... from itertools import cycle
>>> import numpy as np
>>> import pandas as pd
>>> import matplotlib.pyplot as plt
>>> from sklearn import datasets
>>> from sklearn.linear_model import lasso_path , LassoCV
>>> from sklearn.model_selection import train_test_split
>>> from sklearn.metrics import mean_squared_error
>>>
>>> # Load the 'diabetes' dataset
... X, y = datasets.load_diabetes(return_X_y=True)
>>>
>>> # Put the features into a dataframe for getting plot legend
... df = pd.DataFrame(X,
...       columns = ['age','sex','bmi','bp','tc','ldl','hdl', 'tch', 'ltg', 'glu'])
...
... # Standardize data
```

```
>>> X = (X - X.mean())/(X.std())
>>>
>>> # Compute the coefficients paths
... print("Computing regularization path using the lasso")
Computing regularization path using the lasso
>>> alphas_lasso, coefs_lasso, _ = lasso_path(X, y)
>>>
>>> # Display paths results
... plt.figure(1)
<Figure size 640x480 with 0 Axes>
>>> colors = cycle(['b', 'r', 'g', 'c', 'k','m','c','y','lime','violet'])
>>>
>>> # Form the logs of lambdas
... log_alphas_lasso = np.log10(alphas_lasso)
>>>
>>> # Iterate over coefficients and color to plot the lasso path
... for coef_l, c in zip(coefs_lasso,colors):
...     lasso_plot = plt.plot(log_alphas_lasso, coef_l, c=c)
>>> plt.xlabel('Log(alpha)')
Text(0.5, 0, 'Log(alpha)')
>>> plt.ylabel('Coefficients')
Text(0, 0.5, 'Coefficients')
>>> plt.title('Lasso coefficients paths')
Text(0.5, 1.0, 'Lasso coefficients paths')
>>> plt.legend(df.columns)
<matplotlib.legend.Legend object at 0x1393eb8e0>
>>> plt.show()
```

After importing all the dependencies, we load the `diabetes` dataset and put the features into a dataframe for getting later a proper plot legend. Then, we standardize the variables and compute the coefficients paths. Observe that the method `lasso_path()` returns the lambdas with name `alphas_lasso`. This may confuse the user accustomed to indicate the Lasso penalization parameter as "lambda"; it is thus wise to bear in mind this peculiarity.

Once obtained the grid of lambdas (`alphas_lasso`) and the coefficients at each lambda (`coefs_lasso`), we can display the paths results. First, we form the logs of the lambdas, and then we iterate over coefficients and colors to plot the Lasso coefficients' path using the `plot()` method of the package `matplotlib`. Figure 3.18 sets out the results. We can immediately observe that two variables are of outstanding importance, that is, `bmi` (body mass index) and `ltg` (log of serum triglycerides) as they are the last ones to be dropped out from the model and have also (standardized) coefficients larger than all the other features.

Now, we go on by carrying out cross-validation, using the function `LassoCV()`. The following code performs this task:

Fig. 3.18 Lasso
coefficients' plot as function
of log of lambda. Dataset:
diabetes. Target variable:
diabprog

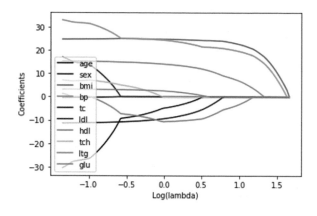

──────────────── Python code and output ────────────────

```
>>> # Carrying out Lasso cross-validation
>>> # Apply "LassoCV": cross-validation method to find out the best Lasso model
>>> # Set multiple alphas values in a list
... alphas = [0.1,0.3, 0.5, 0.8, 1]
>>>
>>> # Fit LassoCV
... lassocv = LassoCV(alphas=alphas, cv=10).fit(X,y)
>>>
>>> # Split the data into training and testing datasets
... xtrain, xtest, ytrain, ytest = train_test_split(X, y, test_size=0.30)
>>>
>>> # Show the optimal lambda
... lassocv.alpha_
1.0
>>>
>>> # Show the optimal coefficients
... lassocv.coef_
array([ -0.       , -9.31956371, 24.83103108, 14.0889837 ,
        -4.83897138, -0.       , -10.62289335, 0.       ,
        24.42088664, 2.56207084])
>>>
>>> # Compute the score (R-squared) at optimal lambda
... score = lassocv.score(X,y)
>>>
>>> # Predict test data
... ypred = lassocv.predict(xtest)
>>>
>>> # Compute the MSE
... mse = mean_squared_error(ytest,ypred)
>>>
>>> # Print the results
... print("Alpha:{0:.2f}, R2:{1:.3f}, MSE:{2:.2f}, RMSE:{3:.2f}"
...      .format(lassocv.alpha_, score, mse, np.sqrt(mse)))
...
... # Visualize the result in a plot
Alpha:1.00, R2:0.513, MSE:3058.63, RMSE:55.30
>>> x_ax = range(len(xtest))
>>> plt.scatter(x_ax, ytest, s=5, color="blue", label="Actual")
<matplotlib.collections.PathCollection object at 0x138f9a340>
>>> plt.plot(x_ax, ypred,lw=0.8, color="red", label="Predicted")
[<matplotlib.lines.Line2D object at 0x138fb8ca0>]
>>> plt.legend()
<matplotlib.legend.Legend object at 0x17de01340>
>>> plt.show()
```

First, we set multiple lambda (`alphas`) values and put them into a list. This will be the grid over which performing the optimal tuning. We fit the `LassoCV()` over the grid using a tenfold cross-validation, and obtain that the optimal lambda is equal to 1. At this lambda, we show the retained coefficients (`lassocv_coef_`).

Then, we split the data into a training and a testing dataset, and compute the R-squared at the optimal lambda, predict the target variable over the testing dataset, compute the test mean squared error, and print all these results. Finally, we visualize the results in a plot where actual and predicted values of the target variable are jointly plotted. Figure 3.19 sets out this plot.

3.14.2 Application P2: Multinomial Regularized Classification in Python

In this application, we fit a multinomial Elastic-net in Python and show how to tune the parameters using the popular `GridSearchCV()` method of Scikit-learn. We use again the `nlsw88` dataset, collecting information on a sample of American women.

We formed a new variable called `dwage` that is a categorical variable taking thee values: 1 = low salary, 2 = medium salary, and 3 = High salary. We are interested in predicting the level of the salary a woman could earn in the job market, based on a series of woman characteristics.

We have already built the training and testing datasets that are stored respectively in the files `data_train_nlsw88class.csv`, including the 70% of the original `nlsw88class.csv` dataset, and `data_test_nlsw88class.csv`, including the remaining 30%.

We start by fitting a multinomial Elastic-net at pre-fixed parameters using the function `LogisticRegression()`. The following code carries it out:

Fig. 3.19 Plot of actual and predicted values. Dataset: `diabetes`; Target variable: `diabprog`

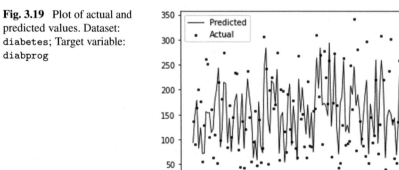

```
──────────────────── Python code and output ────────────────────
>>> # Fitting a multinomial elasticnet (at pre-fixed parameters)
>>> # Import the needed packages
... from sklearn.linear_model import LogisticRegression
>>> from sklearn.model_selection import GridSearchCV
>>> import pandas as pd
>>> import numpy as np
>>> import matplotlib.pyplot as plt
>>> import os
>>>
>>> # Set the directory
... os.chdir("/Users/giocer/Dropbox/Giovanni/Book_Stata_ML_GCerulli/lasso/sjlogs")
>>>
>>> # Set the train and test datasets
... dataset="data_train_nlsw88class.dta" # 70% from original nlsw88class
>>> dataset_new="data_test_nlsw88class.dta" # 30% from original nlsw88class
>>>
>>> # Load the training dataset as pandas dataframe
... df = pd.read_stata(dataset)
>>>
>>> # Define "y", the target variable
... y=df.iloc[:,0].astype(int)
>>>
>>> # Define "x", the features
... X=df.iloc[:,1::]
>>>
>>> # Fit a multinomial elasticnet with all the main parameters
... mlrc = LogisticRegression(penalty='elasticnet',
...                           C=0.5,
...                           solver='saga',
...                           multi_class='multinomial',
...                           l1_ratio = 0.2)
...
```

In the first part of the code, we import the needed packages that we will use for the subsequent bunches of code.

After setting the directory, and the training and testing datasets, we load in the memory the training dataset as a pandas dataframe and define y, the target variable, and x the features.

Then, we fit a multinomial Elastic-net with all the main parameters fixed at specific values. In particular, we set the Elastic-net parameter l1_ratio equal to 0.2, and the inverse of the penalizing parameter (1/lambda) C equal to 0.5.

We can show now how to tune l1_ratio and C over a grid using a tenfold cross-validation based on the GridSearchCV() method. The core syntax of this method is as follows:

GridSearchCV(estimator, param_grid, cv = *num_folds*, scoring =

    scoring_type, return_train_score = TRUE *or* FALSE, ...)

where:

- estimator: a Scikit-learn estimator object;
- param_grid: dictionary with parameter names as keys and lists of parameter settings to try as values;
- *num_folds*: number of cross-validation folds to create. Default is 5.
- *scoring_type*: one of 'accuracy', 'roc_auc', ... See documentation.

- return_train_score: TRUE if the training scores has to be reported, FALSE if not;
- ... : other options.

The following code implements the search of the optimal parameters over the grid:

```
――――――――――――――――――――――――――――――――――――― Python code and output ―――――――――――――
>>> # Tuning a regularized elasticnet multinomial classification
>>> # Generate the two grids (for "C" and for "l1_ratio") as lists
... # 1. "C": inverse of regularization strength
... #    (smaller values specify stronger regularization)
... # 2. "l1_ratio": the elastic-net mixing parameter
... #    (with 0 <= l1_ratio <= 1)
>>> # Form the grid for "C"
... grid_tau = [0.1, 1, 5, 10, 40, 60, 100 , 120, 200]
>>>
>>> # Form the grid for "l1_ratio"
... grid_elastic=[0, 0.5, 1] # 1 = LASSO ; 0 = RIDGE
>>>
>>> # Put the generated grids into a Python dictionary
... param_grid = {´C´: grid_tau, ´l1_ratio´: grid_elastic}
>>>
>>> # Build a "grid search classifier"
... grid = GridSearchCV(LogisticRegression(),
...                      param_grid,
...                      cv=10,
...                      scoring=´accuracy´,
...                      return_train_score=True,
...                      verbose=0)
...
... # Train the classifier over the grids
>>> grid.fit(X, y)
>>>
>>> # View the results: ´mean train error´, ´mean test error´, ´std. test error´
... pd.DataFrame(grid.cv_results_)[[´mean_train_score´,´mean_test_score´,´std_test_score´]]
    mean_train_score  mean_test_score  std_test_score
0           0.649324         0.618378        0.050752
1           0.649324         0.618378        0.050752
2           0.649324         0.618378        0.050752
3           0.647262         0.620668        0.054352
4           0.647262         0.620668        0.054352
5           0.647262         0.620668        0.054352
6           0.647520         0.616064        0.059069
7           0.647520         0.616064        0.059069
8           0.647520         0.616064        0.059069
9           0.646148         0.614514        0.057236
10          0.646148         0.614514        0.057236
11          0.646148         0.614514        0.057236
12          0.646233         0.620710        0.062176
13          0.646233         0.620710        0.062176
14          0.646233         0.620710        0.062176
15          0.646061         0.618396        0.059405
16          0.646061         0.618396        0.059405
17          0.646061         0.618396        0.059405
18          0.647863         0.619153        0.059623
19          0.647863         0.619153        0.059623
20          0.647863         0.619153        0.059623
21          0.645115         0.622999        0.056432
22          0.645115         0.622999        0.056432
23          0.645115         0.622999        0.056432
24          0.648980         0.622224        0.059011
25          0.648980         0.622224        0.059011
26          0.648980         0.622224        0.059011
```

First, we generate two grids as lists: one for `C`, the inverse of regularization strength (remember that smaller values specify stronger regularization); and one for `l1_ratio`, the Elastic-net mixing parameter (ranging from 0 to 1).

We put then the generated grids into a Python dictionary, and build a grid search classifier. We thus train the classifier over the grid using `GridSearchCV()` and display the mean training error, the mean testing error, and the mean standard deviation of the testing error obtained over the 10 folds at each point of the grid. Subsequently, we explore the results, both numerically and graphically. The following code carries out this task:

```
────────────────────────────────────── Python code and output ──────────────
>>> # Examine the best model and plot the results: training vs. test error
>>> # Examine the best model
... print(grid.best_score_)
0.6229994036970782
>>> print(grid.best_params_)
{'C': 120, 'l1_ratio': 0}
>>> print(grid.best_estimator_)
LogisticRegression(C=120, l1_ratio=0)
>>> print(grid.best_index_)
21
>>>
>>> # Select 'mean train score' and 'mean test score'
... xx=pd.DataFrame(grid.cv_results_)[['mean_test_score']]
>>> yy=pd.DataFrame(grid.cv_results_)[['mean_train_score']]
>>>
>>> # Plot xx and yy separately
... plt.plot(xx)
[<matplotlib.lines.Line2D object at 0x137ef50d0>]
>>> plt.plot(yy)
[<matplotlib.lines.Line2D object at 0x138cfa850>]
>>>
>>> # Name the x axis
... plt.xlabel("Index")
Text(0.5, 0, 'Index')
>>>
>>> # Name the y axis
... plt.ylabel('Accuracy')
Text(0, 0.5, 'Accuracy')
>>>
>>> # Give a title to the graph
... plt.title('Test vs. Train accuracy')
Text(0.5, 1.0, 'Test vs. Train accuracy')
>>> plt.axvline(x=grid.best_index_ , color='black',
...             linestyle='--', linewidth=3)
...
... # Function to show the plot
<matplotlib.lines.Line2D object at 0x151f15e10>
>>> plt.show()
```

The main results of implementing the function `GridSearchCV()` are contained in a series of returns objects. Specifically, `grid.best_score_` is a Python variable containing the average testing accuracy reached that in this case is equal to 62%; `grid.best_params_` is a dictionary containing the best parameters that are 120 for the parameter `C` and 0 for the `l1_ratio`; printing `grid.best_estimator_` allows to show the fitted estimator at its optimal parameters; finally, `grid.best_index_` contains the best index of the considered grid that in this example is equal to 21.

Fig. 3.20 Plot of training
and test error as a function of
the grid index. Dataset:
`diabetes`. Target variable:
`diabprog`

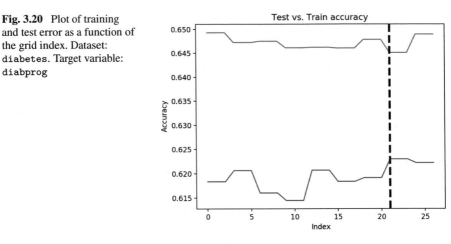

The code goes on by plotting the mean testing accuracy and the mean train-ing accuracy over the grid index. To perform this task, we first put the main out-put of `GridSearchCV()`, i.e., `grid.cv_results_`, into a dataframe where we extract the two columns xx (i.e., `mean_test_score`) and yy (i.e., `mean_train_score`) to then plot them using the `plt.plot()` method.

The plot is displayed in Fig. 3.20. As expected, the training accuracy is more optimistic compared to the test accuracy. The graph also confirms that the maximum of the testing accuracy is reached at an index value of the grid equal to 21.

Finally, after tuning optimally the model, we want to predict both in-sample and out-of-sample at optimal parameters. The following code carries out this task:

```
────────────────────────────────  Python code and output  ────────────────────────────────
>>> # Predicting at optimal parameters
>>> # Store the two best parameters into two Python variables
... opt_c = grid.best_params_.get(´C´)
>>> opt_lambda = grid.best_params_.get(´l1_ratio´)
>>>
>>> # Use the best parameters to make predictions
... # train your model using all data and the best parameters
... mlrc = LogisticRegression(penalty = ´elasticnet´,
...                           C = opt_c,
...                           solver = ´saga´,
...                           multi_class = ´multinomial´,
...                           l1_ratio = opt_lambda)
...
... # Fit the model
>>> mlrc.fit(X, y)
/Library/Frameworks/Python.framework/Versions/3.7/lib/python3.7/site-packages/sklearn/linear_model/_s
> ag.py:330: ConvergenceWarning: The max_iter was reached which means the coef_ did not converge
  "the coef_ did not converge", ConvergenceWarning)
LogisticRegression(C=120, l1_ratio=0, multi_class=´multinomial´,
                   penalty=´elasticnet´, solver=´saga´)
>>>
>>> # Put coefficients in a dataframe
... mlrc_coefs = pd.DataFrame(mlrc.coef_)
>>>
>>> # Rename the columns of "mlrc_coefs" with the variables names
... #mlrc_coefs.info()
```

```
... vars_names = X.columns.tolist()
>>> mlrc_coefs.columns = vars_names
>>> mlrc_coefs.rows = [´white´,´black´,´hispanic´]
__main__:1: UserWarning: Pandas doesn´t allow columns to be created via a new attribute name - see ht
> tps://pandas.pydata.org/pandas-docs/stable/indexing.html#attribute-access
>>>
>>> # Compute in-sample prediction for "y"
... y_hat = mlrc.predict(X)
>>>
>>> # Compute in-sample probability prediction
... prob = mlrc.predict_proba(X)
>>>
>>> # Compute out-of-sample "label" prediction for "y" using the test dataset
... Xnew = pd.read_stata(dataset_new)
>>> ynew = mlrc.predict(Xnew)
>>>
>>> # Compute out-of-sample "probability" prediction for "y" using a test dataset
... prob_new = mlrc.predict_proba(Xnew)
>>>
>>> # Merge label and probability prediction for "y"
... # Use "numpy" to stack by column ´ynew´ and ´prob_new´
... out=np.column_stack((ynew,prob_new))
>>>
>>> # Visualize out-of-sample predictions
... out = pd.DataFrame(out,columns=[´label_pred´, ´prob_low´, ´prob_medium´,´prob_high´] )
>>> print(out)
     label_pred  prob_low  prob_medium  prob_high
0           2.0  0.389730     0.511351   0.098919
1           3.0  0.021346     0.079959   0.898695
2           2.0  0.253804     0.434330   0.311866
3           3.0  0.130126     0.268092   0.601782
4           3.0  0.062127     0.254501   0.683372
..          ...       ...          ...        ...
549         2.0  0.329119     0.392301   0.278580
550         1.0  0.841018     0.152364   0.006618
551         1.0  0.763355     0.211558   0.025087
552         3.0  0.091048     0.288833   0.620119
553         1.0  0.703154     0.247265   0.049581

[554 rows x 4 columns]
```

First, we store the two best parameters for C and l1_ratio into two Python variables called respectively opt_c and opt_lambda. Then, we use the best parameters to train our model using all data and the best parameters. We fit it, and obtain the coefficients (mlrc_coefs). We can then compute both in-sample and out-of-sample predictions for our target variable, as well as for target's label probabilities. Finally, we visualize the results thus obtained for the out-of-sample predictions.

3.15 Conclusion

In this chapter, we presented theory and applications with Stata, R, and Python of regularized (or penalized) regression and classification methods. These methods are particularly relevant as they allow for optimal (parametric) prediction and inference in high-dimensional data settings, i.e., datasets where the number of features may be much larger than the number of observations.

We mainly focused on the Stata implementation, although the R and Python sections provide extensive illustrative examples for learning how to apply these methods within these two platforms.

We have not covered inferential Lasso with R and Python. However, the hdm package of R can perform high-dimensional inference as well. For the sake of conciseness and because it is very close to the implementation provided by the LASSO package of Stata, we have not presented an application of the hdm package in the R section of this chapter. In this case, the reader can refer directly to Chernozhukov et al. (2016). As for Python, to our knowledge, no packages have been so far developed for inferential regularized methods.

References

Ahrens, A., Hansen, C., & Schaffer, M. (2020). Lassopack: Model selection and prediction with regularized regression in Stata. *Stata Journal, 20*(1), 176–235.

Angrist, J. D., & Krueger, A. B. (1991). Does compulsory school attendance affect schooling and earnings? *The Quarterly Journal of Economics, 106*(4), 979–1014.

Belloni, A., & Chernozhukov, V. (2014b). Inference on treatment effects after selection among high-dimensional controls. *The Review of Economic Studies, 81*(2(287)), 608–650.

Belloni, A., Chen, D., Chernozhukov, V., & Hansen, C. (2012). Sparse models and methods for optimal instruments with an application to eminent domain. *Econometrica, 80*(6), 2369–2429.

Belloni, A., & Chernozhukov, V. (2011). High dimensional sparse econometric models: An introduction. In P. Alquier, E. Gautier, & G. Stoltz (Eds.), *Inverse problems and high-dimensional estimation: Stats in the chteau summer school, August 31-September 4, 2009* (pp. 121–156). Lecture notes in statistics. Berlin, Heidelberg: Springer.

Belloni, A., Chernozhukov, V., & Hansen, C. (2014). High-dimensional methods and inference on structural and treatment effects. *Journal of Economic Perspectives, 28*(2), 29–50.

Belloni, A., Chernozhukov, V., & Wei, Y. (2016). Post-selection inference for generalized linear models with many controls. *Journal of Business & Economic Statistics, 34*(4), 606–619.

Bergmeir, C., Hyndman, R. J., & Koo, B. (2016). hdm: High-dimensional metrics. *The R Journal, 8*(2), 185–199.

Bergmeir, C., Hyndman, R. J., & Koo, B. (2018). A note on the validity of cross-validation for evaluating autoregressive time series prediction. *Computational Statistics & Data Analysis, 120*, 70–83.

Bühlmann, P., & van de Geer, S. (2011). Theory for l1/l2-penalty procedures. Springer Series in StatisticsIn P. Bühlmann & S. van de Geer (Eds.), *Statistics for high-dimensional data: Methods, theory and applications* (pp. 249–291). Berlin, Heidelberg: Springer.

Cerulli, G. (2020). SUBSET: Stata module to implement best covariates and stepwise subset selection. https://econpapers.repec.org/software/bocbocode/s458647.htm

Chernozhukov, V., Chetverikov, D., Demirer, M., Duflo, E., Hansen, C., Newey, W., Robins, J. (2018). Double/debiased machine learning for treatment and structural parameters. *The Econometrics Journal, 21*(1), C1–C68. _eprint: https://onlinelibrary.wiley.com/doi/pdf/10.1111/ectj. 12097

Chernozhukov, V., Hansen, C., & Spindler, M. (2015). Post-selection and post-regularization inference in linear models with many controls and instruments. *The American Economic Review, 105*(5), 486–490. Publisher: American Economic Association. https://www.jstor.org/stable/43821933

Daubechies, I., Defrise, M., & Mol, C. D. (2004). An iterative thresholding algorithm for linear inverse problems with a sparsity constraint. *Communications on Pure and Applied Mathematics, 57*(11), 1413–1457.

Daubechies, I., Defrise, M., & Mol, C. D. (2010). Regularization paths for generalized linear models via coordinate descent. *Journal of Statistical Software, 33*(1), 1–22.

Eaton, J. P., & Haas, C. A. (1995). *Titanic: Triumph and tragedy* (2nd ed.). New York: W. W. Norton & Company.

Efron, B., Hastie, T., Johnstone, I., & Tibshirani, R. (2004). Least angle regression. *The Annals of Statistics, 32*(2), 407–499. Publisher: Institute of Mathematical Statistics.

Efron, B., Hastie, T., Johnstone, I., & Tibshirani, R. (2011). Regression shrinkage and selection via the lasso: a retrospective. *Journal of the Royal Statistical Society. Series B (Statistical Methodology), 73*(3), 273–282. Publisher: [Royal Statistical Society, Wiley]. https://www.jstor.org/stable/41262671

Fan, J., & Li, R. (2001). Variable selection via nonconcave penalized likelihood and its oracle properties. *Journal of the American Statistical Association, 96*(456), 1348–1360.

Foster, D. P., & George, E. I. (1994). The risk inflation criterion for multiple regression. *The Annals of Statistics, 22*(4), 1947–1975.

Friedman, J., Hastie, T., Höfling, H., & Tibshirani, R. (2007). Pathwise coordinate optimization. *The Annals of Applied Statistics, 1*(2), 302–332.

Fu, W. J. (1998). Penalized regressions: The bridge versus the lasso. *Journal of Computational and Graphical Statistics, 7*(3), 397–416.

Geer, S. A. V. D., & Bühlmann, P. (2009). On the conditions used to prove oracle results for the Lasso. *Electronic Journal of Statistics, 3*(none), 1360–1392.

Gorman, J. W., & Toman, R. J. (1966). Selection of variables for fitting equations to data. *Technometrics, 8*(1):27–51. Publisher: Taylor & Francis. https://amstat.tandfonline.com/doi/abs/10.1080/00401706.1966.10490322

Harrison, D. J., & Rubinfeld, D. L. (1978). Hedonic housing prices and the demand for clean air. *Journal of Environmental Economics and Management, 5*(1), 81–102. Publisher: Elsevier. https://ideas.repec.org/a/eee/jeeman/v5y1978i1p81-102.html

Hastie, T., Tibshirani, R., & Friedman, J. (2009). *The elements of statistical learning: Data mining, inference, and prediction* (2nd ed.). Springer series in statistics. New York: Springer. https://www.springer.com/gp/book/9780387848570

Hastie, T., Tibshirani, R., & Wainwright, M. (2015). *Statistical learning with sparsity: The lasso and generalizations*. Chapman and Hall/CRC.

Hocking, R. R., & Leslie, R. N. (1967). Selection of the best subset in regression analysis. *Technometrics, 9*(4), 531–540.

Hoerl, A. E., & Kennard, R. W. (1970). Ridge regression: Applications to nonorthogonal problems. *Technometrics, 12*(1), 69–82. Publisher: Taylor & Francis. https://amstat.tandfonline.com/doi/abs/10.1080/00401706.1970.10488635

Hoerl, A. E., Kannard, R. W., & Baldwin, K. F. (2007). Ridge regression: Some simulations. *Communications in Statistics—Theory and Methods*. Publisher: Marcel Dekker, Inc. https://www.tandfonline.com/doi/abs/10.1080/03610927508827232

Leeb, H., & Pötscher, B. M. (2008). Can one estimate the unconditional distribution of post-model-selection estimators? *Econometric Theory, 24*(2), 338–376. Publisher: Cambridge University Press. https://www.jstor.org/stable/20142496

Loh, P. -L., & Wainwright, M. J. (2017). Support recovery without incoherence: A case for non-convex regularization. *The Annals of Statistics, 45*(6), 2455–2482. Publisher: Institute of Mathematical Statistics.

Ramaswamy, S., Tamayo, P., Rifkin, R., Mukherjee, S., Yeang, C.-H., Angelo, M., Ladd, C., Reich, M., Latulippe, E., Mesirov, J. P., Poggio, T., Gerald, W., Loda, M., Lander, E. S., & Golub, T. R. (2001). Multiclass cancer diagnosis using tumor gene expression signatures. *Proceedings of the National Academy of Sciences, 98*(26), 15149–15154.

Robinson, P. M. (1988). Root-N-consistent semiparametric regression. *Econometrica, 56*(4), 931–954. Publisher: [Wiley, Econometric Society]. https://www.jstor.org/stable/1912705

Roecker, E. B. (1991). Prediction error and its estimation for subset-selected models. *Technometrics, 33*(4), 459–468.

Theil, H. (1957). Specification errors and the estimation of economic relationships. *Revue de l'Institut International de Statistique/Review of the International Statistical Institute, 25*(1/3), 41–51. Publisher: [International Statistical Institute (ISI), Wiley]. https://www.jstor.org/stable/1401673

Tibshirani, R. (1996). Regression shrinkage and selection via the lasso. *Journal of the Royal Statistical Society. Series B (Methodological), 58*(1), 267–288. Publisher: [Royal Statistical Society, Wiley]. https://www.jstor.org/stable/2346178

van Wieringen, W. N. (2020). Lecture notes on ridge regression. arXiv:1509.09169 [stat].

Wang, Z., Liu, H., & Zhang, T. (2014). Optimal computational and statistical rates of convergence for sparse nonconvex learning problems. *The Annals of Statistics, 42*(6), 2164–2201.

Wold, H., & Faxer, P. (1957). On the specification error in regression analysis. *The Annals of Mathematical Statistics, 28*(1), 265–267. Publisher: Institute of Mathematical Statistics. https://www.jstor.org/stable/2237040

Zou, H. (2006). The adaptive lasso and its oracle properties. *Journal of the American Statistical Association, 101*(476), 1418–1429.

Chapter 4
Discriminant Analysis, Nearest Neighbor, and Support Vector Machine

4.1 Introduction

In this chapter, we cover three related Machine Learning techniques: discriminant analysis (DA), support vector machine (SVM), and k-nearest neighbor (KNN) algorithms. We will focus mainly on classification, but we will show also how to extend SVM and KNN to a regression setup.

Discriminant analysis is a Bayesian approach to classification, that is, it allocates unknown class membership using the Bayes rule. Although DA can be used for descriptive purposes to identify, for example, what features are more relevant to discriminate among groups, we will focus on DA mainly as a predictive technique, in line with our supervised learning perspective. We will discuss both linear and quadratic DA, lending attention to the concept of *decision boundary*.

The chapter goes on by introducing the k-nearest neighbors (KNN) algorithm, which can be seen as an extension of the discriminate analysis to a nonparametric estimation of the decision boundary. The KNN is a *local* imputation technique, in that it allows to classify an unlabeled observation using the *closeness* principle, that is, exploiting the knowledge of the labels of the first k nearest neighbors of the considered observation. The same closeness principle can be adapted to a numerical target variable, thus entailing that the KNN can be extended also to a regression setting.

As the KNN algorithm, the support vector machine (SVM) can be used both in the classification and in the regression setting. We will devote wider attention to the two-class SVM classification, to then extend SVM classification to a multi-class setup, showing how to retrieve class assignment probabilities within this approach. Finally, we will discuss SVM regression.

In the application section, the chapter presents a series of applications of the methods illustrated in the theoretical part using Stata, R, and Python on real datasets. The programming codes and their outputs are presented and extensively explained.

G. Cerulli, *Fundamentals of Supervised Machine Learning*, Statistics and Computing, https://doi.org/10.1007/978-3-031-41337-7_4

4.2 Classification

4.2.1 Linear and Logistic Classifiers

In the introductory chapter, we stated that classifying a given unit i within a certain class k can be carried out using the conditional probability that this unit belongs to class k, conditional on unit's characteristics $\mathbf{x}_i$:

$$p_k(\mathbf{x}_i) = \text{Prob}(y_i = k|\mathbf{x}_i) \tag{4.1}$$

A unit, either in-sample or out-of-sample, will be assigned to the class having the largest probability (Bayes criterion). The conditional probability is inherently unknown from observation, therefore it can be modeled in various ways, including both parametric and nonparametric approaches. Among the family of parametric classifiers, the linear probability model, and the logistic probability model are among the most popular. They classify units using the following formulas of the conditional probabilities:

$$\text{Linear: } p_k(\mathbf{x}_i) = \mathbf{x}_i \beta \tag{4.2}$$

$$\text{Logistic: } p_k(\mathbf{x}_i) = \frac{e^{\mathbf{x}_i \beta}}{1 + e^{-\mathbf{x}_i \beta}} \tag{4.3}$$

where $\mathbf{x}_i \beta = (\beta_0 + x_{1i} \beta_1 + \cdots + x_{pi} \beta_p)$. The linear probability is a simple way to model the conditional probability of a unit to belong to class k. One important limitation of the linear probability model is that it generates probability predictions that may be either smaller than zero, or larger than one. This drawback is prevented by the logistic conditional probability which has a sigmoid shape with fitted probability predictions always ranging between zero and one. Figure 4.1 shows a comparison between linear and logistic probability predictions on simulated data. These two approaches are also different in the way they estimate the unknown parameters β, with the linear probability model using least squares, and the logistic probability model using maximum likelihood. These models can be also adapted to the case of a multi-class target variable, that is, a categorical variable presenting more than two categories. The logistic model, in particular, encompasses both binomial and multinomial maximum likelihood estimation.

Although the linear and logistic probability models are different, the final classification they provide is similar. This depends on the fact that the so-called *decision* or *classification boundary* is linear in both cases. To show this, take a two-class two-feature example. Consider a binary variable y taking just two categories, either *green* or *red*, ad suppose to have only two features, x_1 and x_2. The linear probability model entails that:

$$p_1(x_1, x_2) = \beta_0 + \beta_1 x_1 + \beta_2 x_2 \tag{4.4}$$

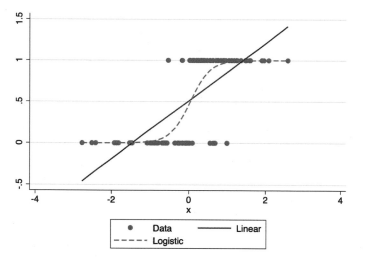

Fig. 4.1 Example of a linear and logistic probability fit

so that each individual i is assigned a color according to this classification rule:

- if $p_1(x_1, x_2) \geq 0.5$, then i is classified as *green*
- if $p_1(x_1, x_2) < 0.5$, then i is classified as *red*

The decision boundary identifies all the pairs $\{x_{i1}, x_{i2}\}$ making one indifferent whether to classify a unit i as *red* or *green*. This is obtained by equalizing the linear probability to 0.5, the decision threshold, thus obtaining the following decision boundary:

$$\text{Linear decision boundary: } x_1 = -\frac{\beta_2}{\beta_1}x_2 - \frac{\beta_0 - 0.5}{\beta_1} \tag{4.5}$$

For a logistic classifier the decision boundary is similar:

$$\text{Logistic decision boundary: } x_1 = -\frac{\beta_2}{\beta_1}x_2 - \frac{\beta_0}{\beta_1} \tag{4.6}$$

This is obtained by equalizing:

$$p_1(x_1, x_2) = \frac{1}{1 + e^{-(\beta_0 + \beta_1 x_1 + \beta_2 x_2)}} = 0.5 \tag{4.7}$$

whose solution entails that $(\beta_0 + \beta_1 x_1 + \beta_2 x_2) = 0$, which is equivalent to Eq. (4.6). Figure 4.2 illustrates a comparison between linear and logistic decision boundaries. As expected, the slope of the decision line is the same, but the intercept is slightly different. This can bring to a few different classifications especially at the center of the cloud, where discrimination is less severe.

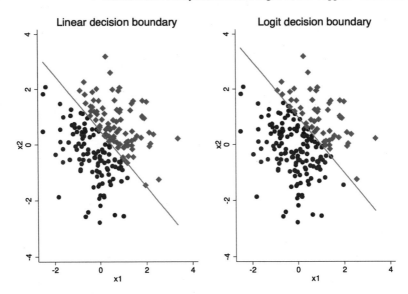

Fig. 4.2 Example of linear and logistic decision boundaries

4.2.2 Discriminant Analysis

A related but different way to classify units, still in a two- or multi-class setting, is the discriminant analysis (DA) (McLachlan, 1992). Instead of directly modeling the class conditional probability, the discriminant analysis—using the Bayes theorem (and formula)—models the probability distribution of each feature x within each class separately. When one assumes a Normal (Gaussian) distribution of each feature x within each class, this produces the so-called linear DA (LDA), or quadratic DA (QDA). Nonetheless, the DA approach is more general, and can in theory encompass other distributions than the Normal one. Typically, however, DA applications are based on Normal distributions, and we will rely on this assumption along the presentation of the DA classifier of this chapter (Huberty, 1994).

For the sake of clarity, consider a classification based on only one feature x. Using the Bayes theorem, one can rewrite the conditional probability of an observation i to belong to class k as follows:

$$\text{Prob}(y_i = k|x_i) = \frac{q_k \text{Prob}(x_i|y_i = k)}{\sum_{k=1}^{K} q_k \text{Prob}(x_i|y_i = k)} \tag{4.8}$$

where K is the total number of classes, $q_k = \text{Prob}(y_i = k)$ is the prior probability to belong to class k, and $\text{Prob}(x_i|y_i = k)$ is the density function of variable x in class k, that we can more compactly indicate as $f_k(x_i)$.

The denominator of Eq. (4.8) does not depend on k. If we assume that q_k is constant, we can classify observation i based on the value of $f_k(x_i)$. Figure 4.3 shows

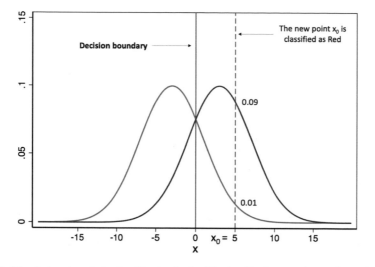

Fig. 4.3 Discriminant analysis classification logic. We classify the new observation with $x = x_0$ as *red*, as $f_{red}(x_0) > f_{green}(x_0)$. Observe that, in this case, we assume same priors, that is, $q_{red} = q_{green}$

the DA classification logic when only two classes are considered, *red* versus *green*, with equal priors. Looking at the figure, we observe that the decision boundary corresponds to a singleton point $x = 0$. The DA classifies the new observation with $x = x_0$ as *red*, because $f_{red}(x_0) = 0.09 > f_{green}(x_0) = 0.01$.

If we make the priors no longer equal, by increasing for example the prior of class *green* (that is, $q_{green} > q_{red}$), we modify the decision boundary by increasing the likelihood to classify a new observation as *green*. This is visible in Fig. 4.4, where we can easily see that the decision boundary has shifted to the right. Estimating the parameters needed to classify observations with the DA is quite straightforward. If the density function in each class is normal, we have

$$f_k(x_i) = \frac{1}{\sqrt{2\pi}\sigma_k} e^{-\frac{1}{2}(\frac{x_i - \mu_k}{\sigma_k})^2} \tag{4.9}$$

By assuming that the variance of x in each class is constant (that is $\sigma_k = \sigma$), and substituting Eq. (4.9) into Eq. (4.8), we have

$$\text{Prob}(y_i = k | x_i) = \frac{q_k \frac{1}{\sqrt{2\pi}\sigma} e^{-\frac{1}{2}(\frac{x_i - \mu_k}{\sigma})^2}}{\sum_{k=1}^{K} q_k \frac{1}{\sqrt{2\pi}\sigma} e^{-\frac{1}{2}(\frac{x_i - \mu_k}{\sigma})^2}} \tag{4.10}$$

By taking the log of Eq. (4.10) and eliminating all the terms that do not depend on k, we obtain that

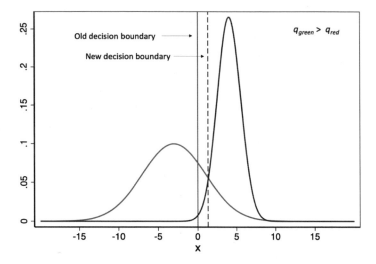

Fig. 4.4 Discriminant analysis classification logic. By increasing the prior of class *green*, we change the decision boundary by increasing the likelihood to classify a new observation as *green*

$$\delta_k(x_i) = x_i \frac{\mu_k}{\sigma^2} - \frac{\mu_k^2}{2\sigma^2} + \log(q_k) \qquad (4.11)$$

Equation (4.11) is a linear function of x. This is called *discriminant score*, and we can use this score to classify units instead of the class conditional probability. We can assign a unit i to the class having the largest discriminant score.

Estimating the discriminant score requires to estimate the sample mean in each class and the overall variance, as well as the prior q_k, generally as the proportion of observations in each class (that is, $\widehat{q}_k = N_k/N$).

When the number of features $p > 1$, the discriminant score becomes

$$\delta_k(\mathbf{x}_i) = \mathbf{x}_i' \, \mathbf{\Sigma}^{-1} \boldsymbol{\mu}_k - \frac{1}{2} \boldsymbol{\mu}_k' \mathbf{\Sigma}^{-1} \boldsymbol{\mu}_k + \log(q_k) \qquad (4.12)$$

where $\mathbf{\Sigma}$ is the covariance matrix of the features. It is important to stress that the discriminant score of Eq. (4.12) is still a linear combination of the column vector of the features $\mathbf{x}_i = [x_{i1}, \ldots, x_{ip}]$. This makes the decision boundary a linear function of the features. Such linearity of the discriminant score, leading to what is called in the literature as *linear discriminant analysis* (LDA), relies on the assumption that $\mathbf{\Sigma}_k = \mathbf{\Sigma}$, that is, the feature covariance matrix is assumed to be the same in each class.

Using LDA for classification purposes can have some advantages compared to the logistic classifier. For example, when the classes are well-separated, the parameter estimates for the logistic model might be fairly unstable, while linear DA does not show this drawback. Also, when the sample size is small and the distribution of the features is close to a normal in every class, the LDA model is again more stable than the logistic one. Finally, LDA is easier to estimate in settings with more than two classes, where multinomial logistic models can have convergence problems (as using maximum likelihood).

This assumption of equal covariance matrix in each class may be however rather strong, as in the real world it is very likely that features covariances are different within different classes. For example, if classes are different levels of wage, and we have two features—let's say age and education—it is unlikely that people with higher wages have the same correlation between age and education than people with lower wages.

When we relax the assumption that $\Sigma_k = \Sigma$, the discriminant score changes becoming:

$$\delta_k(\mathbf{x}_i) = -\frac{1}{2}\log|\Sigma_k| - \frac{1}{2}(\mathbf{x}_i - \boldsymbol{\mu}_k)'\Sigma^{-1}(\mathbf{x}_i - \boldsymbol{\mu}_k) + \log(q_k) \qquad (4.13)$$

where quadratic terms emerge, thus making the decision boundary a quadratic equation. This is called *quadratic discriminant analysis* (QDA).

A special discriminant score case is when one assumes that within each class the features are independent. This leads to obtain a diagonal covariance matrix. In this case, the discriminant score takes on this simplified form:

$$\delta_k(\mathbf{x}_i) \propto \log \prod_{j=1}^{p} f_{kj}(x_{ij}) = \frac{1}{2} \sum_{j=1}^{p} \frac{(x_j - \mu_{kj})^2}{\sigma_{kj}^2} + \log(q_k) \qquad (4.14)$$

The classifier based on Eq. (4.14) is known as Naïve Bayes. When p is very large, this classifier can have considerable computational advantages compared to the logistic regression, the LDA, and the QDA. Also, although the strong assumption of feature independence, the Naïve Bayes classifier has proved to have remarkable prediction accuracy in many real data settings.

In a nutshell, we may say that the logistic classifier is particularly suitable when we have two classes; the LDA and QDA are popular when the sample size is small, when we have a multi-class setting, and when the normality assumption is reasonable; the Naïve Bayes is useful when p, the number of features, is very large.

4.2.3 Nearest Neighbor

Linear, logistic, LDA, and QDA are highly parametric classifiers. This means that their ability to correctly classifying out-of-sample observations when training observations belong to not well-separated groups may be poor.

Nonparametric classifiers, based on more data-driven approaches, may improve classification accuracy at expense however of a higher computational burden. Modern

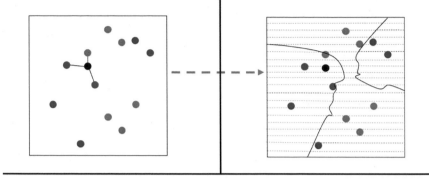

We want to give a label to the test observation **black point**. The 3 closest points are identified, and predict that the test observation belongs to the *most commonly-occurring class*, in this case the blue one.

The M-Nearest-neighbor decision boundary shown in **black**. Blue grid: region in which a test observation will be assigned to the blue class. Orange grid: region in which it will be assigned to the orange class.

Fig. 4.5 Nearest-neighbor classifier with neighborhood of size $M = 3$. The decision boundary is built based on the class obtaining—within the neighborhood of the point at stake—the largest frequency. This approach uses a pure data-driven approach to classify units, based on the *closeness* principle

computers can nonetheless easily handle this issue, thereby explaining the today's large use of these techniques in applications.

Among nonparametric classifiers, the Nearest-neighbor is a popular one. Instead of relaying on a mathematical formula for the conditional probability to belong to a specific class k, the Nearest-neighbor classifier uses a procedure based on the principle of *closeness* (Fix & Hodges, 1951, 1989; Cover & Hart, 1967).

Specifically, consider a unit i and define a neighborhood of it made of those M units j that are the first M nearest neighbors of unit i. Call this neighborhood $\mathcal{N}_i$. The Nearest-neighbor classifier computes class k conditional probability using this formula:

$$\text{Prob}(y_i = k | \mathbf{x}_i) = \frac{1}{M} \sum_{j \in \mathcal{N}_i} I(y_j = k) \tag{4.15}$$

This probability is the frequency of class k within the neighborhood of i. Thus, i will be assigned to the class having the largest frequency within its neighborhood. Figure 4.5 shows an example of Nearest-neighbor classification with a neighborhood of size $M = 3$ (i.e., 3-Nearest-neighbor classification). We consider twelve observations in our training dataset, six belonging to the class *blue*, and six belonging to the class *orange*. Our aim is to give a label (i.e., assign a class) to the test observation represented by the black point. The three closest points are identified, and we predict the label of our test observation as the most commonly-occurring class. In this case, this class is the *blue* one. The right-hand panel shows in black the M-nearest neighbor decision boundary. The blue grid indicates the region in which a test observation

LDA

M-NN, M =1

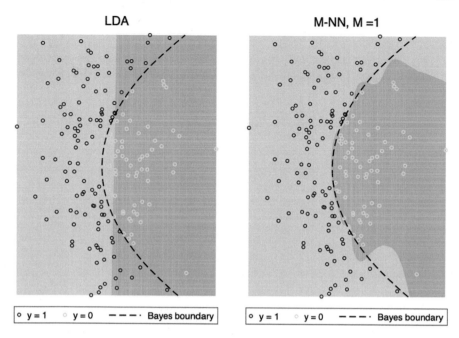

Fig. 4.6 Simulation of decision boundary adaptation: comparison between a linear discriminant classifier (LDA) and a Nearest-neighbor classifier with neighborhood of size $M = 3$ (3-NN). The true decision boundary is a quadratic function

will be assigned to the blue class, while the orange grid is the region in which a test observation will be assigned to the orange class. It is worth noting the ability of the M-nearest neighbor decision boundary to adapt to the data, as it is not determined by a specific mathematical formula but instead based on a pure data-driven procedure.

To better appreciate such ability, Fig. 4.6 shows a simulation comparing the decision boundary generated by a LDA (which is, as said, linear), and that generated by a M-nearest-neighbor (M-NN). In this simulation, the *true* decision boundary (sometimes called "Bayes boundary") is a quadratic function. The adaptation of the M-NN is pretty perfect compared to the adaptation of the LDA that generates—on the contrary—a number of misclassifications also in the training data.

The classification performance of the M-NN depends highly on the size of the neighborhood, that is, on the number of nearest neighbors to consider in the local class assignment procedure. It is easy to prove that the parameter M is a tuning parameter, as it entails a trade-off between prediction bias and prediction variance. Indeed, when M is large, the neighborhood is larger and the sample size over which computing the class frequencies is larger. This reduces the level of the variance, as we have larger information within the neighborhood. However, variance reduction takes place at expense of an increasing bias. Indeed, the bias increases because within the neighborhood we are considering points (i.e., observations) that are very far from the point at which we want to impute the class. This makes such imputation spurious.

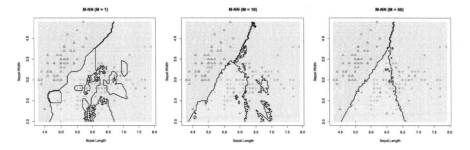

Fig. 4.7 M-nearest neighbor decision boundary at different number of nearest neighbors. The largest model complexity is reached when $M = 1$, the smallest when $M = 50$. The parameter M is a tuning parameter, entailing a trade-off between prediction bias and prediction variance. As such, M can be optimally tuned, for example, by cross-validation

A good compromise between variance increase and bias decrease can be found by optimally tuning M, for example, by cross-validation. Figure 4.7 illustrates the M-NN decision boundary at three levels of M, that is 1, 10, and 50. Model complexity is largest when $M = 1$ while it becomes smaller as soon as M increases. Therefore, the complexity parameter is measured by $1/M$. From the figure, we can see that when $M = 1$, the decision boundary tends to be too flexible, indicating a too high prediction variance with a minimal bias. On the contrary, when $M = 50$, the decision boundary is highly smooth indicating a smaller prediction variance but possibly paired with a larger bias. Intermediate values of M might be more balanced as happens, for example, at $M = 10$.

The M-nearest neighbor method outlined above directly models the conditional probability to belong to a class k. We can use a different perspective using the Bayes formula, and modeling $\text{Prob}(y_i = k|x_i)$ nonparametrically using the nearest neighbor algorithm. In this case, the formula becomes

$$\text{Prob}(x_i|y_i = k) = \frac{n_{i,k}}{N_k} \tag{4.16}$$

where $n_{i,k}$ is the number of the nearest neighbors of i coming from class k. Thus, the Bayes formula of the conditional probability to belong to class k for unit i becomes:

$$\text{Prob}(y_i = k|x_i) = \frac{q_k \frac{n_{i,k}}{N_k}}{\sum_{k=1}^{K} q_k \frac{n_{i,k}}{N_k}} \tag{4.17}$$

To appreciate how we can classify a new instance using formula (4.17), let's consider an example based on Fig. 4.8. In this example, we have $N = 24$ training observations, and two classes—*blue* and *red*—with respectively $N_{blue} = 14$ and $N_{red} = 10$ (so that, $q_{blue} = 14/24$, and $q_{red} = 10/24$). We want to impute the class for a new instance $i=x$. For this purpose, we use a 5-Nearest neighbor algorithm, which generates the

Fig. 4.8 Example of a 5-Nearest neighbor algorithm with $N = 24$ training observations, and two classes—*blue*. For a new instance $i = $ x, we impute the class using the Bayes conditional probability formula (4.17)

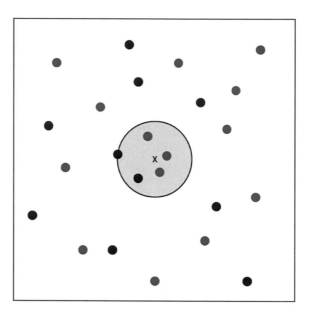

neighborhood of point x colored in gray within the figure. For point i=x, $n_{i,blue} = 3$, and $n_{i,red} = 2$. Therefore, we have that:

- $\text{Prob}(y_i = red|x_i) = \frac{10/24 \cdot 2/10}{(10/24 \cdot 2/10 + 14/24 \cdot 3/14)} = 0.4$
- $\text{Prob}(y_i = blue|x_i) = \frac{14/24 \cdot 3/14}{(10/24 \cdot 2/10 + 14/24 \cdot 3/14)} = 0.6$

implying that point x will be classified as *blue*. To build neighbors, one needs some distance metrics. For continuous variables, the Mahalanobis distance is often used. Specialized metrics for binary, or for a mix of binary and continuous features are also implemented in standard software. Although formulas (4.15) and (4.17) lead to similar classifications, formula (4.17) has the advantage of allowing for manipulating priors in a more direct way. As such, it can be preferable to be used in applications where manipulating priors can be of some usefulness. An example will be provided in the application section.

4.2.4 Support Vector Machine

Support Vector Machine (SVM) is a supervised ML approach one can use for both regression and classification. It was developed in the computer science community in the 1990s (Cortes & Vapnik, 1995). Here, we will mainly focus on SVM for classification. SVM performs rather well in many applied settings, and it is among the most popular learners used in applications. Technically, it can be seen as the

Fig. 4.9 Example of a
separating hyperplane with
two classes

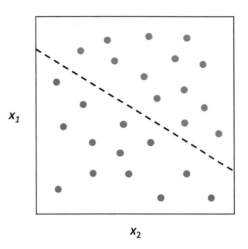

generalization of a more basic method known as *maximal margin classifier*. It is well
suited for two-class classification settings, but it can be easily extended to multi-class
outcomes.

To properly introduce SVM, it is useful to start with the notion of *separating
hyperplane*. If we consider a two-class classification problem, finding a separating
hyperplane means trying to find—in the feature space—a plane that separates the
two classes. Sometimes, the nature of the data does not allow for finding a perfect
separation, and we have to deal with non-perfect (fuzzier) separation.

A separating hyperplane is a linear decision boundary, that is, a linear combination
of the features that makes it possible to perfectly separate the two classes. Figure 4.9
shows a linear decision boundary which is a separating hyperplane, defined as pairs
$\{x_1, x_2\}$ such that $\beta_0 + \beta_1 x_1 + \beta_2 x_2 = 0$. Of course, we can extend this figure to
multi-class and multi-feature settings. Suppose to have p features, define $h(\mathbf{x}) =
(\beta_0 + \beta_1 x_1 + \cdots + \beta_p x_p)$, and code the class *green* as $+1$ and the class *orange* as
-1. Then, we can see that $y_i \times h(\mathbf{x}_i) > 0$ for every i in the dataset, with the separating
hyperplane obtained when $h(\mathbf{x}) = 0$.

Figure 4.10 shows that a separating hyperplane cannot be unique. One can define
many separating hyperplanes within the same dataset, leading to the problem of
finding the "best" one. How can we define the "best" in this specific setting? A
simple procedure can be based on finding the so-called *maximal margin hyperplane*,
which can be found through a two-step procedure. It entails that for each separating
hyperplane:

1. we compute the perpendicular distance from each training observation
2. we retain the smallest distance (called "margin").

The maximal margin hyperplane is the separating hyperplane for which the margin
is largest, that is, the hyperplane that has the largest minimum distance to the training
observations.

Fig. 4.10 Non-uniqueness of a separating hyperplane

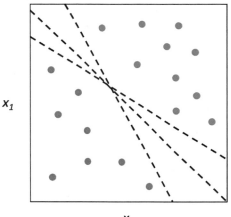

Fig. 4.11 Computation of the *margin*

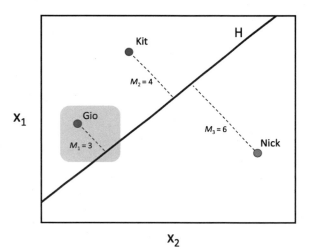

Consider first how to compute the margin. Figure 4.11 shows a simple case with three people (our observations) named "Gio", "Kit", and "Nick", and one hyperplane called H. We define by M_i the perpendicular distance of unit i to the hyperplane H.

Given the hyperplane H, we see that "Gio" is the winner, as he owns the *smallest* perpendicular distance ($M_{Gio} = 3$) from the hyperplane H among the three people.

Once we know how to compute the margins, the last step is to consider all the possible hyperplane, compute the margin, and choose the hyperplane H^* having the largest margin. This will be the optimal separating hyperplane, known as the maximal margin hyperplane. Figure 4.12 shows an example that considers three people— "Sam", "Bill", and "Frank"—and looks for the best hyperplane between H_1, H_2, and H_3. Among H_1, H_2, and H_3, the maximal margin hyperplane is H_2, as its minimal

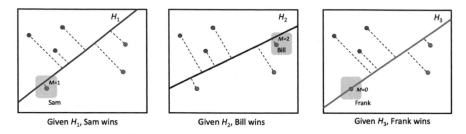

Given H_1, Sam wins   Given H_2, Bill wins   Given H_3, Frank wins

Fig. 4.12 Computation of the *maximal margin hyperplane*

margin is the largest. This is because "Bill" has the largest margin $M_{Bill} = 2$, when compared to "Sam" ($M_{Sam} = 1$) and "Frank" ($M_{Frank} = 0$).

It can be proved, that finding the maximal margin hyperplane entails to find an optimal set of parameters $\{\beta_0^*, \beta_1^*, \ldots, \beta_p^*\}$ that maximizes the margin M under the following constrains:

$$
\begin{cases}
\max_{\beta_0, \ldots, \beta_p} M \\[4pt]
\text{subject to:} \\[4pt]
(1) \; \sum_{j=1}^{p} \beta_j^2 = 1 \\[8pt]
(2) \; y_i(\beta_0 + \beta_1 x_{1i} + \cdots + \beta_p x_{pi}) \geq M
\end{cases}
\tag{4.18}
$$

Under constraint (1), the product between y_i and $(\beta_0 + \beta_1 x_{1i} + \cdots + \beta_p x_{pi})$ measures the perpendicular distance between each observation i (with $i = 1, \ldots, N$) and the hyperplane. Condition (2) implies that each observation must have a distance equal or larger than the margin M. Figure 4.13 provides a representation of the solution. We can see that no observation is allowed to lie between the optimal hyperplane (solid line) and the marginal hyperplane (dashed line)[1]. The observation lying on the marginal hyperplanes are called *support vectors*, and it can be proved that the solution to the optimization problem (4.18) depends only on these observations.

Unfortunately, the case of perfectly separable classes is not very common in real applications. If classes are *not separable*, a maximal margin hyperplane able to perfectly separate them does not exist. As a consequence, it would be useful to find a way to separate classes with the aim of (i) providing larger robustness to specific configurations of the observations, and (ii) improving classification accuracy for the training observations.

When classes are not perfectly separable, we would like to allow an observation to cross the boundaries provided by the marginal hyperplanes, and possibly allow also for misclassifications (that is, lying in the wrong side of the optimal hyperplane). In other words, an observation i might be (1) on the right side of the margin, (2) in the

---

[1] Observe, also, that we have two marginal hyperplanes, one for each class.

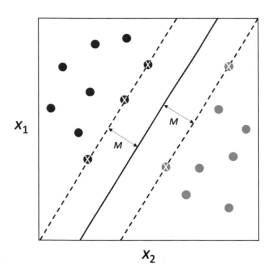

Fig. 4.13 Maximal margin hyperplane with perfectly separable classes. No observation is allowed to lie between the hyperplane and the marginal hyperplane

wrong side of the margin, but on the right side of the optimal hyperplane, (3) in the wrong side of the optimal hyperplane.

A possible way to deal with non-separable classes is to rely on a Support Vector Classifier (SVC). The SVC optimization problem takes on this form:

$$
\begin{cases}
\displaystyle\max_{\beta_0,\dots,\beta_p,\epsilon_1,\dots,\epsilon_N} \quad M \\[2mm]
\text{subject to:} \\[2mm]
(1) \ \displaystyle\sum_{j=1}^{p} \beta_j^2 = 1 \\[2mm]
(2) \ y_i(\beta_0 + \beta_1 x_{1i} + \cdots + \beta_p x_{pi}) \geq M(1 - \epsilon_i) \\[2mm]
(3) \ \epsilon_i \geq 0 \\[2mm]
(4) \ \displaystyle\sum_{i=1}^{N} \epsilon_i \leq \tau
\end{cases}
\tag{4.19}
$$

where M is the width of the margin (as above); τ is a non-negative constant parameter; ϵ_i are known as *slack variables*, and they allow observations to be on the wrong side of the margin or of the hyperplane.

The value ϵ_i influences the location of unit i. We can interpret ϵ_i as the proportional amount by which unit i is on the wrong side of its margin of reference. When an observation has $\epsilon_i = 0$, its distance from the separating hyperplane H—namely, $d(i; H) = y_i(\beta_0 + \beta_1 x_{1i} + \cdots + \beta_p x_{pi})$—must be equal or larger than M, as $M(1 - \epsilon_i) = M$; when $0 < \epsilon_i < 1$, observation i may be located within the

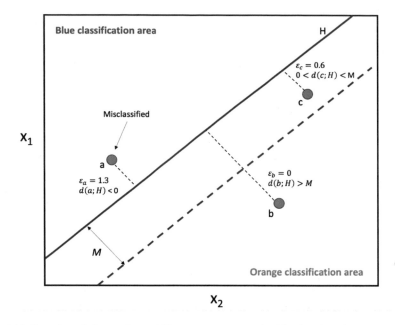

Fig. 4.14 Location of observations within a support vector classification, based on system (4.19)

separating hyperplane and the marginal hyperplane, as $0 < M(1 - \epsilon_i) < M$; finally, if $\epsilon_i > 1$, then observation i may be located in the wrong side of the separating hyperplane, as $M(1 - \epsilon_i) < 0$.

Figure 4.14 shows a two-class example where only the *orange* class is considered (while the other class is *blue*). We have three points with three different values of ϵ_i, point a, point b, and point c. Point b has $\epsilon_i = 0$; thus, by constraint (2) of system (4.19), a point like this must be located either in the margin or on the right side of the margin. In this example, b lies on the right side of the margin. Point c has $\epsilon_i = 0.6$, thus it is surely located on the right side of the separating hyperplane, but possibly also located between the separating hyperplane and the (orange) marginal hyperplane. In this example, c is located exactly between the separating hyperplane and the marginal hyperplane. Finally, point a has $\epsilon_i = 1.3$, thus an observation like this might be located in the wrong side of the separating hyperplane, as we placed this point. Point a is also misclassified, as it is an orange point located in the decision area where points are classified as blue. It is important to notice that—according to constraint (2) of system (4.19)—having an $\epsilon_i > 1$ does not assure that unit i will be misclassified, has its distance $d(i; H)$ may be positive and thus larger than the negative $M(1 - \epsilon_i)$. Therefore, an $\epsilon_i > 1$ is only an indication that a unit i *might be misclassified*.

It is important to notice that—according to constraint (4) of system (4.19)—if we interpret ϵ_i as an error measure, the sum of such errors cannot be larger than the

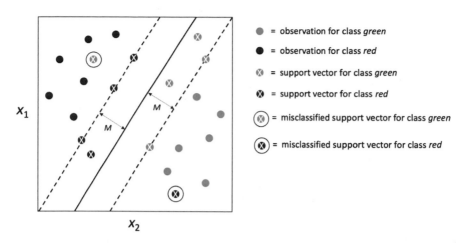

Fig. 4.15 Support vector classifier for non-separable classes. Observations are allowed to lie both in the wrong side of the margin or of the hyperplane. These observations—known as support vectors—are those responsible for classification

constant parameter τ, sometimes called *total cost* parameter. Therefore, τ poses an upper bound to the possible number of misclassifications that the SVC can commit.

The optimization entailed by (4.19) owns an important property: only observations that either lie on the margin or that violate the margin will affect the optimal separating hyperplane, and thus the classifier obtained; therefore, an observation that lies strictly on the correct side of the margin does not affect the support vector classification. The observations that lie directly on the margin, or on the wrong side of the margin for their class are known as *support vectors*. These are the observations that affect classification performance also in non-separable settings.

Figure 4.15 shows a representation of the solution provided by optimization (4.19). As we can see, for being a support vector, it is sufficient that an observation lies on the wrong side of the marginal hyperplane (according to its reference class). This observation can also be placed in the wrong size of the optimal separating hyperplane, thus being misclassified.

Although system (4.19) clarifies well how the optimal hyperplane parameters are obtained, it is not the standard way in which the solution is computed. It can be proved that the solution of system (4.19) is more easily computed using this equivalent quadratic programming optimization:

$$
\begin{cases}
\displaystyle \min_{\beta_0,\ldots,\beta_p,\epsilon_1,\ldots,\epsilon_N} \left[\frac{1}{2} \sum_{j=1}^{p} \beta_j^2 + C \sum_{i=1}^{N} \epsilon_i \right] \\
\text{subject to:} \\
y_i (\beta_0 + \beta_1 x_{1i} + \cdots + \beta_p x_{pi}) \geq (1 - \epsilon_i) \\
\epsilon_i \geq 0
\end{cases}
\tag{4.20}
$$

where the main objective function takes the form of a penalized loss. It is important to notice that the penalization parameter C is equal to $1/\tau$, and has a central role in the classification performance. By enlarging or reducing C, we modify the set of support vectors, and thus we generate different classification outcomes.

Larger values of C (i.e., smaller values of τ) produce tighter margins. On the contrary, smaller values of C (i.e., larger values of τ) produce wider margins. It is rather clear that C (or τ) is a tuning parameter, as it engenders a trade-off between prediction variance and prediction bias (Chapelle et al., 2002). Let's focus on this point.

First, when $C = \infty$ (i.e., when $\tau = 0$), we are in the setting provided by the maximal margin hyperplane optimization problem, with no violations of the margins allowed and perfect separability of the classes. When C increases (and τ decreases), we are narrowing the margin. In this case, the SVC solution depends on a smaller set of observations (smaller number of support vectors), implying an increase in the prediction variance due to a smaller sample size employed for classifying. At the same time, however, a smaller τ severely bounds the sum of the ϵ_i (see constrain (4) in system 4.19), thereby reducing the number and the size of the violations of either the marginal hyperplane or the separating hyperplane. In this case, only a few misclassifications are allowed, so that prediction bias decreases. By contrast, when C decreases (and τ increases), prediction variance goes down as classification is now made by using a larger set of units (that is, a larger number of support vectors). But with a larger τ, we are also allowing for more violations of the marginal and possibly also of the separating hyperplanes, thereby increasing prediction bias. Figure 4.16 sets out an example of SVC for non-separable classes at different levels of the cost parameter C (or τ).

For specific configurations of the data, a linear boundary can work very poorly, although the value of C has been optimally tuned. The example in Fig. 4.17 shows the case in which a linear boundary performs poorly in classifying correctly the training observations, as both orange and green points are in both sides of the hyperplane. How can we deal with situations like this?

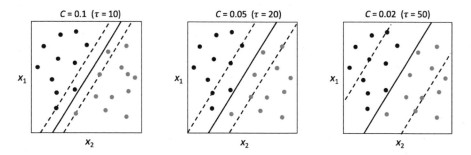

Fig. 4.16 Support vector classifier for non-separable classes at different levels of the cost parameter C (or τ). Larger values of C (i.e., smaller values of τ) produce tighter margins. Variations of C induce a trade-off between prediction bias and prediction variance

Fig. 4.17 Support vector classifier for non-separable classes where a linear boundary is unable to classify well

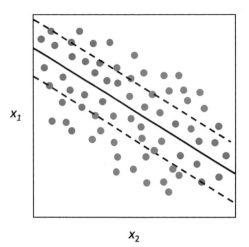

x_1

x_2

A possible solution would be that of generating hyper-curves instead of hyperplanes, thereby producing nonlinear decision boundaries. There are two ways to do this: (1) feature polynomial expansion, and (2) Support Vector Machine (SVM), based on either polynomial kernel or radial basis function (RBF) kernel (Schölkopf et al., 1997).

With feature polynomial expansion, we expand the space of features by a polynomial series transformation. This transformation allows to pass from p to $P > p$ features obtained by the polynomial expansion. For example, with two features, x_1 and x_2, a polynomial of degree 2 contains the following 6 terms: $x_1, x_2, x_1x_2, x_1^2, x_2^2, x_1^2x_2^2$. Given this enrichment of the feature space, we can fit an SCV to the P new features, thereby obtaining nonlinear decision boundaries in the original feature space defined by the pairs $\{x_{1i}, x_{2i} : i = 1, \ldots, N\}$. In particular, if we consider a 2-degree polynomial expansion with two-features, the equation of the decision boundary will be

$$\beta_0 + \beta_1 x_1 + \beta_2 x_2 + \beta_3 x_1 x_2 + \beta_4 x_1^2 + \beta_5 x_2^2 + \beta_6 x_1^2 x_2^2 = 0$$

which corresponds to a quadratic function. Of course, we can also consider larger polynomial degree expansions.

The second approach to deal with non-separable classes is the SVM based on Kernel adaptation. To address this approach, we first show an important characterization of the solution of the Support Vector Classifier optimization problem. Although its apparent complexity, the solution of the Support Vector Classifier's optimization takes on this form:

$$G(\mathbf{x}) = \beta_0 + \sum_{i=1}^{N} \alpha_i \langle \mathbf{x}, \mathbf{x}_i \rangle \tag{4.21}$$

where $\langle \mathbf{x}, \mathbf{x}_i \rangle$ is the *inner product* between vector $\mathbf{x}$ and vector $\mathbf{x}_i$, and α_i is a parameter specific to every observation $i = 1, \ldots, N$. For instance, if we have two observations, i and i', their inner product is defined as

$$\langle \mathbf{x}_i, \mathbf{x}_{i'} \rangle = \sum_{j=1}^{p} x_{ij} x_{i'j} \tag{4.22}$$

Therefore, in order to estimate all the parameters—that is, $\beta_0, \alpha_1, \ldots, \alpha_N$—it is sufficient to compute the $N(N-1)/2$ inner products. However, for the solution to be obtained, one does not need to compute *all* the inner products, but only the subset including the support vectors. This is why the SCV depends only on the support vectors. Indicating with S the set of support vectors, we can rewrite Eq. (4.21) as

$$G(\mathbf{x}) = \beta_0 + \sum_{i \in S} \alpha_i \langle \mathbf{x}, \mathbf{x}_i \rangle \tag{4.23}$$

which classifies whatever observation having features equal to $\mathbf{x}$. We can generalize the previous formula by considering nonlinear transformation of the inner products using Kernel transformation as follows:

$$G(\mathbf{x}) = \beta_0 + \sum_{i \in S} \alpha_i K(\mathbf{x}, \mathbf{x}_i) \tag{4.24}$$

where $K(\cdot)$ is an appropriate kernel function. Two Kernel transformations are generally used: *polynomial* kernel and *radial* kernel.

The formula of a polynomial kernel of degree d is

$$K(\mathbf{x}_i, \mathbf{x}_{i'}) = \left[1 + \sum_{j=1}^{p} x_{ij} x_{i'j}\right]^d \tag{4.25}$$

where, for $d = 1$, we obtain the linear Support Vector Classifier. In this case the metric used to measure the distance between point i and point i' is the Pearson correlation.

Differently, the formula of the radial kernel is

$$K(\mathbf{x}_i, \mathbf{x}_{i'}) = \exp\left[-\gamma \sum_{j=1}^{p} (x_{ij} - x_{i'j})^2\right] \tag{4.26}$$

where $K(\cdot)$ is a function of the Euclidean distance between the pair i and i' instead of the inner product. The interpretation of the radial kernel is straightforward when we consider the exponential function therein used rewritten in this way:

$$K = e^{-\gamma d} \tag{4.27}$$

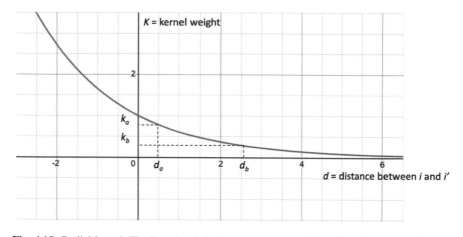

Fig. 4.18 Radial kernel. The kernel weight decreases exponentially with an increasing distance between point i and point i'

where d is the distance between i and i'. Figure 4.18 shows the Radial kernel function, where it is clear that the kernel weight decreases exponentially with an increasing distance between point i and point i'. If a given test observation x^* is far from a training observation x_i in terms of Euclidean distance, then $K(\cdot)$ will be very tiny. This means that x_i will play no role in contributing to $G(x^*)$. The predicted class label for the test observation x^* is based on the sign of $G(x^*)$. Thus, training observations far from x^* will play no role in the predicted class label for x^*. Radial kernel has very local behavior, as only nearby training observations have an effect on the class label of a test observation. This makes SVM similar to the nearest-neighbor classifier.

This analogy with the nearest-neighbor classifier can be put forward also through the role played by the γ parameter of the RBF, which is similar to the number of nearest neighbors to consider in KNN. In fact, when γ is low, the same distance d between a pair of observations will produce a larger $K(\cdot)$ (see Fig. 4.19), implying that classification becomes more sensitive to the distance and even large distances can count a lot for the classification of a given test point. This is equivalent to consider a large number of nearest neighbors in the KNN. The decision boundary thus becomes flatter, entailing smaller variance and larger bias. On the contrary, when γ is high, the same distance d between a pair of observations will produce a smaller $K(\cdot)$ (see Fig. 4.19), thus classification becomes less sensitive to the distance and larger distances can count much less for classification. This is equivalent to consider a small number of nearest neighbors in the KNN. The decision boundary becomes very irregular, forming like "islands" around data points. This entails larger variance, but smaller bias. It is clear that γ is a tuning parameter entailing a variance-bias trade-off. As such, it can be optimally tuned via cross-validation.

The advantage of SVM over feature expansion is mainly computational, as in many applications the enlargement of the feature space makes the computation of the solution no longer achievable. SVM only requires the calculation of the

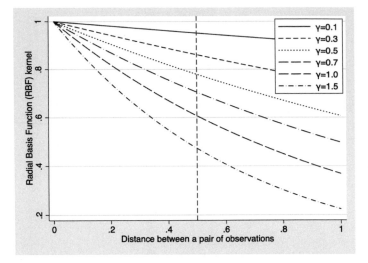

Fig. 4.19 Radial Basis Function (RBF) kernel as function of the γ parameter

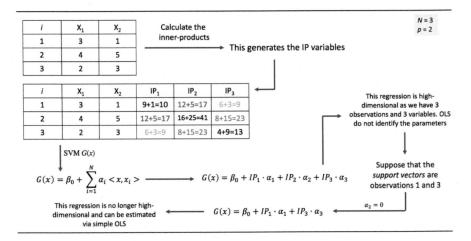

Fig. 4.20 Numerical example of computation of the Support Vector Machine (SVM) classifier with 3 training observations and 2 features

$N \times N$ matrix of training observations' inner-products or inner-distances. Figure 4.20 provides a simple numerical example of the computation of the Support Vector Machine (SVM) classifier with 3 training observations and 2 features.

It is important to notice that SVM do not compute probabilities. Instead, they output a score:

$$\widehat{G}(\mathbf{x}) = \widehat{\beta}_0 + \sum_{i \in \mathcal{S}} \widehat{\alpha}_i \langle \mathbf{x}, \mathbf{x}_i \rangle \tag{4.28}$$

Therefore, a positive score means that $\widehat{y} = +1$ is more likely; a negative score means that $\widehat{y} = -1$ is more likely; and a score equal to 0 means that both classes are equally likely. How can we obtain class probabilities from the scores? Generally, we can map the score into a pseudo-probability using, in the training dataset, a logistic regression with y, the 0/1 true class membership, as the dependent variable and the score as the only independent variable. For each training unit i, the probability p_i is thus obtained as:

$$p_i = E[y_i|\widehat{G}(\mathbf{x}_i)] = \Lambda[\beta_0 + \beta_1\widehat{G}(\mathbf{x}_i)]$$

Using the estimated parameters from this logit regression, we obtain class probabilities also for new unlabeled instances.

So far, we have considered SVM classification in a two-class setting. We can extend SVM classification to a multi-class setting as well. We have two different approaches that we can follow when we have $J > 2$ classes: (i) *one versus all* (OVA), and (ii) *one versus one* (OVO). In the OVA approach, one fits $J > 2$ different two-class classifiers $\widehat{C}_j(\mathbf{x})$, with $j = 1, \ldots, J$. In other words, every class j against the rest of the classes. In the OVO approach, we fit $\binom{J}{2}$ pairwise classifiers $\widehat{C}_{jm}(\mathbf{x})$. In this case, suppose that we have to classify a new observation $\mathbf{x}_0$. According to OVO, this new observation will be classified within the class that wins the largest number of pairwise competitions. For example, suppose to have $J = 3$ classes, 1, 2, and 3, and suppose that we have that $\widehat{C}_{12}(\mathbf{x}_0) = 1$, $\widehat{C}_{13}(\mathbf{x}_0) = 1$, $\widehat{C}_{23}(\mathbf{x}_0) = 3$. In this case, $\mathbf{x}_0$ will be classified within class 1, as this class is the one with the largest numbers of wins (i.e., 2). Choosing between OVA and OVO mostly depends on how large is J. If J is small, OVO is preferable, and OVA otherwise.

4.3 Regression

Both the k-nearest neighbor (KNN) and support vector machine (SVM) learners can be extended to predict numerical outcomes. In other words, they can be used for regression purposes. In what follows, we present both approaches.

4.3.1 SVM Regression

Support vector machine can be also extended to the regression case, that is, when the target variable is numerical instead of categorical. In the regression case, however, SVM has a different setting that we have to investigate more in depth.

First of all, it is important to define what is the loss function adopted by SVM when a continuous target variable is considered. We know that ordinary least squares

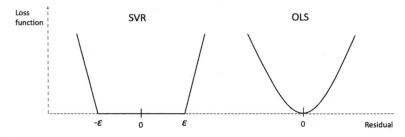

Fig. 4.21 OLS and SVM regression loss functions: a comparison

(OLS) use the quadratic loss function. In this case, if we consider a linear conditional mean—that is $E(y|\mathbf{x}) = \mathbf{x}\beta$—we have that, for a generic observation i:

$$L_i^{ols} = (y_i - \mathbf{x}_i\beta)^2 \tag{4.29}$$

that must be minimized in the vector β over the sample. SVM, instead, adopts the so-called ϵ-insensitive loss function, that is

$$L_i^{svm} = \begin{cases} 0 \text{ if } |y_i - \mathbf{x}_i\beta| < \epsilon \\ |y_i - \mathbf{x}_i\beta| - \epsilon \text{ if } |y_i - \mathbf{x}_i\beta| \geq \epsilon \end{cases} \tag{4.30}$$

This loss function entails that any residual smaller than ϵ is set equal to zero, while for any residual larger than ϵ a linear loss is used.

It is clear, that there is some similarity with the support vector classifier: in the classification setting, we saw that observations located on the correct side of the decision boundary and distant from it are not used in the optimization procedure. In the SVM regression setting, similarly, observations with small residuals are not relevant in affecting the loss, and are thus not accounted for in the optimization procedure. Figure 4.21 shows a comparison between the OLS and SVM regression loss functions, where it is clear how they weigh differently the residuals. Observe that the parameter ϵ is a tuning parameter and thus can be optimally selected, for example, via cross-validation.

It is possible to show that minimizing over the training sample the SVM loss function of Eq. (4.30) is equivalent to find the best linear-SVM regression parameters β^* that solve the following minimization:

$$\min_{\beta} \frac{1}{2}\|\beta\|^2 + C \sum_{i=1}^{n} \max(0, |\mathbf{x}_i\beta - y_i| - \epsilon) \tag{4.31}$$

Unfortunately, if there are observations outside of the ϵ-band around the SVM regression line, the previous problem does not provide any solution. For taking into account also observations that are located outside of the ϵ-band, an equivalent expression can be formed using slack variables:

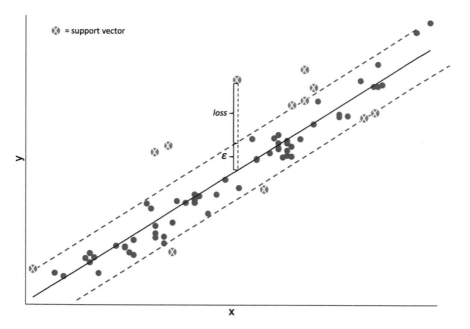

Fig. 4.22 Example of a SVM regression fit

$$\min_{\beta, \xi} \frac{1}{2} \|\beta\|^2 + C \sum_{i=1}^{n} \xi_i$$

$$\text{s.t. } \xi_i \geq \mathbf{x}_i \beta - y_i - \epsilon, \qquad\qquad (4.32)$$

$$\xi_i \geq -\mathbf{x}_i \beta + y_i - \epsilon,$$

$$\xi_i \geq 0$$

Figure 4.22 shows a possible solution of the previous minimization. We see that the observations that affect the loss are those laying outside the ϵ-band, and they are the support vectors. Also, the figure displays the loss for one specific support vector.

Smola & Schölkopf (2004) provides a tutorial to show how to solve problems like the one in (4.32). As in the case of SVM classification, also for SVM regression nonlinear (polynomial and kernel) extensions are available.

4.3.2 KNN Regression

The k-nearest neighbor regression is a natural extension of the KNN classifier. In this case, the target variable is numerical, often continuous. The KNN regression algorithm estimates the conditional mean $E(y|\mathbf{x})$ *locally* in every specific point $\mathbf{x}_0$

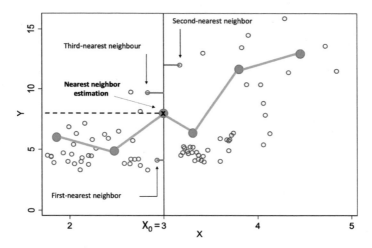

Fig. 4.23 Illustrative example of the k-nearest neighbor regression algorithm with only one feature, and $k = 3$

by imputing $E(y|\mathbf{x}_0)$ using the average of the outcomes of the first k nearest neighbors of $\mathbf{x}_0$. By repeating the same procedure pointwise (that is, at every observation values, or at given grid values), we obtain the KNN regression curve.

Fig. 4.23 provides an illustrative example of how the KNN regression algorithm works with only one feature, and $k = 3$ nearest neighbors.

In some implementations of the KNN regression algorithm, a kernel weighting scheme is also considered. This is a weighting scheme giving larger weight to points that are closer to the point of imputation $\mathbf{x}_0$. It means that the imputation of the conditional mean at $\mathbf{x}_0$ is carried out using a weighted average of the k nearest neighbors' outcomes, with weights summing up to one. It goes without saying that also for the KNN regression the main tuning parameter is k, the number of nearest neighbors to consider. It is also common to tune the type of kernel weighting scheme to use when different options are available.

4.4   Applications in Stata, R, and Python

4.4.1   Application S1: Linear Discriminant Analysis Classification in Stata

In Stata, discriminant analysis and nearest neighbor classification can be carried out using the `discrim <subcommand>` command, where `<subcommand>` can be one of: `knn` (kth-nearest-neighbor discriminant analysis), `lda` (linear discriminant

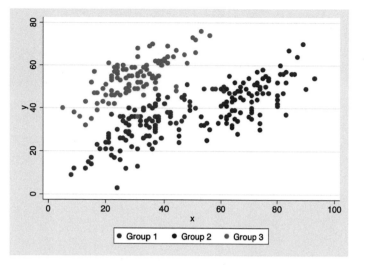

Fig. 4.24 Scatterlplot of the three groups defined within the feature space

analysis), `logistic` (logistic discriminant analysis), or `qda` (quadratic discriminant analysis).

To start with, consider the dataset `threegroup`, which contains three classes (or groups)—1, 2, and 3—and two features, `y` and `x`. The below Stata code implements an LDA over these data (Fig. 4.24).

```
─────────────────────────── Stata code and output ───────────────────────────
. * Linear discriminant analysis (LDA)

. * - Three groups ---> j = 1,2,3
. * - Two features: {y,x}

. * Open the dataset
. use http://www.stata-press.com/data/r15/threegroup , clear
(Three Groups)

. * Plot the three groups in a scatterplot
. graph twoway ///
> (scatter y x if group==1)   ///
> (scatter y x if group==2)   ///
> (scatter y x if group==3) , ///
> legend(label(1 "Group 1") label(2 "Group 2") label(3 "Group 3") rows(1)) ///
> scheme(s2color)

. * Run "lda"
. discrim lda y x, group(group)
Linear discriminant analysis
Resubstitution classification summary
```

Key
Number
Percent

	Classified			
True group	1	2	3	Total
1	93	4	3	100
	93.00	4.00	3.00	100.00
2	3	97	0	100
	3.00	97.00	0.00	100.00
3	3	0	97	100
	3.00	0.00	97.00	100.00
Total	99	101	100	300
	33.00	33.67	33.33	100.00
Priors	0.3333	0.3333	0.3333	

* Show the "Leave-one-out classification table"
. estat classtable , looclass

Leave-one-out classification table

Key
Number
Percent

	LOO Classified			
True group	1	2	3	Total
1	93	4	3	100
	93.00	4.00	3.00	100.00
2	3	97	0	100
	3.00	97.00	0.00	100.00
3	3	0	97	100
	3.00	0.00	97.00	100.00
Total	99	101	100	300
	33.00	33.67	33.33	100.00
Priors	0.3333	0.3333	0.3333	

. * List the misclassified observations
. estat list, varlist misclassified

	Data		Classification		Probabilities		
Obs	y	x	True	Class.	1	2	3
19	49	37	1	3 *	0.2559	0.0000	0.7441
29	49	57	1	2 *	0.4245	0.5750	0.0005
47	49	37	1	3 *	0.2559	0.0000	0.7441
55	24	45	1	2 *	0.4428	0.5572	0.0000
70	48	61	1	2 *	0.0661	0.9339	0.0000
74	49	58	1	2 *	0.3041	0.6957	0.0003
92	37	22	1	3 *	0.3969	0.0000	0.6031
143	27	45	2	1 *	0.6262	0.3738	0.0000
161	39	49	2	1 *	0.8026	0.1973	0.0001
185	49	54	2	1 *	0.7782	0.2187	0.0030
238	48	44	3	1 *	0.8982	0.0017	0.1001
268	50	44	3	1 *	0.7523	0.0009	0.2469
278	36	31	3	1 *	0.9739	0.0000	0.0261

* indicates misclassified observations

```
. * Calculate the "training error rate" and the "test error rate"
. discrim lda y x, group(group) notable

. estat errorrate , class     // training error rate
Error rate estimated by error count
```

	group			
	1	2	3	Total
Error rate	.07	.03	.03	.0433333
Priors	.3333333	.3333333	.3333333	

```
. estat errorrate , looclass // test error rate
Error rate estimated by leave-one-out error count
```

	group			
	1	2	3	Total
Error rate	.07	.03	.03	.0433333
Priors	.3333333	.3333333	.3333333	

Let's comment on the code and the outputs. After opening the dataset, we plot the three groups in a scatterplot defined within the feature space. Then, we run the discrim lda command, where the option group() takes as argument the target variable that in this case is called group. The output provides the resubstitution classification table, which is the in-sample classification table (sometimes also called *confusion matrix*) generated by the LDA. However, to assess the real out-of-sample prediction ability of the model, one possibility is to estimate the leave-one-out classification table. This is achieved by typing estat classtable with option looclass. This table looks like the previous one, but it has happened only by chance. Finally, after listing the misclassified observations, we calculate the training error rate and the test error rate, which in this specific case is the same and equal to 0.043.

We provide now a second example where we compute also predictions. For this purpose, we use the riding-mower data from Johnson and Wichern (2007). We have 2 groups (12 riding-lawnmower owners, and 12 nonowners) and 2 features: income (income in thousands of dollars), and lotsize (lot size in thousands of square feet). Suppose that a riding-mower manufacturer wants to know whether these two variables adequately separate owners from nonowners, and if so to then direct their marketing on the basis of the separation of owners from nonowners. Specifically, we want to predict whether a new individual showing up in the city's neighborhood, will be a owner or a nonowner. To perform this analysis, we run the below Stata code.

```
─────────────────────────── Stata code and output ───────────────────────────
. * Open the dataset
. use http://www.stata-press.com/data/r15/lawnmower2 , clear
(Johnson and Wichern (2007) Table 11.1)

. * RUN "LDA"
. discrim lda lotsize income , group(owner) notable

. * Out-of-sample predictions

. * See how the LDA classifies observations with:
. * - income of $90,000, $110,000, and $130,000,
```

```
. * - lot size of 20,000 square feet for all.
.
. * Procedure
. * Add 3 observations to the bottom of the dataset containing these values:
. input
        owner  income  lots~e
25. owner income lotsize
'owner' cannot be read as a number
25.    .    90     20
26.    .   110     20
27.    .   130     20
28. end
```

```
. * Use "predict" to obtain "classifications" and "probabilities"
. * Predicted classification
. predict grp in 25/L, class
(24 missing values generated)
. * Predicted probabilities
. predict pr* in 25/L, pr
(24 missing values generated)
. * List the results
. . list in 25/L
```

	owner	income	lotsize	grp	pr1	pr2
25.	.	90.0	20.0	0	.5053121	.4946879
26.	.	110.0	20.0	1	.1209615	.8790385
27.	.	130.0	20.0	1	.0182001	.9818

```
. * Predicting "training" and "test" classification
. use http://www.stata-press.com/data/r15/lawnmower2 , clear
(Johnson and Wichern (2007) Table 11.1)
. qui discrim lda lotsize income, group(owner) notable
. predict gr_class , class   // training classification
. predict gr_loo , looclass  // test classification (leave-one-out group
membership classification)
```

```
. * Predicting "training" and "test" error
. estat errorrate , class      // training error rate
Error rate estimated by error count
```

	owner nonowner	owner	Total
Error rate	.1666667	.0833333	.125
Priors	.5	.5	

```
. estat errorrate , looclass  // test error rate
Error rate estimated by leave-one-out error count
```

	owner nonowner	owner	Total
Error rate	.25	.1666667	.2083333
Priors	.5	.5	

Let's comment on the code and the outputs. We start by opening the dataset
lawnmower2, and then we run the lda using the same command seen above. After
this step, we perform some out-of-sample predictions to see how the LDA classifies

new incoming observations with an income of $90,000, $110,000, and $130,000, and a lot size of 20,000 square feet for all. For this purpose, we add 3 observations to the bottom of the dataset containing these values using the input command, to then use predict to obtain both classifications and probabilities. As a last step, we predict both train and test classification (that is, leave-one-out group membership classification). Finally, we predict train and test errors. In this case, the estimated in-sample error rate is equal to 0.125, while the error rate estimated by leave-one-out error count is more pessimistic (as expected) with a value of around 0.21.

4.4.2   Application S2: Quadratic Discriminant Analysis Classification in Stata

In this application, we use quadratic discriminant analysis (QDA) over the dataset rootstock by Rencher and Christensen (2012). This dataset contains information about 8 trees from each of six apple tree rootstocks. The features are trunk girth at 4 years (y1), extension growth at 4 years (y2); trunk girth at 15 years (y3), and weight of tree above ground at 15 years (y4). The target (i.e., grouping) variable is rootstock. Below, we can find the Stata code.

```
───────────────────────────── Stata code and output ─────────────────────────────
. * Quadratic discriminant analysis (qda)

. * Open the dataset
. use http://www.stata-press.com/data/r15/rootstock , clear
(Table 6.2 Rootstock Data, Rencher and Christensen (2012))

. * Run QDA
. discrim qda y1 y2 y3 y4, group(rootstock) lootable
Quadratic discriminant analysis
Resubstitution classification summary
```

Key
Number
Percent

True	Classified						
rootstock	1	2	3	4	5	6	Total
1	8	0	0	0	0	0	8
	100.00	0.00	0.00	0.00	0.00	0.00	100.00
2	0	7	0	1	0	0	8
	0.00	87.50	0.00	12.50	0.00	0.00	100.00
3	1	0	6	0	1	0	8
	12.50	0.00	75.00	0.00	12.50	0.00	100.00
4	0	0	1	7	0	0	8
	0.00	0.00	12.50	87.50	0.00	0.00	100.00
5	0	3	0	0	4	1	8

	0.00	37.50	0.00	0.00	50.00	12.50	100.00
6	2	0	0	0	1	5	8
	25.00	0.00	0.00	0.00	12.50	62.50	100.00
Total	11	10	7	8	6	6	48
	22.92	20.83	14.58	16.67	12.50	12.50	100.00
Priors	0.1667	0.1667	0.1667	0.1667	0.1667	0.1667	

Leave-one-out classification summary

```
Key

Number
Percent
```

True rootstock	Classified 1	2	3	4	5	6	Total
1	2	0	0	3	1	2	8
	25.00	0.00	0.00	37.50	12.50	25.00	100.00
2	0	3	0	2	2	1	8
	0.00	37.50	0.00	25.00	25.00	12.50	100.00
3	1	2	4	0	1	0	8
	12.50	25.00	50.00	0.00	12.50	0.00	100.00
4	1	1	3	2	0	1	8
	12.50	12.50	37.50	25.00	0.00	12.50	100.00
5	0	4	1	0	2	1	8
	0.00	50.00	12.50	0.00	25.00	12.50	100.00
6	3	1	0	0	2	2	8
	37.50	12.50	0.00	0.00	25.00	25.00	100.00
Total	7	11	8	7	8	7	48
	14.58	22.92	16.67	14.58	16.67	14.58	100.00
Priors	0.1667	0.1667	0.1667	0.1667	0.1667	0.1667	

```
. * Calculate the "training" and "test" error rate
. estat errorrate, nopriors class
Error rate estimated by error count
```

	rootstock 1	2	3	4	5	6	Total
Error rate	0	.125	.25	.125	.5	.375	.2291667

```
. estat errorrate, nopriors looclass
Error rate estimated by leave-one-out error count
```

	rootstock 1	2	3	4	5	6	Total
Error rate	.75	.625	.5	.75	.75	.75	.6875

Let's comment on the code and its outputs. We start by opening the dataset rootstock, and then we run the QDA using discrim with the subcommand qda and the lootable option to obtain as a result the leave-one-out confusion matrix. We then compute the train and test error rates, obtaining a training error rate of 0.23

against a test error rate of 0.69, which is around three times larger. Notice that with option `nopriors` we have suppressed the display of prior probabilities.

4.4.3 Application S3: Nearest Neighbor Classification in Stata

In this application, we apply the k-nearest neighbor classifier using `discrim` with the subcommand `knn`. We consider the `head` dataset (Rencher and Christensen, 2012) investigating head measurements and the degree of separability of three groups: (1) High school football players, (2) College football players, (3) Non football players. The target variable containing these groups is called `group`. The number of features is 6, corresponding to six head dimensions measured for each subject. They are `wdim`, head width at its widest dimension; `circum`, head circumference; `fbeye`, front-to-back measurement at eye level; `eyehd`, eye to top of head measurement; `earhd`, ear to top of head measurement; `jaw`, jaw width. These measurements do not have same ranges, and Mahalanobis distance is used to standardize them. Below, we set out the Stata code.

```
————————————————————— Stata code and output ——————————————
. * k-nearest neighbor discriminant analysis (knn)

. * Open the dataset
. use http://www.stata-press.com/data/r15/head, clear
(Table 8.3 Head measurements, Rencher and Christensen (2012))

. * Run "knn" with "mahalanobis" option
. discrim knn wdim-jaw, k(5) group(group) mahalanobis
Kth-nearest-neighbor discriminant analysis
Resubstitution classification summary
```

```
┌──────────┐
│   Key    │
├──────────┤
│  Number  │
│ Percent  │
└──────────┘
```

True group	Classified high school	college	nonplayer	Unclassified	Total
high school	26	0	1	3	30
	86.67	0.00	3.33	10.00	100.00
college	1	19	9	1	30
	3.33	63.33	30.00	3.33	100.00
nonplayer	1	4	22	3	30
	3.33	13.33	73.33	10.00	100.00
Total	28	23	32	7	90
	31.11	25.56	35.56	7.78	100.00
Priors	0.3333	0.3333	0.3333		

```
. * Run "knn" with "mahalanobis" and "ties(nearest)" option
. discrim knn wdim-jaw, k(5) group(group) mahalanobis ties(nearest)
```

```
Kth-nearest-neighbor discriminant analysis
Resubstitution classification summary
```

```
┌─────────┐
│ Key     │
├─────────┤
│ Number  │
│ Percent │
└─────────┘
```

True group	Classified high school	college	nonplayer	Total
high school	28	0	2	30
	93.33	0.00	6.67	100.00
college	1	20	9	30
	3.33	66.67	30.00	100.00
nonplayer	1	4	25	30
	3.33	13.33	83.33	100.00
Total	30	24	36	90
	33.33	26.67	40.00	100.00
Priors	0.3333	0.3333	0.3333	

Let's comment on the code and its outputs. First, we open the head dataset. Then, we run a 5-nearest neighbor classifier with the mahalanobis option, thus obtaining the related resubstitution classification table. We can observe that the unclassified observations are those that resulted in being "ties". Thus, we can force ties to be classified by changing the ties() option. The default is ties(missing), which classifies ties as missing values. The option ties(nearest) breaks the tie by classifying to the group of the nearest tied observation. We thus run the discrim knn with this option showing that, in this case, all units become classified.

We consider now a second application using the k-nearest neighbor classifier in Stata when we have to deal with *binary* features. Indeed, when data is binary (i.e., taking values either 0 or 1), the concept of distance between observations must be re-considered and cannot be the same as the one used for continuous variables. In the literature, besides distance measures for continuous data (such as the Euclidean and the Mahalanobis), a variety of binary measures are available such as, for example, the Jaccard binary similarity coefficient, which reports the proportion of matches of ones when at least one of the observations contains a one; or the Dice binary similarity measure, that weighs matches of ones twice as heavily as the Jaccard measure.

In this second application, we consider the mushroom dataset with the aim of classifying "edible" versus "poisonous" mushrooms. The number of groups is thus

2 (edible and poisonous), and the features are 23 variables that describe mushrooms characteristics. The Stata code implementing this application is below.

```
──────────────────────── Stata code and output ────────────────────────
. * Open the dataset
. use http://www.stata-press.com/data/r15/mushroom , clear
(Lincoff (1981) Audubon Guide; http://archive.ics.uci.edu/ml/datasets/Mushroom)

. * Consider just 2,000 mushrooms at random
. set seed 12345678

. cap generate u = runiform()

. sort u

. * Run knn (with many binary features and "dice" option)
. xi, noomit: discrim knn i.population i.habitat i.bruises i.capshape ///
> i.capsurface i.capcolor in 1/2000, k(15) group(poison) measure(dice)
Kth-nearest-neighbor discriminant analysis
Resubstitution classification summary
```

Key
Number
Percent

True poison	Classified edible	poisonous	Total
edible	848	65	913
	92.88	7.12	100.00
poisonous	29	1,058	1,087
	2.67	97.33	100.00
Total	877	1,123	2,000
	43.85	56.15	100.00
Priors	0.5000	0.5000	

```
. * Run the previous "knn" (with larger prior probability to poisonous mushrooms)
. estat classtable in 1/2000 , priors(.09, .91)
Resubstitution classification table
```

Key
Number
Percent

True poison	Classified edible	poisonous	Total
edible	728	185	913
	79.74	20.26	100.00
poisonous	0	1,087	1,087
	0.00	100.00	100.00
Total	728	1,272	2,000
	36.40	63.60	100.00
Priors	0.0900	0.9100	

```
. * Check prediction of poison status on all dataset
. predict cpoison, classification priors(.09, .91)
```

```
. label values cpoison poison
. tabulate poison cpoison
```

| | classification | | |
poison	edible	poisonous	Total
edible	2,450	718	3,168
poisonous	0	3,789	3,789
Total	2,450	4,507	6,957

Let's comment on the output. First of all, just for the sake of exposition, we consider only 2,000 mushrooms at random, as the dataset is very huge. Then, we run a 15-nearest neighbor classifier with the binary features and dice option. We can see that only 29 poisonous mushrooms have been misclassified as edible. The result is worrying, as we want zero poisonous mushrooms misclassified as edible. In order to have this result, we thus change the priors by setting the prior probability of poisonous mushrooms ten times higher than that of the edible mushrooms. Thus, we run the previous KNN with a larger prior probability for poisonous mushrooms, finding that the number of poisonous mushrooms now classified as edible is zero, as we wanted. Of course, this happens at expense of a larger number of edible mushrooms classified as poisonous: before (that is, with uniform priors) they were 65, but now they are 185. Finally, we check prediction of poisonous status over all the dataset finding, again, that no poisonous mushrooms are misclassified.

4.4.4 Application S4: Tuning the Number of Nearest Neighbors in Stata

In this application, we use the mushroom dataset and the command gridsearch for tuning k, the number of nearest neighbors, in a k-nearest neighbor classification. The code below does this procedure providing the final output.

```
————————————————————————————————————— Stata code and output ———————————————
. * Tuning "k" in the k-nearest neighbor classification

. * Open the dataset
. use http://www.stata-press.com/data/r15/mushroom , clear
(Lincoff (1981) Audubon Guide; http://archive.ics.uci.edu/ml/datasets/Mushroom)

. * Shuffle the data for generating a random order of the observations
. set seed 793742
. generate u = runiform()
. sort u

. * Keep only the first 1000 observations
. keep in 1/1000
(5,957 observations deleted)

. * Run the command "gridsearch" to find the best "k"
. global y "poison"
. global X "i.population i.habitat i.bruises i.capshape i.capsurface i.capcolor"
```

```
. global greed "1 2 3 4 5 6 7 8 9 10 11 12 13 14 15"
. qui xi: gridsearch discrim knn $X , ///
> par1name(k) par1list($greed) group($y) criterion(accuracy) method(trainfraction .8)

. * Output the best parameter combination
. matrix gridsearch= r(gridsearch)
. svmat gridsearch , names(col)
. list accuracy k seconds in 1/15
```

	accuracy	k	seconds
1.	.95	1	.729
2.	.92	2	.793
3.	.89	3	.738
4.	.885	4	.738
5.	.885	5	.744
6.	.875	11	.819
7.	.875	12	.841
8.	.875	13	.837
9.	.875	15	.838
10.	.87	6	.779
11.	.87	8	.79
12.	.87	9	.875
13.	.87	10	.812
14.	.87	14	.832
15.	.865	7	.777

Let's comment on the code and its output. After opening the dataset, we reshuffle the data for generating a random order of the observations. This is necessary, as we will use a training-testing validation to tune our parameter k. For the sake of exposition, we only keep the first 1,000 observations. Then, we go on by setting the group target variable ("poison"), the features, and the greed of nearest neighbors values into specific macros. We thus have all the ingredients to run the command gridsearch discrim knn. Let's describe the used options. Option par1name() takes as argument the name of the parameter that we want to tune. In our case it is k, the number of nearest neighbors. Option par1list() takes as argument a list of candidate values for being the optimal number of nearest neighbors. In our case this list coincides with greed. Option criterion() identifies the predictive criterion to use. In our case, because we are considering a classification setting, we use the accuracy, that is "one minus the classification error rate". Finally, the option method() identifies the out-of-sample method used to tune the parameter of interest. We could use K-fold cross-validation obtained via setting method(cv #) (with # indicating the number of folds), but we opted to use method(trainfraction .8), which generates randomly only one fold with a fraction of training observations of 80%, and a fraction of testing observations of 20%.

After running this command, we can output the best parameter combination (although, in this case, we have only one parameter), and list the accuracy associated to every value of the greed. We can immediately see that the largest accuracy is reached when only one nearest neighbor is used. This accuracy is equal to 0.95

(calculation took only 0.73 s). One can does use this optimal number of nearest neighbors (i.e., k = 1) to draw optimal predictions.

4.4.5   Application R1: K-Nearest Neighbor Classification in R

In R, k-nearest neighbor classification can be carried out using the knn() function. This function automatically provides class and probability predictions for the test observations from the fit made on the training set. The usage of this function is rather easy as we show in the code below, where we apply the knn() function to the Iris dataset.

──────────────────────────── R code and output ────────────────────────────

```
# K-Nearest Neighbors on the "Iris" dataset

# Create training and test datasets

# Split of iris data as 70% train and 30% test datasets
> set.seed(1234)
> index <- sample(2, nrow(iris), replace=TRUE, prob=c(0.7, 0.3))
> trainData <- iris[index==1,]
> testData <- iris[index==2,]

# Remove the target factor variable "Species" from training and test datasets
> trainData1 <- trainData[-5]
> testData1 <- testData[-5]

# Check the dimensions of train and test datasets
> dim(trainData)
[1] 112   5

> # Store target variable for testing and training data
> iris_train_labels <- trainData$Species
> iris_test_labels  <- testData$Species

> # Train "knn" on the train data
> library(class)
> iris_test_pred1 <- knn(train = trainData1,
+                        test = testData1,
+                        cl= iris_train_labels,
+                        k = 3,
+                        prob=TRUE)

# See the attributes of the last function run
> attributes(.Last.value)
$levels
[1] "setosa"     "versicolor" "virginica"

$class
[1] "factor"

$prob
 [1] 1.0000000 1.0000000 1.0000000 1.0000000 1.0000000 1.0000000 1.0000000 1.0000000 1.0000000
[10] 1.0000000 1.0000000 1.0000000 1.0000000 1.0000000 1.0000000 1.0000000 1.0000000 1.0000000
[19] 1.0000000 1.0000000 1.0000000 1.0000000 0.6666667 1.0000000 1.0000000 1.0000000 1.0000000
[28] 1.0000000 1.0000000 1.0000000 0.6666667 1.0000000 0.6666667 1.0000000 1.0000000 1.0000000
[37] 0.6666667 1.0000000

# Create confusion matrix
> confMatrix <- table(iris_test_labels,iris_test_pred1)
```

```
# This function takes as argument the confusion matrix and return the accuracy
> accuracy <- function(x){sum(diag(x)/(sum(rowSums(x)))) * 100}
> accuracy(confMatrix)
[1] 97.36842
```

Let's comment on the code and its output. We set out by creating the train and test sets. For this purpose, we use the `sample()` function where 70% of data will be used for training and 30% for testing. This function returns an index for the observations, equal to 1 for training and 2 for testing observations. Therefore, using `index`, we can generate the two datasets that we need, by removing the target factor variable `Species` from training and test datasets, thus obtained only datasets made of the features. Of course, we store separately the class target variable `Species` for the testing and the training data, thus generating the two target variables: `iris_train_labels` and `iris_test_labels`.

We have now all the ingredients to run `knn()`. The first argument taken by this function is the training data (that is, the training feature dataset); the second argument is the testing set; the third argument, `cl`, is the training target variable; the forth argument, `k`, is the number of nearest neighbors to use; finally, the fifth argument requests to provide the proportion of the votes for the winning class. We can also see the main attributes of this function, that are: `$levels`, `$class`, and `$prob`.

After running `knn()`, we obtain the label predictions on the test dataset, and use them to form the confusion matrix. Finally, from the confusion matrix, we compute the accuracy which is, in this case, equal to 97.4%.

4.4.6 Application R2: Tuning the Number of Nearest Neighbors in R

In this application, we use the `cancer` dataset, a copy of the popular "Breast Cancer Wisconsin (Diagnostic) Data Set" to predict whether a cancer is benign or malignant based on ten real-valued features computed for each cell nucleus: (1) `radius` (mean of distances from center to points on the perimeter); (2) `texture` (standard deviation of gray-scale values); (3) `perimeter`; (4) `area`; (5) `smoothness` (local variation in radius lengths); (6) `compactness` ($perimeter^2/area - 1.0$); (7) `concavity` (severity of concave portions of the contour); (8) `concave points` (number of concave portions of the contour); (9) `symmetry`; (10) `fractal dimension` ("coastline approximation" $- 1$). These features are computed from a digitized image of a fine needle aspirate (FNA) of a breast mass. They describe characteristics of the cell nuclei present in the image.

The mean, standard error and "worst" or largest (mean of the three largest values) of these features were computed for each image, resulting in 30 features. For instance, field 3 is Mean Radius, field 13 is Radius SE, field 23 is Worst Radius.

The target variable is `diagnosis`, taking value "M" (= malignant) or "B" (= benign). Class distribution entails 357 benign, and 212 malignant cancers, for a total 569 observations, and 31 final features. Also, no missing values are present in the dataset.

We want to predict whether a cancer is benign or malignant using the k-nearest neighbor classifier by optimally tuning the number of nearest neighbors. For this purpose, we will use the CARET package. The R script for the whole computation is displayed below.

──────────────── R code and output ────────────────

```
# Tuning k-Nearest Neighbors on Breast Cancer Wisconsin Data Set
# Data from: https://www.kaggle.com/datasets/uciml/breast-cancer-wisconsin-data

# Clear all
> rm(list = ls())

# Set the directory
> WD <- "~/Dropbox/Giovanni/Book_Stata_ML_GCerulli/discrim/sjlogs/Dataset_cancer"
> setwd(WD)

# Load data into R
> library(readr)
> cancer <- read_csv("cancer.csv")

# Make data as a DataFrame
> df <- as.data.frame(cancer)

# Explore the structure of the dataset
> dim(df)
[1] 569  31

# Import the needed packages for this application
> library(mclust)
> library(dplyr)
> library(ggplot2)
> library(caret)
> library(pROC)

# Create training and testing datasets, preserving the 50/50 class split in each
> training_index <- createDataPartition(df$diagnosis,
+                                        p = 0.8,
+                                        list = FALSE)
>

# Generate the train and test datasets
> training_set <- df[training_index, ]
> testing_set <- df[-training_index, ]

# Confirm the class split in the training set
> table(training_set$diagnosis)

  B   M
286 170


# Hyper-parameter tuning via cross-validation:
# - 10-fold cross-validation repeated 10 times
# - "summaryFunction" force to use a CV metrics: "accuracy" and "kappa" statistics
> training_control <- trainControl(method = "repeatedcv",
+                                  summaryFunction = defaultSummary,
```

```
+                               classProbs = TRUE,
+                               number = 10,
+                               repeats = 10)
>

# Usee the "train()" function to tune the "k" hyperparameter.
# The range of "k" is from 1 to 81 in steps of 2 (odd distances only)
# We set a random seed for reproducibility
> set.seed(2)
> knn_cv <- train(diagnosis ~ .,
+                 data = training_set,
+                 method = "knn",
+                 trControl = training_control,
+                 metric = "Accuracy",
+                 tuneGrid = data.frame(k = seq(1,81,by = 2)))
>

# Display the results
> print(knn_cv)
k-Nearest Neighbors

456 samples
 30 predictor
  2 classes: 'B', 'M'

No pre-processing
Resampling: Cross-Validated (10 fold, repeated 10 times)
Summary of sample sizes: 410, 411, 410, 410, 410, 410, ...
Resampling results across tuning parameters:

  k    Accuracy   Kappa
   1   0.9245024  0.8372550
   3   0.9461981  0.8840705
   5   0.9429227  0.8772882
   7   0.9440000  0.8793759
   9   0.9402850  0.8708354
  11   0.9378696  0.8654590
  13   0.9363285  0.8614982
  15   0.9334783  0.8547546
  17   0.9325942  0.8526175
  19   0.9325942  0.8526555
  21   0.9319372  0.8512073
  23   0.9317198  0.8506852
  25   0.9310628  0.8490212
  27   0.9306425  0.8477386
  29   0.9304251  0.8470289
  31   0.9304251  0.8468534
  33   0.9306473  0.8474714
  35   0.9308599  0.8478356
  37   0.9304300  0.8466904
  39   0.9306473  0.8472375
  41   0.9289034  0.8432418
  43   0.9269227  0.8387328
  45   0.9258261  0.8362151
  47   0.9253913  0.8352629
  49   0.9240725  0.8323913
  51   0.9232029  0.8303985
  53   0.9234155  0.8309034
  55   0.9242899  0.8328003
  57   0.9229662  0.8299134
  59   0.9220870  0.8279041
  61   0.9227440  0.8294228
  63   0.9231787  0.8304861
  65   0.9229517  0.8299565
  67   0.9236135  0.8314452
  69   0.9234010  0.8308100
  71   0.9218696  0.8273072
```

```
73   0.9214348   0.8261622
75   0.9201304   0.8231780
77   0.9207729   0.8245947
79   0.9209952   0.8251794
81   0.9212174   0.8255511
```

Accuracy was used to select the optimal model using the largest value.
The final value used for the model was k = 3.

```
# Show the best value of "k"
> knn_cv$bestTune$k
[1] 3

# Predict at the best tuning parameter

# Store the target variable for testing and training data
> diagnosis_train_labels <- training_set$diagnosis
> diagnosis_test_labels <- testing_set$diagnosis

# Consider only the matrix of features
> trainData1 <- training_set[-1]
> testData1 <- testing_set[-1]

# Apply knn() at the optimal k
> library(class)
> opt_k <- knn_cv$bestTune$k
> diagnosis_test_pred <- knn(train = trainData1,
+                            test = testData1,
+                            cl= diagnosis_train_labels,
+                            k = opt_k,
+                            prob=TRUE)
>

# See the attributes of the last function run
> attributes(.Last.value)
$levels
[1] "B" "M"

$class
[1] "factor"

$prob
  [1] 1.0000000 1.0000000 1.0000000 0.6666667 1.0000000 0.6666667 1.0000000 1.0000000 1.0000000
 [10] 1.0000000 1.0000000 1.0000000 1.0000000 0.6666667 1.0000000 1.0000000 1.0000000 1.0000000
 [19] 1.0000000 1.0000000 1.0000000 1.0000000 1.0000000 1.0000000 1.0000000 0.6666667 1.0000000
 [28] 1.0000000 1.0000000 1.0000000 1.0000000 0.6666667 1.0000000 1.0000000 1.0000000 1.0000000
 [37] 1.0000000 1.0000000 1.0000000 1.0000000 1.0000000 1.0000000 1.0000000 0.6666667 1.0000000
 [46] 1.0000000 0.6666667 1.0000000 1.0000000 1.0000000 1.0000000 1.0000000 1.0000000 1.0000000
 [55] 1.0000000 1.0000000 0.6666667 1.0000000 1.0000000 1.0000000 1.0000000 1.0000000 1.0000000
 [64] 1.0000000 1.0000000 1.0000000 1.0000000 1.0000000 1.0000000 1.0000000 1.0000000 1.0000000
 [73] 1.0000000 1.0000000 1.0000000 0.6666667 1.0000000 1.0000000 1.0000000 1.0000000 1.0000000
 [82] 0.6666667 1.0000000 1.0000000 0.6666667 1.0000000 1.0000000 1.0000000 1.0000000 1.0000000
 [91] 1.0000000 0.6666667 0.6666667 1.0000000 0.6666667 1.0000000 1.0000000 1.0000000 1.0000000
[100] 1.0000000 1.0000000 1.0000000 1.0000000 1.0000000 1.0000000 1.0000000 1.0000000 1.0000000
[109] 1.0000000 1.0000000 1.0000000 1.0000000 0.6666667

# Create confusion matrix
> confMatrix <- table(diagnosis_test_labels,diagnosis_test_pred)
> confMatrix
                     diagnosis_test_pred
diagnosis_test_labels  B   M
                   B  68   3
                   M  10  32

# Estimate the accuracy
> accuracy <- function(x){sum(diag(x)/(sum(rowSums(x)))) * 100}
> accuracy(confMatrix)
```

```
[1] 88.49558

# Provide a larger set of fitting statistcs using "confusionMatrix()"

# Transform test labels into a "factor" (as "confusionMatrix()" wants factors)
> diagnosis_test_labels <- as.factor(diagnosis_test_labels)
> class(diagnosis_test_labels)
[1] "factor"

# Fit the confusion matrix
> knn_cm <- confusionMatrix(diagnosis_test_labels, diagnosis_test_pred, mode = "everything")

# Obtain the results
> print(knn_cm)
Confusion Matrix and Statistics

          Reference
Prediction  B  M
         B 68  3
         M 10 32

               Accuracy : 0.885
                 95% CI : (0.8113, 0.9373)
    No Information Rate : 0.6903
    P-Value [Acc > NIR] : 1.023e-06

                  Kappa : 0.745

 Mcnemar's Test P-Value : 0.09609

            Sensitivity : 0.8718
            Specificity : 0.9143
         Pos Pred Value : 0.9577
         Neg Pred Value : 0.7619
              Precision : 0.9577
                 Recall : 0.8718
                     F1 : 0.9128
             Prevalence : 0.6903
         Detection Rate : 0.6018
   Detection Prevalence : 0.6283
      Balanced Accuracy : 0.8930

       'Positive' Class : B
```

Let's comment on the code and its output. After cleaning the environment and loading the dataset as data frame, we explore the structure of the dataset showing that it is correctly made of 569 cancers, and 31 features. We import the needed packages for this application, to then create training and testing datasets, preserving the 50/50 class split in each, using the function createDataPartition(). We go on by generating the train and test datasets, thus having all the ingredients to run the CARET package. As usual with CARET, we first set up the function trainControl(), and then we use the train() function to tune the "k" hyper-parameter. The range of "k" goes from 1 to 81 in steps of 2 (only odd values are considered). Results are included in the generated knn_cv object. We print it and find that the optimal number of nearest neighbors to use is equal to 3, corresponding to an accuracy of 0.946.

We are now ready to generate prediction at the optimal tuning parameter. To do this, we first store the target variable for testing and training data in two new variables and consider only the matrix of features. Then, we apply the function knn() at the

Statistics	Identifier	Target	Classifiers		
	ID	Y	C-NIR	C-1	C-2
	1	1	0	1	0
	2	1	0	1	0
	3	1	0	0	0
	4	1	0	0	1
	5	0	0	0	0
	6	0	0	0	0
	7	0	0	0	1
	8	0	0	0	0
	9	0	0	0	1
	10	0	0	0	1
N. of matches			6	8	4
Accuracy			6/10	8/10	4/10

Fig. 4.25 Comparison of two classifiers, C-1 and C-2, performing respectively better and worse than the "No Information Rate" (NIR) classifier

optimal "k", create the confusion matrix, and estimate the accuracy, which is equal to 88.5%.

However, to provide a larger set of fitting statistics, we can use the `confusionMatrix()` function. We can see that this function provides a series of accuracy statistics of interest. Besides the classical accuracy—confirmed to be equal to 88.5%—this function provides also the accuracy's 95% confidence interval, and the "No Information Rate" (NIR), with a test to establish whether the accuracy is larger than the NIR. The NIR is computed as the largest proportion of the observed classes. In this case "benign" is the class with largest proportion.

The NIR corresponds to the accuracy of a trivial classifier (let's call it "C-NIR") that would assign a new unlabeled instance to the class having the largest frequency in the initial dataset. In other words, the NIR classifier classifies any observation within the same class, the one with largest probability.

If a classifier C obtains an accuracy smaller than the accuracy of C-NIR, it means that classifying randomly—that is, based on the initial frequency distributions of the classes—is better than using any sophisticated classification approach using features, as entailed by performing C. A hypothesis test is also computed to evaluate whether the accuracy rate of C is greater than the NIR.

To better catch the interpretation of the NIR, Fig. 4.25 shows a comparison of two classifiers, C-1 and C-2, performing respectively better and worse than the "No Information Rate" classifier (C-NIR). We see that the accuracy of C-NIR is exactly equal to the proportion of the largest class of the dataset (in this case, the class "0"). This accuracy is equal to $6/10 = 0.6$. The accuracy of C-1 is larger, and equal to $8/10 = 0.8$, while the accuracy of C-2 is smaller and equal to $4/10 = 0.4$. We conclude that only C-1 provides additional non-trivial information compared to a classification based on random guesses. It is also clear that performing better than C-NIR also means that the information brought about by the features is relevant for prediction purposes.

In our example, the KNN accuracy is larger than the NIR's, as also confirmed by accepting the null Acc $\geq$ NIR of the test. We can conclude that our classification is informative.

4.4.7 Application P1: Support Vector Machine Classification in Python

In this section, we provide an application of support vector machine (SVM) classification in Python, using the Scikit-learn platform. We consider the datasets `dnlw_train` and `dnlw_test`, that we extracted from the National Longitudinal Surveys of Young Women and Mature Women (NLSW), a survey designed to gather information on labor market activities and other significant life events for American women in 1988.

We generated a categorical variable `dwage` indicating three levels of wage (1 = low, 2 = medium, and 3 = high) based on equal percentiles. We would like to predict the woman's wage level based on this set of features: `age` (age at current year), `race` (white, black, or other), `married` (whether married), `never_married` (whether never married), `grade` (whether the current grade was completed), `collgrad` (college graduated), `south` (if living in south), `smsa` (if living in SMSA), `c_city` (if living in central city), `union` (if having a union membership), `hours` (usual hours worked), `ttl_exp` (total work experience), `tenure` (job tenure in years). Finally 12 industry dummies are included.

The Python code for the whole SVM computation is displayed below.

```
──────────────────────────────────────── Python code and output ────────────────────────────
>>> # SUPPORT VECTOR MACHINE CLASSIFICATION USING PYTHON

>>> # IMPORT THE NEEDED PYTHON PACKAGES
... from sklearn import svm
>>> from sklearn.model_selection import GridSearchCV
>>> import pandas as pd
>>> import os
>>> from sklearn.metrics import accuracy_score
>>> from sklearn.metrics import confusion_matrix
>>> from sklearn.metrics import ConfusionMatrixDisplay
>>>
>>> # SET THE DIRECTORY
... os.chdir("/Users/cerulli/Dropbox/Giovanni/Book_Stata_ML_GCerulli/discrim/sjlogs/Dataset_dnlw")
>>>
>>> # SET THE TRAIN/TEST DATASET
... data_train="dnlw_train.dta"
>>> data_test="dnlw_test.dta"
>>>
>>> # LOAD TRAIN AND TEST DATASETS AS PANDAS DATAFRAME
... df_train = pd.read_stata(data_train)
>>> df_test = pd.read_stata(data_test)
>>>
>>> # DESCRIBE THE DATA
... df_train.info()
<class 'pandas.core.frame.DataFrame'>
Int64Index: 1748 entries, 0 to 1747
Data columns (total 26 columns):
dwage           1748 non-null float32
race            1748 non-null int8
```

```
collgrad          1748 non-null int8
age               1748 non-null int8
married           1748 non-null int8
never_married     1748 non-null int8
grade             1748 non-null int8
south             1748 non-null int8
smsa              1748 non-null int8
c_city            1748 non-null int8
union             1748 non-null int8
hours             1748 non-null int8
ttl_exp           1748 non-null float32
tenure            1748 non-null float32
ind1              1748 non-null int8
ind2              1748 non-null int8
ind3              1748 non-null int8
ind4              1748 non-null int8
ind5              1748 non-null int8
ind6              1748 non-null int8
ind7              1748 non-null int8
ind8              1748 non-null int8
ind9              1748 non-null int8
ind10             1748 non-null int8
ind11             1748 non-null int8
ind12             1748 non-null int8
dtypes: float32(3), int8(23)
memory usage: 73.4 KB
>>>
>>> # DEFINE y_train AND y_test (THE TARGET VARIABLES)
... y_train = df_train.iloc[:,0].astype(int)
>>> y_test = df_test.iloc[:,0].astype(int)
>>>
>>> # DEFINE X_train AND X_test (THE FEATURES)
... X_train = df_train.iloc[:,1::]
>>> X_test = df_test.iloc[:,1::]
>>>
>>> # SELECT THE BEST TUNING PARAMETERS
>>> # SEARCHING MULTIPLE PARAMETERS SIMULTANEOUSLY
>>> # INITIALIZE A SVM (with parameters: kernel='rbf', C = 10.0, gamma=0.1)
... svmc = svm.SVC(kernel='rbf', C = 10.0, gamma=0.1, probability=True)
>>>
>>> # SVMC "CROSS-VALIDATION" FOR "C" AND "GAMMA" BY PRODUCING A "GRID SEARCH"
... # GENERATE THE TWO PARAMETERS' GRID AS A "LIST"
... gridC = [30 , 50 , 70]
>>> gridG=[0.3, 0.5 , 0.7]
>>>
>>> # PUT THE GENERATED GRIDS INTO A PYTHON DICTIONARY
... param_grid = {'C': gridC, 'gamma': gridG}
>>>
>>> # BUILD A "GRID SEARCH CLASSIFIER"
... grid = GridSearchCV(svm.SVC(),param_grid,cv=10,scoring='accuracy',return_train_score=True)
>>>
>>> # TRAIN THE CLASSIFIER OVER THE GRIDS
... grid.fit(X_train, y_train)
GridSearchCV(cv=10, estimator=SVC(),
             param_grid={'C': [30, 50, 70], 'gamma': [0.3, 0.5, 0.7]},
             return_train_score=True, scoring='accuracy')
>>>
>>> # VIEW THE RESULTS: ONLY TRAIN AND TEST ACCURACY
... pd.DataFrame(grid.cv_results_)[['mean_train_score','mean_test_score']]
   mean_train_score  mean_test_score
0               1.0         0.509110
1               1.0         0.472509
2               1.0         0.412453
3               1.0         0.509110
4               1.0         0.472509
5               1.0         0.412453
6               1.0         0.509110
```

```
7                1.0          0.472509
8                1.0          0.412453
>>>
>>> # EXAMINE THE BEST MODEL
... print(grid.best_score_)
0.5091100164203612
>>> print(grid.best_params_)
{'C': 30, 'gamma': 0.3}
>>> print(grid.best_estimator_)
SVC(C=30, gamma=0.3)
>>> print(grid.best_index_)
0
>>>
>>> # STORE THE TWO BEST PARAMETERS INTO TWO VARIABLES
... opt_c=grid.best_params_.get('C')
>>> opt_gamma=grid.best_params_.get('gamma')
>>>
>>> # USING THE BEST PARAMETERS TO MAKE PREDICTIONS
... # TRAIN YOUR MODEL USING ALL DATA AND THE BEST PARAMETERS
... svmc = svm.SVC(kernel='rbf', C=opt_c, gamma=opt_gamma, probability=True)
>>>
>>> # FIT THE MODEL
... svmc.fit(X_train,y_train)
SVC(C=30, gamma=0.3, probability=True)
>>>
>>> # MAKE TRAIN PREDICTION FOR y, AND PUT IT INTO A DATAFRAME
... y_train_hat = svmc.predict(X_train)
>>> y_df_train_pred=pd.DataFrame(y_train_hat)
>>> print(y_df_train_pred)
      0
0     3
1     1
2     2
3     1
4     2
...  ..
1743  1
1744  3
1745  1
1746  1
1747  2

[1748 rows x 1 columns]
>>>
>>> # MAKE TRAIN PROBABILITY PREDICTION, AND PUT IT INTO A DATAFRAME
... prob_train = svmc.predict_proba(X_train)
>>> prob_df_train=pd.DataFrame(prob_train)
>>> print(prob_df_train)
             0         1         2
0     0.018269  0.075475  0.906256
1     0.856300  0.116068  0.027632
2     0.070379  0.828432  0.101189
3     0.847476  0.128474  0.024050
4     0.125865  0.781778  0.092357
...        ...       ...       ...
1743  0.845557  0.134593  0.019850
1744  0.018263  0.076734  0.905003
1745  0.862517  0.109683  0.027800
1746  0.854583  0.118061  0.027356
1747  0.135632  0.782141  0.082227

[1748 rows x 3 columns]
>>>
>>> # MAKE TEST PREDICTION FOR y, AND PUT IT INTO A DATAFRAME
... y_test_hat = svmc.predict(X_test)
>>> y_df_test_pred=pd.DataFrame(y_test_hat)
>>> print(y_df_test_pred)
   0
```

```
0    2
1    3
2    3
3    2
4    1
..  ..
95   2
96   3
97   2
98   1
99   1
[100 rows x 1 columns]
>>>
>>> # MAKE OUT-OF-SAMPLE "PROBABILITY" PREDICTION FOR y USING A PREPARED DATASET
... prob_test = svmc.predict_proba(X_test)
>>> prob_df_test=pd.DataFrame(prob_test)
>>> print(prob_df_test)
            0         1         2
0    0.316517  0.361416  0.322067
1    0.329950  0.326881  0.343169
2    0.229596  0.283875  0.486529
3    0.213931  0.584514  0.201555
4    0.608635  0.313221  0.078144
..       ...       ...       ...
95   0.158170  0.679312  0.162518
96   0.270291  0.297439  0.432270
97   0.163218  0.405669  0.431113
98   0.543281  0.401873  0.054846
99   0.649473  0.294982  0.055545
[100 rows x 3 columns]
>>>
>>> # COMPUTE TRAIN ACCURACY
... accuracy_score(y_df_train_pred , y_train)
1.0
>>>
>>> # COMPUTE TEST ACCURACY
... accuracy_score(y_df_test_pred , y_test)
0.5
>>>
>>> # SHOW THE CONFUSION MATRIX
... cm = confusion_matrix(y_df_test_pred , y_test)
>>>
>>> # DISPLAY THE CONFUSION MATRIX
... cm_display = ConfusionMatrixDisplay(cm).plot()
```

Let's comment on the code and its output.

First we import the needed Python packages, and in particular SVR() and GridSearchCV() from Scikit-learn. After setting the directory, we load the train and the test datasets as pandas data frames, and describe the training data. We see that we have 1,748 observation and 26 variables (1 target variable, and 25 features). We go on by extracting the train and test target variables (y_train and y_test), as well as the matrix of features (x_train and x_test). Then, we search for the best SVM tuning parameters, using the radial basis function (RBF) kernel. As stressed in the theoretical part of this chapter, the two SVM hyper-parameters are c and gamma that we initialize respectively equal to 19 and 0.1. We are thus ready to provide a k-fold cross-validation for c and gamma, by producing a grid search over two lists for respectively the two parameters (gridC and gridG). We consider only three values for each grid, thereby having a grid space of $3 \times 3 = 9$ combinations. We put the

Fig. 4.26 Graphical
representation of the SVM
confusion matrix

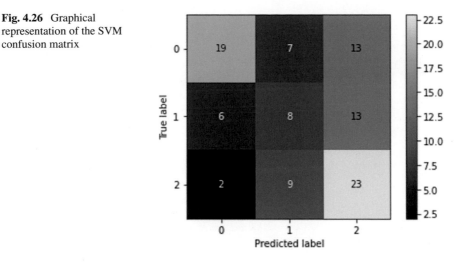

generated grids into a Python dictionary and build a grid-search classifier using the
GridSearchCV() function. We train this classifier over the grid and view the results
only for the train and test accuracy obtained from a tenfold cross-validation.

By examining the best model, we see that the best accuracy is reached at a level
of 0.5, corresponding to c = 30, and gamma = 0.3. Therefore, we store the two best
parameters into two variables, to then use these best parameters to make (optimal)
class and probability predictions, both in-sample and out-of-sample. Finally, using
the test dataset, we compute the test accuracy which confirms to be equal to 0.5,
a rather low level. The application ends by showing and displaying graphically the
confusion matrix (Fig. 4.26).

4.4.8 Application P2: Support Vector Machine Regression in Python

In this section, we provide an application of support vector machine (SVM) regression
in Python, using the Scikit-learn platform. We consider the classical boston dataset
collecting by the U.S. Census Service concerning housing in the area of Boston
Mass. The whole dataset contains 506 cases, but we form two related datasets: a
training dataset (with 405 observations), and a test dataset (with 101 observations).
We are interested in predicting the median value of the houses in the neighborhood
measured in thousands of dollars. Several features explaining the price of the houses
are considered. As we used this dataset often in this book, we do not describe again
these features.

We fit the SVM regression by optimally tuning the three main parameters that characterize this algorithm, that is, C, gamma, and epsilon using Scikit-learn, and estimate test accuracy, along with plotting the actual versus the fitted outcome values. The Python code for the whole SVM computation is displayed below.

```
──────────────────────────────────── Python code and output ────────────────────────────────
>>> # SUPPORT VECTOR MACHINE REGRESSION USING PYTHON

>>> # IMPORT NEEDED PACKAGES
... from sklearn.svm import SVR
>>> from sklearn.model_selection import GridSearchCV
>>> from sklearn.metrics import mean_absolute_percentage_error
>>> import pandas as pd
>>> import matplotlib.pyplot as plt
>>> import os
>>> import random
>>>
>>> # SET THE DIRECTORY
... os.chdir("/Users/cerulli/Dropbox/Giovanni/Book_Stata_ML_GCerulli/discrim/sjlogs/Dataset_boston")
>>>
>>> # SET RANDOM SEED
... random.seed(10)
>>>
>>> # FORM THE TRAIN DATASET
... df_train = pd.read_stata("boston_train.dta")
>>>
>>> # DEFINE y_train THE TARGET VARIABLE
... y_train=df_train.iloc[:,0]
>>>
>>> # DEFINE X_train THE FEATURES
... X_train=df_train.iloc[:,1::]
>>>
>>> # FORM THE TEST DATASET
... df_test = pd.read_stata("boston_test.dta")
>>>
>>> # DEFINE y_test THE TARGET VARIABLE
... y_test=df_test.iloc[:,0]
>>>
>>> # DEFINE X_test THE FEATURES
... X_test=df_test.iloc[:,1::]
>>>
>>> # INITIALIZE A SVM (with parameters: kernel='rbf', C = 10.0, gamma=0.1, epsilon=0.1)
... model = SVR(kernel='rbf', C = 10.0, gamma=0.1 , epsilon=0.1)
>>>
>>> # SVM "CROSS-VALIDATION" FOR "C", "GAMMA", AND "EPSILON" USING "GRID SEARCH"
... # GENERATE THE THREE PARAMETERS' GRID AS A "LIST"
... gridC=[10, 50, 70, 100]
>>> gridG=[0.1, 0.5, 0.9]
>>> gridE=[0.1, 0.5, 0.9]
>>>
>>> # PUT THE GENERATED GRIDS INTO A PYTHON DICTIONARY
... param_grid = {'C': gridC, 'gamma': gridG, 'epsilon': gridE}
>>>
>>> # SET THE NUMBER OF CROSS-VALIDATION FOLDS
... n_folds = 10
>>>
>>> # INSTANTIATE THE GRID
... grid = GridSearchCV(model, param_grid, cv=n_folds, scoring='explained_variance', return_train_score=
True
> )
>>>
>>> # FIT THE GRID
... grid.fit(X_train, y_train)
GridSearchCV(cv=10, estimator=SVR(C=10.0, gamma=0.1),
            param_grid={'C': [10, 50, 70, 100], 'epsilon': [0.1, 0.5, 0.9],
```

```
                           ´gamma´: [0.1, 0.5, 0.9]},
            return_train_score=True, scoring=´explained_variance´)
>>>
>>> # VIEW THE RESULTS
... CV_RES=pd.DataFrame(grid.cv_results_)[[´mean_train_score´,´mean_test_score´,´std_test_score´]]
>>> print(CV_RES)
    mean_train_score  mean_test_score  std_test_score
0           0.821821         0.000483        0.038480
1           0.800872         0.000237        0.000728
2           0.799404         0.000012        0.000036
3           0.819321         0.000757        0.037107
4           0.798809         0.000223        0.000684
5           0.797363         0.000011        0.000033
6           0.814428         0.001057        0.035830
7           0.794346         0.000209        0.000642
8           0.792919         0.000010        0.000031
9           0.999889         0.004638        0.051195
10          0.999888         0.000311        0.000880
11          0.999888         0.000015        0.000044
12          0.997349         0.004825        0.049530
13          0.997325         0.000296        0.000837
14          0.997323         0.000014        0.000042
15          0.991862         0.004940        0.047940
16          0.991718         0.000281        0.000794
17          0.991707         0.000013        0.000039
18          0.999889         0.004638        0.051195
19          0.999888         0.000311        0.000880
20          0.999888         0.000015        0.000044
21          0.997349         0.004825        0.049530
22          0.997325         0.000296        0.000837
23          0.997323         0.000014        0.000042
24          0.991862         0.004940        0.047940
25          0.991718         0.000281        0.000794
26          0.991707         0.000013        0.000039
27          0.999889         0.004638        0.051195
28          0.999888         0.000311        0.000880
29          0.999888         0.000015        0.000044
30          0.997349         0.004825        0.049530
31          0.997325         0.000296        0.000837
32          0.997323         0.000014        0.000042
33          0.991862         0.004940        0.047940
34          0.991718         0.000281        0.000794
35          0.991707         0.000013        0.000039
>>>
>>> # EXAMINE THE BEST MODEL
... print("                                               ")

>>> print("                                               ")

>>> print("-----------------------------------------------")
-----------------------------------------------------
>>> print("CROSS-VALIDATION RESULTS TABLE")
CROSS-VALIDATION RESULTS TABLE
>>> print("-----------------------------------------------")
-----------------------------------------------------
>>> print("The best score is:")
The best score is:
>>> print(grid.best_score_)
0.004939684910073849
>>> print("-----------------------------------------------")
-----------------------------------------------------
>>> print("The best parameters are:")
The best parameters are:
>>> print(grid.best_params_)
{´C´: 50, ´epsilon´: 0.9, ´gamma´: 0.1}
>>>
>>> # STORE THE BEST HYPER-PARAMETERS OBTAINED
```

```
... opt_c=grid.best_params_.get('C')
>>> opt_gamma=grid.best_params_.get('gamma')
>>> opt_epsilon=grid.best_params_.get('epsilon')
>>> print("-----------------------------------------------------")
-----------------------------------------------------
>>> print("The best estimator is:")
The best estimator is:
>>> print(grid.best_estimator_)
SVR(C=50, epsilon=0.9, gamma=0.1)
>>> print("-----------------------------------------------------")
-----------------------------------------------------
>>> print("The best index is:")
The best index is:
>>> print(grid.best_index_)
15
>>> print("-----------------------------------------------------")
-----------------------------------------------------
>>>
>>> # TRAIN THE MODEL USING TRAIN DATA AND BEST PARAMETERS
... model = SVR(kernel='rbf', C=opt_c, gamma=opt_gamma , epsilon=opt_epsilon)
>>>
>>> # FIT THE OPTIMAL MODEL ON TRAIN DATA
... model.fit(X_train, y_train)
SVR(C=50, epsilon=0.9, gamma=0.1)
>>>
>>> # MAKE OPTIMAL TRAIN PREDICTION
... y_hat_train = model.predict(X_train)
>>>
>>> # FIT THE OPTIMAL MODEL ON TEST DATA
... model.fit(X_test, y_test)
SVR(C=50, epsilon=0.9, gamma=0.1)
>>>
>>> # MAKE OPTIMAL TEST PREDICTION
... y_hat_test = model.predict(X_test)
>>>
>>> # COMPUTE THE TRAIN and TEST MAPE (mean absolute percentage error)
... MAPE_TRAIN = mean_absolute_percentage_error(y_train, y_hat_train)
>>> print(MAPE_TRAIN)
0.04458781822453716
>>> MAPE_TEST = mean_absolute_percentage_error(y_test, y_hat_test)
>>> print(MAPE_TEST)
0.04513263906038097
>>>
>>> # PLOT THE FIT ON TEST DATA
... plt.plot(y_test, color='black', linestyle='dashed',label="Actual test data")
[<matplotlib.lines.Line2D object at 0x17e04fd30>]
>>> plt.plot(y_hat_test, color = 'green',label="Fitted test data")
[<matplotlib.lines.Line2D object at 0x17e0173d0>]
>>> plt.xlabel("House identifier")
Text(0.5, 0, 'House identifier')
>>> plt.ylabel('Median value of homes in $1000')
Text(0, 0.5, 'Median value of homes in $1000')
>>> plt.legend(loc="upper left")
<matplotlib.legend.Legend object at 0x17e0171f0>
>>> plt.show()
```

Let's comment on the code and the outputs. After importing the needed packages, setting the current directory and the random seed for reproducibility, we form the train and test datasets, with related targets and features. Then, we initialize an SVM regression with parameters: `kernel='rbf'`, $C = 10.0$, gamma=0.1, and epsilon=0.1. For cross-validation we generate three grids for the three parameters' as lists. For the sake of fast computation, we only consider a few values within the grid, so we only

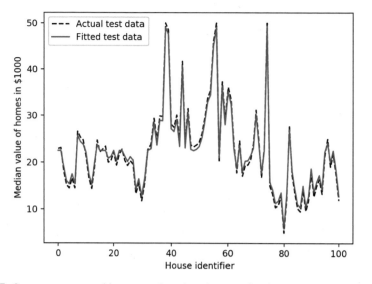

Fig. 4.27 Support vector machine regression. Actual versus fitted outcomes on the test dataset. Dataset: `Boston`. Target variables: Median value of homes in 1000s dollars

have $4 \times 3 \times 3 = 36$ parameters combinations. To go on, we put the generated grid into a Python dictionary, set the number of cross-validation folds to 10, and instantiate the grid search using, as usual, the `GridSearchCV()` function. We fit the grid, and look at the results contained in the object `grid.cv_results_`. We examine the best model, that is, the best score reached by cross-validation, and the related best parameters (`grid.best_params_`). After storing the best hyper-parameters thus obtained into three Python variables, we fit the model using train data and the best parameters, and obtain optimal training prediction. We then fit the optimal model on test data, and make optimal test prediction as well.

As final test accuracy measure, we compute the train and test MAPE (mean absolute percentage error), obtaining a train error of 4.4% and an only slightly larger test error of 4.5%. Both signal a very good fit, as also showed by Fig. 4.27 plotting the actual versus the fitted outcome on the test dataset.

4.5 Conclusions

In this chapter, we have presented the discriminant analysis (DA), the support vector machine (SVM), and the k-nearest neighbor (KNN) algorithms. We mainly focused on classification, but we discussed also SVM and KNN regression. The applications in Stata, R, and Python have proved the large flexibility of these approaches to deal with both predicting labels and predicting numerical outcomes (only for SVM and KNN, as the DA is only a classification method). Also, the centrality of the decision boundary has been widely discussed. Of course, the array of applications of these methods is much larger that what presented in this chapter; nonetheless, the reader can take these applications as useful templates for applications in different fields of study and larger datasets.

References

Chapelle, O., Vapnik, V., Bousquet, O., & Mukherjee, S. (2002). Choosing multiple parameters for support vector machines. *Machine Learning, 46*(1–3), 131–159. https://doi.org/10.1023/A:1012450327387

Cortes, C., & Vapnik, V. (1995). Support-vector networks. *Machine Learning, 20*(3), 273–297.

Cover, T. M., & Hart, P. E. (1967). Nearest neighbor pattern classification. *IEEE Transactions on Information Theory, 13*(1), 21–27. https://doi.org/10.1109/TIT.1967.1053964

Fix, E., & Hodges, J. (1951). Discriminatory analysis—Nonparametric discrimination: Consistency properties. Technical report, California University Berkeley. Section: Technical reports. https://apps.dtic.mil/sti/citations/ADA800276

Fix, E., & Hodges, J. L. (1989). Discriminatory analysis. Nonparametric discrimination: Consistency properties. *International Statistical Review/Revue Internationale de Statistique, 57*(3), 238–247. Publisher: [Wiley, International Statistical Institute (ISI)]. https://www.jstor.org/stable/1403797

Huberty, C. J. (1994). *Applied discriminant analysis/book and disk.* New York: Wiley-Interscience.

Johnson, R. A., & D. W. Wichern. (2007). *Applied Multivariate Statistical Analysis.* 6th ed. Englewood Cliffs, NJ: Prentice Hall.

McLachlan, G. J. (1992). *Discriminant analysis and statistical pattern recognition.* New York: Wiley-Interscience.

Rencher, A. C., & W. F. Christensen. (2012). *Methods of Multivariate Analysis.* 3rd ed. Hoboken, NJ: Wiley

Schölkopf, B., Sung, K. K., Burges, C. J. C., Girosi, F., Niyogi, P., Poggio, T. A., & Vapnik, V. (1997). Comparing support vector machines with Gaussian kernels to radial basis function classifiers. *IEEE Transactions on Signal Processing, 45*(11), 2758–2765. https://doi.org/10.1109/78.650102

Smola, A. J., & B. Schölkopf. (2004). A tutorial on support vector regression. *Statistics and Computing, 14*(3), 199–222.

Chapter 5
Tree Modeling

5.1   Introduction

Tree-based regression and classification are greedy supervised Machine Learning methods involving a sequential stratification of the initial sample by splitting over the feature space (Breiman et al., 1984; Sheppard, 2017; Maimon & Rokach, 2014). The logic of growing a tree mimics a decision process where, at each step, one has to decide whether or not to go in one direction according to a specific rule that follows an *if–then* logic (Quinlan, 1986, 1990). Decision trees are easy to interpret and results straightforward to communicate but, as a greedy approach, tree prediction power can be poor (Quinlan, 1999). For this reason, tree-related ensemble methods, specifically *bagging*, *random–forests*, and *boosting*, have been proposed in the literature. They are ensemble methods in that they aggregate several different trees to decrease the prediction variance, thus increasing prediction quality. As we will see, the cost of obtaining higher precision comes at expense of lower interpretability. Indeed, unlike growing one single tree, ensemble methods are sometimes black boxes whose contextual interpretation may be poor. To tackle this limitation, ensemble methods allow for computing feature importance measures that weigh the relative contribution of the features to increase prediction performance (Hastie et al., 2009).

Growing decision trees entails a bias–variance trade-off as more complex trees increase prediction variance although reducing the bias (James et al., 2013; Frank et al., 1998). Therefore, identifying the optimal tree size becomes central in this respect. As these methods are highly nonparametric, we cannot use information criteria to cherry-pick the *best* model (i.e. the best tree). Cross-validation is, however, a suitable and more general alternative in this case, as well as in the case of ensemble tree methods (EML, 2010).

5.2   The Tree Approach

To familiarize with decision trees it is worth starting from an illustrative example. Figure 5.1 sets out a regression tree for predicting an outcome y on the basis of two features, X_1 and X_2. Let's first comment on the anatomy of a decision tree. A decision tree is a graph object made of four elements: *nodes*, *branches*, *leaves* (or *terminal nodes*), and a *binary splitting criterion*. Figure 5.1 is a representation of a tree built from the top (*root* or *initial node*) to the bottom (*leaves*), where the tree construction starts from the root. The label above an internal node is the splitting criterion indicating the splitting rule of the sample at that node. For example, in the initial node of the tree of Fig. 5.1, the split is made using the criterion $X_1 < s_1$, where the value of s_1 is the *threshold* (or *cutoff*). By convention, the left-hand branch corresponds to a positive response to the question "is $X_1 < s_1$?", while the right-hand branch corresponds to a negative one. The partition of the dataset is thus made by sequential binary splitting, and no overlap takes place among the sample units. The splitting rule $X_1 < s_1$ forms two non-overlapping subsamples, where only the one on the right-hand is split again according to a new splitting criterion, $X_2 < s_2$, that generates in turn two new subsamples. At the terminal nodes we finally have the tree leaves made, in this case, of three final subgroups of units (also called *regions*), R_1, R_2, and R_3. The prediction of the outcome is made by taking the average of the value of y over the units belonging to the specific region R_j that in the figure are indicated by $\bar{y}_1$, $\bar{y}_2$, and $\bar{y}_3$.

   Using the `boston_house` dataset, Fig. 5.2 sets out a regression tree for predicting the median value of owner-occupied housing units (variable `medv`), as a function of the average number of rooms per dwelling (variable `rm`), and the percentage of population that is lower status (variable `lstat`).

   Looking at the tree, we observe that, at initial node, the feature `rm` is chosen for the split with a cutoff equal to 6.94. The right-hand branch ends at a terminal

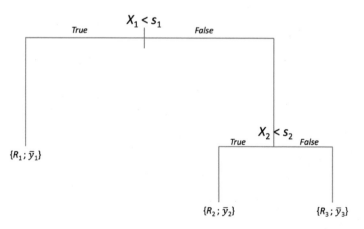

Fig. 5.1 An illustrative tree plot based on binary splitting

Fig. 5.2 Regression tree plot predicting the median value of owner-occupied housing units (variable `medv`), as a function of the average number of rooms per dwelling (variable `rm`), and the percentage of population that is lower status (variable `lstat`). Dataset: `boston_house`

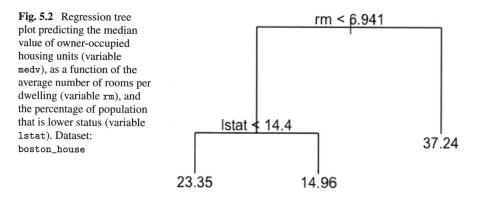

node with a predicted value of the house median price equal to 37.24. The left-end branch terminates with a new split over the variable `lstat`, generating in turn two new subgroups with predicted house median value respectively equal to 23.35 and 14.96. From an interpretative point of view, this simple example suggests that the most important variable explaining the house value is the average number of rooms per dwelling. However, when this number is below a certain threshold, what matters is the percentage of lower status population within the neighborhood, that reduces the predicted value of the house when `lstat` is above a certain threshold.

5.3 Fitting a Tree via Recursive Binary Splitting

As for any other Machine Learning approach, a good starting point to fit a tree may be that of minimizing the training RSS. Assuming to have split the observations into J non-overlapping regions $R_1, \ldots, R_J$, we define the total RSS as:

$$\text{RSS} = \sum_{j=1}^{J} \sum_{i \in R_j} (y_i - \widehat{y}_{R_j})^2 \tag{5.1}$$

where $\widehat{y}_{R_j}$ is the average value of the target variable for the training observations within the j-th region. Minimizing the RSS of Eq. (5.1) over $(R_1, \ldots, R_J)$ is however computationally prohibitive, as would entail to explore an enormous set of alternatives. Because of this, a top-down greedy approach, as the *recursive binary splitting*, may be a feasible alternative.

Fitting a tree by recursive binary splitting entails a series of steps. At each node, starting from the initial one, in fact, one has to take these decisions: (i) whether or not to split the node and, in the event of splitting, (ii) which variable X_j to split, and at which cutoff s. The fit of the tree through recursive binary splitting takes place sequentially, and the logic is straightforward. Suppose to be located on a given node, and want to carry out a binary splitting of the current observations. By defining the two splitting regions as:

$$R_1(j, s) = \{X | X_j < s\} \text{ and } R_2(j, s) = \{X | Xj \geq s\} \tag{5.2}$$

we would like to find the j^* and the s^* that minimize the sum of the RSS of the two regions, that is:

$$\sum_{i:x_i \in R_1(j,s)} (y_i - \widehat{y}_{R_1})^2 + \sum_{i:x_i \in R_2(j,s)} (y_i - \widehat{y}_{R_2})^2 \tag{5.3}$$

where $\widehat{y}_{R_1}$ is the average value of the target variable for the training observations belonging in region $R_1(j, s)$, and $\widehat{y}_{R_2}$ is the the average value of the target variable for the training observations in region $R_2(j, s)$.

Operationally, for each considered feature, we form a grid of possible splitting cutoffs $s_1, s_2, \ldots, s_K$, and compute over this grid the values taken by (5.3). The pair $(j; s)$ providing the smallest value represents the optimal splitting criterion. Once the split is carried out, we obtain two new nodes, and we have to decide which one to split. For this purpose, the literature suggests two alternative approaches: *depth–first* or *best–first*.

In the depth-first approach, decision trees expand nodes by first completing each depth level (first, second, third level, etc.). In this approach, a node is no longer split when the node-specific MSE becomes zero (regression case), or when all observations in the node have the same class label (classification case). Alternative stopping criteria can be considered as, for example, stopping the split if the number of observations of the node is smaller than a certain amount. In classification, when all observations in a node present the same class label, we say that the node is *pure*. Clearly, node purity entails a zero training error rate.

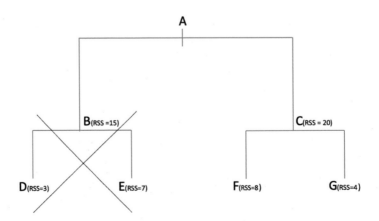

Fig. 5.3 Growing a tree by recursive binary splitting and best–first approach

In the best-first approach, decision trees expand nodes by splitting first the node that allows for the largest reduction in the RSS (regression) or error rate (classification). For example, suppose to have two nodes, B and C with a related RSS of respectively 15 and 20 (see Fig. 5.3). We must decide which node to split first. Assume that splitting the node B yields an RSS, measured as the sum of the two sub-nodes' RSS computed by (5.3), equal to 10, whereas splitting the node C yields an RSS of 12. This entails that the RSS reduction produced by splitting node B is $5(= 15 - 10)$, and that obtained by splitting node C is $8(= 20 - 12)$, thus implying that node C will be split first. Once we split node C, we generate a tree made of three terminal nodes, B (that is still pending), and D and E (the two nodes generated by splitting C). We have now to establish again which node to split first among B, D, and E, and for this purpose, we replicate the same procedure, by splitting first the one producing the largest RSS reduction. Even in this case, one goes on splitting until either the RSS (or error rate) are zero, or the number of observations within the node drops below a certain pre-fixed amount.

By letting the nodes to compete for the split, the best-first approach induces a natural hierarchy among the nodes, whereas the depth-first approach provides a hierarchy only among the tree depth levels. Both approaches result, however, in the same terminal fully-grown tree, while differences emerge when one decides to prune the tree at a specific size. Indeed, while the depth-first approach allows to prune the tree only at a certain depth, the first-best approach is more flexible, as it allows to prune a tree at a certain number of terminal nodes (final leaves). It means, as we will discuss later on, that the two algorithms may generate different optimal pruned trees depending on two different pruning parameters, the tree's depth on the one hand, and the number tree's terminal nodes, on the other hand.

5.4 Tree Overfitting

Is growing a tree based on recursive binary splitting affected by overfitting? The response to this question is unfortunately affirmative and depends on the bias–variance trade-off induced by expanding a tree further and further. To better catch this aspect, consider a small tree characterized by a few splits (few terminal nodes or regions). This tree is fairly simple, with high interpretative power, and every terminal node contains a substantially large number of observations. It is clear that the simpler the tree, the lower the prediction variance (due to a larger sample size to compute terminal node's mean), but at expense of a the larger bias (due to a too poor exploration of features interactions). A larger tree, characterized by many terminal nodes, is conceivably more complex, containing in each terminal node a smaller amount of observations; however, it explores lots of interactions (i.e, conditional splits) among the features, thus reducing the prediction bias at expenses of a larger variance. Moreover, if we expand a tree until the end using the procedure set out in the previous section, the terminal nodes will contain only one observation, generating by construction a perfect fit. The overfitting on a training dataset, however, does not measure the real fit of the tree model, as we know that the test MSE (or the error rate) is not minimized at the largest possible tree one can build.

In a tree, the main tuning parameter implying the bias–variance trade-off is thus either the tree depth (if the tree is built by the depth-first procedure), or the number of terminal nodes (if the tree is built by the best-first procedure). It makes thus sense to seek for an *optimal tree* based on an appropriate tuning of the tree size. In the tree context, we refer to *tree optimal pruning*.

5.5   Tree Optimal Pruning

Pruning a tree refers to a procedure able to cut some nodes in a proper way with the aim of increasing tree prediction accuracy by minimizing prediction error (Zhou & Yan, 2019; Esposito et al., 1997). This procedure seeks for an optimal tree size, that is, an optimal tree depth (depth-first) or an optimal number of leaves (best-first). Observe that "optimal" refers, as usual, to finding the tree size that minimizes the test prediction error. Given the highly nonparametric fitting procedure entailed by constructing a tree (void of any use of functional forms), K-fold cross-validation is the appropriate computational procedure to prune a tree. Before presenting the K-fold cross-validation procedure for optimal tree pruning, nonetheless, it is necessary to restate the problem of optimal pruning in the light of a RSS penalization setting. In what follows, we will focus on optimal pruning using the best-first as a procedure to grow a tree; optimal pruning using the depth-first approach follows, however, a fairly similar protocol.

As growing a tree faces a bias–variance prediction trade–off, optimal pruning aims at balancing bias reduction and variance increase within the largest possible tree T_0 one can fit to the training dataset. This "balancing" can be achieved by *penalizing* too complex trees in favor of more parsimonious ones, avoiding at the same time to produce a too large bias.

The largest tree, T_0, contains a sequence of subtrees that we can index through a (nonnegative) parameter α controlling for terminal nodes' penalization. By considering such a sequence of trees, at each value of α, it corresponds a specific subtree T_α of T_0 such that this quantity is minimized:

$$\sum_{j=1}^{|T_\alpha|} \sum_{x_i \in R_j} (y_i - \widehat{y}_{R_j})^2 + \alpha |T_\alpha| \tag{5.4}$$

where $|T_\alpha|$ is the number of leaves of the tree T_α, R_j is the region corresponding to the j-th terminal node, and $\widehat{y}_{R_j}$ the prediction of the target variable estimated as the average of the training observations belonging in R_j.

Figure 5.4 shows the sequence of trees generated by increasing the penalization parameter α. When $\alpha = 0$, we obtain the largest tree possible T_0 with five terminal

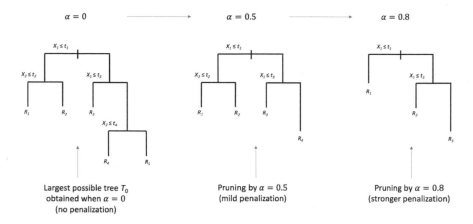

Fig. 5.4 Sequence of trees indexed by the penalization parameter α

Table 5.1 Algorithm for optimal tree pruning via K–fold cross-validation

Algorithm. *Optimal tree pruning via K–fold cross–validation*
Split the sample into $k = 1, \ldots, K$ non-overlapping folds, and call the generic fold as s_k
for $k = 1, \ldots, K$:
Consider the training dataset S excluding the k-th fold, and call it S_{-k}
Using observations in S_{-k}, produce the largest possible tree $T_{0,k}$
Prune $T_{0,k}$ by choosing a specific value of α, thus obtaining a specific tree $T_{0,k,\alpha}$
for i_k in s_k:
Consider each observation i_k in s_k and, based on its features, allocate it within the tree $T_{0,k,\alpha}$, where it will be allocated in one specific region $R(i_k)$
Calculate the residual for this unit: $\mathrm{Res}(i_k) = [y_{i_k} - \mathrm{mean}(y \text{ in } R(i_k))]$
end
Calculate the average of all the $\mathrm{Res}(i_k)$ over i_k and call it $\mathrm{MSE}_{k,\alpha}$
Repeat this procedure on various α, and store the $\mathrm{MSE}_{k,\alpha}$ at various α
end

nodes; as long as α increases the tree becomes increasingly pruned, being characterized by four leaves when $\alpha = 0.5$, and three leaves when $\alpha = 0.8$.

Following a procedure similar to the one developed for optimal tuning in a penalized regression setting, the problem we face here is that of finding an optimal α (i.e., an optimal depth or number of leaves) using cross-validation. In this case, however, we need a specific cross-validation algorithm able to accommodate the peculiarities of tree model fitting. We set out this algorithm in Table 5.1.

After splitting the sample in K folds, the algorithm starts by looping over each fold. At every iteration, one of the K folds is left out and the tree built as large as possible using the retained folds. Subsequently, one chooses a specific α to prune the largest tree, so to get the subtree $T_{0,k,\alpha}$. Each left-out observation is then allocated

Table 5.2 Example of a tree cross-validation MSEs calculation with a 5-fold split and a grid of three values of α

	α_1	α_2	α_3
$k = 1$	$MSE_{1,1}$	$MSE_{1,2}$	$MSE_{1,3}$
$k = 2$	$MSE_{2,1}$	$MSE_{2,2}$	$MSE_{2,3}$
$k = 3$	$MSE_{3,1}$	$MSE_{3,2}$	$MSE_{3,3}$
$k = 4$	$MSE_{4,1}$	$MSE_{4,2}$	$MSE_{4,3}$
$k = 5$	$MSE_{5,1}$	$MSE_{5,3}$	$MSE_{5,3}$

within this subtree, and the error on this observation calculated. In so doing, one can compute the MSE (or the error rate for classification) for all the units of the left-out fold. We re-iterate this procedure for all the folds, thus getting K MSE associated to the specific choice of α. By looping on a given grid of αs, we finally obtain a MSE for each left-out fold, and each value of α, as represented in Table 5.2 for the case of a $K = 5$ fold split and three values of α.

According to this table, we obtain 15 MSE estimates. For each level of α (namely, by column), we compute the *integrated* MSE obtained by taking the average of the MSEs over the 5 folds given a value of α:

$$IMSE(\alpha) = \frac{1}{K} \sum_{k=1}^{K} MSE(k, \alpha) \qquad (5.5)$$

The optimal subtree is the one indexed by α^*, the value that minimizes the IMSE(α) of Eq. (5.5).

5.6 Classification Trees

Building a classification tree follows a procedure similar to that for building a regression tree, with the difference that the target variable to predict is in this case categorical (qualitative response). As usual, this categorical y may be either binary (two classes) or multi-valued (several classes). In classification, we have to predict a class (or label), not a numerical value. This entails to find a different prediction principle than the arithmetic mean used in a regression setting. With categorical outcomes, the prediction of each training dataset observation will be in fact "the most commonly class occurring in the region (or terminal node) to which the observation belongs". In the same way, we cannot use the RSS as our loss function in the binary splitting, as summing labels does not make sense. As possible alternatives to RSS, the literature has suggested three popular loss functions: the error rate, the Gini index, and the cross-entropy function.

The classification error rate (ERR) is the share of the training observations in a given terminal node not belonging to the most common class. Its formula is:

$$ERR_j = 1 - \max_k(\widehat{p}_{jk}) \tag{5.6}$$

where p_{jk} is the proportion of training observations in the j-th terminal node belonging to the k-th class.

The Gini index takes on this form:

$$G_j = \sum_{k=1}^{K} \widehat{p}_{jk}(1 - \widehat{p}_{jk}) \tag{5.7}$$

It can be interpreted as the total variance over the K classes. In fact, indicate by D_{ik} the binary variable taking value 1 if observation i belongs to class k. If data are i.i.d., the variance of this variable is $\text{Var}(D_{ik}) = p_k(1 - p_k)$, where p_k is $\text{E}(D_{ik}) = \text{Pr}(D_{ik} = 1)$. It is thus clear, that the total variance across all the K classes, given by the sum of all classes variance, is equal to the Gini index. It is also straightforward to see that the Gini index is zero (thus taking its minimum) when all the observations in a given terminal node belong to the same class, while it is maximum if class frequencies are exactly the same (perfect equality). For a two-class classification, for example, the Gini is maximum if the proportion of observations in class 1 and in class 2 are equal to $1/2$. Therefore, the Gini index is a measure of *node purity*, a measure of the extent to which a node is preponderantly characterized by the presence of a specific class.

The cross-entropy (or deviance) has this formula:

$$D_j = -\sum_{k=1}^{K} \widehat{p}_{jk}\log(\widehat{p}_{jk}) \tag{5.8}$$

The cross-entropy takes numerical values that are close to those of the Gini index. Figure 5.5 shows this occurrence, by comparing the plot of the classification error rate, Gini, and cross-entropy metrics in a two-class setting as a function of the frequency of class 1. We observe that all the three metrics have a maximum at $p = 0.5$, that is, when the two classes have the same frequency. The error rate increases (decreases) linearly, but with an angular vertex at $p = 0.5$, where the Gini index and the cross-entropy are smoother functions. For this reason, they are more suitable to be used as loss functions for numerical optimization problems.

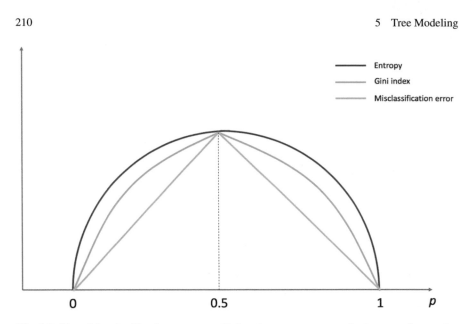

Fig. 5.5 Plot of the classification error rate, Gini, and cross-entropy metrics in a two-class setting as a function of the proportion of class 1

5.7   Tree-Based Ensemble Methods

Regression and classification trees are powerful fitting methods for the conditional expectation of y given $\mathbf{x}$. The conditional mean estimated by a tree of size M is in fact:

$$E(y|\mathbf{x}) = \sum_{j=1}^{M} c_j \cdot 1[\mathbf{x} \in R_j] \tag{5.9}$$

where $c_j = E(y|\mathbf{x} \in R_j)$. When the relationship between y and $\mathbf{x}$ is highly non-linear, fitting a tree may provide a better approximation of the underlying data generating process. Decision trees are also intuitive, easy to interpret and explain, and have a simple graphical representation. Compared to a linear regression, trees do not provide regression coefficients (and thus, partial effects), but have the virtue to allow for identifying critical values (the cutoffs) of the predictors that may be used in procedural decision-making. Moreover, qualitative predictors—having a natural cutoff—can be more easily managed and interpreted in tree modeling than in standard regression. For example, the coefficients of a linear regression with many interactions among either numerical and categorical variables are generally hard to interpret. On the contrary, interactions among the features are the essence of tree-based models.

Despite providing some advantages, trees own the weakness of not generally having a strong predictive power. This depends on the *coarsening* process operated by the cutoff-based construction of the predictions, that makes tree predictions poorly precise, and especially affected by high variance.

Tree-based ensemble methods, specifically *bagging*, *random forests*, and *boosting*, by aggregating many decision trees, can increase substantially the predictive power of both tree regression and classification. Why aggregating trees can provide this improvement? What we mean by the term "aggregating"? The response to these two questions relies on the properties of the sample mean. In fact, if $X_1, \ldots, X_N$ are independent and identically distributed (i.i.d.) variables with mean μ and variance σ^2, their mean, $\bar{X}_N = \sum_{j=1}^{N} X_j$, has mean equal to μ and variance equal to σ^2/N. It means that the variance of the aggregated statistic $\bar{X}_N$ is N times smaller than that of the single X_j. This also explains why aggregated data are generally less variable than dis-aggregated ones. For example, linear regressions on more aggregated unit of analysis (as "countries", for instance) have generally higher R^2 than regressions on less aggregated units (as "companies", for instance). This depends on the fact that the estimation error becomes smaller as aggregating reduces prediction variance. Of course, not all the aggregating approaches provide the same predictive performance. Depending on the context of application, one approach can outperform the other, but the reason why prediction variance shrinks still draws on the properties of the sample mean. In what follows we focus in three popular tree-based aggregating approach, namely, *bagging*, *random forests*, and *boosting*.

5.7.1 Bagging

Bagging (or *bootstrap aggregation*) is a general approach one can use to aggregate statistical methods with the aim of reducing prediction variance (Breiman, 1996). Its application has been particular popular for tree–based regression and classification.

At the heart of bagging, there is the use of the bootstrap re-sampling method (Davison et al., 2003; Efron, 1987; Efron & Tibshirani, 1993). As discussed in Chap. 2, the bootstrap can be used for generating with replacement B new training datasets from the initial training dataset. The idea is to fit B distinct trees, estimate at each $\mathbf{x}$ the corresponding predictions $\widehat{f}_1(\mathbf{x}), \widehat{f}_2(\mathbf{x}), \ldots, \widehat{f}_B(\mathbf{x})$, and then sum them up to obtain:

$$\widehat{f}_{bag}(\mathbf{x}) = \frac{1}{B} \sum_{b=1}^{B} \widehat{f}_b(\mathbf{x}), \tag{5.10}$$

Table 5.3 sets out the bagging algorithm we described above. This algorithm can be suitably applied both to regression and classification. However, while for regression, summing prediction makes sense, as predictions are numerical values, for classification it is not clear how to obtain an aggregated result. As we stressed above when speaking of label prediction in tree terminal nodes, also in this case a

Table 5.3   Algorithm for the computation of the bagging ensemble prediction

Algorithm. *Bagging ensemble prediction.*
1. Generate B different bootstrapped training datasets.
2. Fit the tree on the b–th bootstrapped dataset and obtain:
$$\widehat{f}_b(\mathbf{x})$$
that is, the prediction at $\mathbf{x}$.
3. Take the average all the predictions, thus obtaining:
$$\widehat{f}_{bag}(\mathbf{x}) = \tfrac{1}{B}\sum_{b=1}^{B}\widehat{f}_b(\mathbf{x})$$

majority vote principle can be applied, entailing that the overall prediction will be equal to the most commonly occurring class among the B predicted labels.

The trees built at each bootstrapped sample of the bagging procedure are grown completed, thus no pruning is generally required for bagging to reduce prediction variance. However, an estimate of the bagging test error is relevant to gauge the prediction accuracy of this learner. The next section will focus on this important aspect.

5.7.2   Test-Error Estimation for Bagging

The bootstrap procedure adopted by bagging is not only useful for reducing prediction variance, but it also has an indirect benefit, that of excluding a certain number of observations from each bootstrapped sample. On average, indeed, only 2/3 of observations are sampled for building the tree in each bootstrap. The left-out 1/3 of observations—known as *out-of-bag* (OOB) observations—can be used to compute the test error. How can we practically compute the test error using the OOB observations?

Consider an observation i. For this observation, there are around $B/3$ predictions that we can compute in which this observation is OOB. To obtain a single out-of-sample prediction for observation i, we can average its OOB predictions (regression), or consider the label having the majority vote (classification). Once we have one single prediction for observation i, we can compute its squared error (regression case), or the classification error (classification case). We can re-do the same procedure for each observation $i = 1, \ldots, N$, thus getting N predictions that we can use to estimate the MSE (regression) or the ERR (classification). Table 5.4 illustrates in detail the necessary steps and formulas to estimate the OOB test error for bagging.

When B is enough large, the test MSE (or ERR) obtained by the OOB procedure is equivalent to that obtained using the leave-one-out (LOO) cross-validation. The OOB procedure is particularly appealing when carrying out bagging on large datasets, where the computation burden can be substantial.

Table 5.4 Algorithm for the computation of the out-of-bag (OOB) test error for regression and classification bagging

Algorithm. *Out–of–bag cross–validation for bagging.*
for $i = 1, \ldots, N$:
Compute the $B/3$ predictions in which this observation is OOB
if regression:
Average these OOB predictions obtaining $\widehat{y}_{i,oob}$
Compute the squared error: $(y_i - \widehat{y}_{i,oob})^2$
if classification:
Find the label having the majority vote obtaining $\widehat{y}_{i,oob}$
Compute the mis–classification binary index: $1[y_i \neq \widehat{y}_{i,oob}]$
end
Compute the $\text{MSE}_{\text{OOB}} = 1/N \sum_{i=1}^{N}(y_i - \widehat{y}_{i,oob})^2$ for regression
Compute the $\text{ERR}_{\text{OOB}} = 1/N \sum_{i=1}^{N} 1[y_i \neq \widehat{y}_{i,oob}]$ for classification

5.7.3 Measuring Feature Importance for Bagging

One problem with bagging is that this approach, by aggregating many trees, may be able to increase prediction at expense of a lower interpreatabilty. In particular, it may become difficult to singling out which predictor (or feature) is more relevant than others in driving the phenomenon under scrutiny. Measuring feature important is thus key when applying ensemble learners.

How can we measure feature importance? In general, in statistics and econometrics, there are two ways to measure feature importance: either by *average partial effect* (APE), or by the *contribution in reducing the loss objective function* (RSS and ERR).

Average partial effect

Given a conditional expectation $E(y|\mathbf{x}) = E(y|x_1, \ldots, x_j, \ldots, x_p)$, the average partial effect of predictor x_j on the target variable y is defined as:

$$\text{APE}_{x_j \to y} = E\left[\frac{\partial E(y|\bar{x}_1, \ldots, x_j, \ldots, \bar{x}_p)}{\partial x_j}\right] \qquad (5.11)$$

where other features than x_j are evaluated at their mean value. The average partial effect is an average of the derivative of $E(y|\mathbf{x})$ on x_j, with all the other features held constant at their mean. As for generalized nonlinear models, this derivative varies with different values of x_j, and the mean is taken over the entire support of x_j. In linear regression, for a given predictor, the APE simply corresponds to β_j, the regression coefficient. Therefore, the feature importance measured by the APE has to do with the effect (both direct and indirect) of x_j on y, and increases in absolute value as long as x_j correlates well with y. Observe, finally, that for comparing the

APEs of different predictors, it is necessary to standardize them before estimation, as predictors may have different unit of measurement.

Loss function reduction

Another way to build feature importance is that of computing the contribution of a given feature to reduce the loss objective function of reference. In regression models, this objective function is the RSS, while in a classification setting the Gini index is usually considered.

A simple procedure to obtain an importance measure for feature x_j is that of computing, in each training dataset bootstrapped tree, the amount of RSS (or Gini index) reduction due to all the splits occurred to this feature. By considering all these B reductions, one can finally take their average. The formula of the feature importance (FI) measure for x_j is thus:

$$\text{FI}_j = \frac{1}{B} \sum_{b=1}^{B} \left[\sum_{k=1}^{K_{jb}} \Delta_{jbk}^{RSS} \right] \tag{5.12}$$

where Δ_{jbk}^{RSS} is the reduction in the RSS occurring at the k-th split of variable x_j in the b-th bootstrapped tree. For classification, as usual, one can use the Gini index instead of the RSS.

5.7.4 Random Forests

Aggregating trees via bagging is a powerful ensemble method for increasing prediction accuracy (Ho, 1995; Schonlau & Zou, 2020b). Bagging, however, has an important limitation. Indeed, the trees obtained in each bootstrap are generally highly correlated, implying that they are grown very similarly. This is reasonable, as bootstrap re-samples over the observations, not over the variables. This means that, if splitting over one or two variables leads to a relevant reduction in the RSS (or ERR), all the produced trees will tend to start by splitting over these variables, possibly several times over the subsequent splits. This leads to poorly different trees that, once aggregated, do not allow to explore the whole interaction space among the features. As a consequence of this shortage of new and heterogeneous information in the generated trees, bagging's reduction of variance may be too small.

A possible way to overcome this limitation is to draw a random sample of m out of p predictors at every split, thereby allowing for strong predictive features not to be used in every split (Geurts et al., 2006). This small but essential modification of the bagging algorithm leads to the *random forest* approach, one of the most popular ML methods (Breiman, 2001; Blanquero et al., 2021). The random forest, by drawing at random a subset of features to split on at every split, de-correlates the trees, thus allowing to explore a larger set of features interactions, creating and aggregating different trees, and increasing prediction precision.

Table 5.5 Random forest implementation algorithm

Algorithm: Random forest implementation
for $b = 1, \dots, B$:
Draw a bootstrapped sample
while units in node $<$ minimum number of units:
Draw at random m out of p features
Use these m features to grow a tree by *binary splitting*
end
end
Regression: obtaining the numerical prediction by averaging over all trees
Classification: predict the expected label by majority vote over all trees

The choice of the parameter m is central in random forest implementation. Interestingly, we can think of m as a new tuning parameter of our tree-based prediction exercise. Indeed, if m is small, the degree of de-correlation becomes higher because the chance to generate trees very different one each other is higher. This is a good thing on the one hand, but on the other hand the trees thus grown will be too rougher representations of the tree obtained using all the predictors. Hence, the prediction variance may sharply decrease but at expense of a larger bias. When, on the contrary, we choose a large m (thus closer to p), we use pretty all the available variables at any split, the trees get closer to the correct one, but the benefit of aggregating shrinks sensibly. In this latter case, the bias would be smaller, but the variance larger. Following the above reasoning, one can think of m as a parameter to properly tune. As a rule-of-thumb, the literature firstly proposed to choose $m = \sqrt{p}$. However, a more correct strategy would be to employ cross-validation for setting an appropriate m. This is what we will do in the random forest application section.

Table 5.5 sets out the random forest implementation algorithm. It follows quite strictly the bagging algorithm except for the new step carrying out a sampling of size m from the initial set of p variables. The feature importance index for the random forest can still be computed by formula (5.12), as we did for bagging. It goes without saying, finally, that when $m = p$, random forest and bagging coincide.

5.7.5 Boosting

Boosting is another ensemble approach using many trees to build one single aggregated prediction (Schonlau, 2005b). Differently from bagging, however, boosting does not carry out bootstrap sampling, but trees are grown in a *sequential* basis entailing that the current generated tree exploits information from the previous generated tree (Bühlmann & Hothorn, 2007; Ridgeway, 1999). This means that trees are no longer grown independently. On the contrary, they are sequentially dependent.

Table 5.6 Implementation of the boosting–tree algorithm for regression

Algorithm. Boosting–tree implementation for regression
Start by setting $\widehat{f}(\mathbf{x}_i)=0$ for every unit i
Set the residual $r_i = (y_i - 0) = y_i$ for every unit i of the training dataset
for $b = 1,...,B$:
Fit a regression tree with $d+1$ terminal nodes to the training dataset $(\mathbf{x}, r)$
Update the function $\widehat{f}(\mathbf{x}_i)$ using the last fit available:
$\widehat{f}(\mathbf{x}_i) \longleftarrow \widehat{f}(\mathbf{x}_i) + \lambda \widehat{f}^b(\mathbf{x}_i)$, where λ is a penalization parameter
Update the residuals:
$r_i \longleftarrow r_i - \lambda \widehat{f}^b(\mathbf{x}_i)$
end
Obtain boosting prediction: $\widehat{f}_{\text{boost}}(\mathbf{x}_i) = \sum_{b=1}^{B} \lambda \widehat{f}^b(\mathbf{x}_i)$

Pioneered by Freund and Schapire (1997), one can show that boosting belongs to the larger family of generalized additive models (GAMs), where $f(\mathbf{x})$, i.e. the expectation of y conditional on $\mathbf{x}$, is expressed as the sum of B unknown functions:

$$f(\mathbf{x}) = f_1(\mathbf{x}) + f_1(\mathbf{x}) + \cdots + f_B(\mathbf{x}) \tag{5.13}$$

Generally, GAMs are estimated using the *backfitting* algorithm. In the case of boosting, we estimate the single basis functions sequentially, that is, by estimating first $f_1(\mathbf{x})$, then $f_2(\mathbf{x})$, subsequently $f_3(\mathbf{x})$, and so forth.

Boosting can be used both for regression and classification (Wang & Chen, 2021; Friedman, 2001). Table 5.6 sets out the boosting algorithm for regression.

The algorithm starts by initializing the prediction equal to a constant (zero, for simplicity), thus setting at the beginning the residual r_i equal to the outcome y_i. Then, a loop starts over B iterations. When $b = 1$, we fit a regression tree made of d splits (i.e., $d + 1$ terminal nodes), over the training dataset $(\mathbf{x}, r)$. Compared to bagging, we can immediately observe that boosting regress the residuals (not the outcome) on the features, implying that boosting directly learns from the error. Let us look at the algorithm moving step–by–step. When $b = 1$, given the initialization, the first tree is fitted on the dataset $(\mathbf{x}, y)$. Once we have fitted this tree, we can compute the prediction $\widehat{f}^1(\mathbf{x}_i)$, and thus update the initial prediction obtaining that $\widehat{f}(\mathbf{x}_i) = \lambda \widehat{f}^1(\mathbf{x}_i)$, where $0 < \lambda \leq 1$ is a penalization parameter capturing the speed of the learning process. Given this new prediction, we can update also the residual that becomes equal to $r_i = y_i - \lambda \widehat{f}^1(\mathbf{x}_i)$. When $b = 2$, we fit a new tree on the new (updated) dataset $(\mathbf{x}, r)$, thus obtaining that $\widehat{f}(\mathbf{x}_i) = \lambda \widehat{f}^1(\mathbf{x}_i) + \lambda \widehat{f}^2(\mathbf{x}_i)$, and $r_i = y_i - \lambda \widehat{f}^1(\mathbf{x}_i) - \lambda \widehat{f}^2(\mathbf{x}_i)$. After B iteration, we obtain that the boosting prediction is equal to $\sum_{b=1}^{B} \lambda \widehat{f}^b(\mathbf{x}_i)$.

For classification, the boosting algorithm is similar but with some different parametrization. However, in order to better understand the boosting algorithm for classification, it is necessary to add some further analytic details regarding the fitting rational of boosting.

5.7.6 Generalized Gradient Boosting

We have stated that boosting can be seen as a way to optimally estimate a generalized additive model (Friedman, 2001; Friedman et al., 2000). More specifically, one can think of boosting as fitting a *basis functions* expansion, taking this general form:

$$f(\mathbf{x}) = \sum_{b=1}^{B} \beta_b g(\mathbf{x}, \phi_b) \tag{5.14}$$

where β_b is the b-th expansion coefficient, and $\beta_b g(\mathbf{x}, \phi_b)$ the the b-th basis function. The set of parameters ϕ_b depends on the specific type of basis that we can consider. In the case of a decision tree, ϕ_b contains: (i) the variable over which to split, (ii) the split cutoffs at the internal nodes, and (iii) the predictions obtained at the terminal nodes.

The fit of a model designed as in Eq. (5.14), requires to find optimal values for the pair (β_b, ϕ_b). Technically, this is carried out by minimizing a loss function $L(\cdot)$ averaged over the training data (where squared–error or a likelihood-based loss are popular examples) of this type:

$$\min_{(\beta_b, \phi_b)} \sum_{i=1}^{N} L\left(y_i, \sum_{b=1}^{B} \beta_b g(\mathbf{x}, \phi_b)\right) \tag{5.15}$$

The optimization problem entailed by (5.15) is tackled through specialized numerical optimization techniques, often computationally laborious. Typically, the *forward stagewise* algorithm can be used for this purpose. This algorithm works *sequentially* by adding new basis functions to the series with no change in the parameters and coefficients of the bases already added. Table 5.7 displays the step-by-step implementation of this algorithm.

The forward stagewise procedure starts by initializing the prediction to zero. Subsequently, a loop with B iteration starts. At each iteration b, given the specific loss function considered, one finds the optimal β and the optimal ϕ. Observe that we add $\beta g(\mathbf{x}_i, \phi)$, containing the parameters over which we optimize, to the current (and already optimized) series expansion $f_{b-1}(\mathbf{x}_i)$. Hence, previous parameters are not modified, but taken as optimal from the previous step.

How can the previous machinery be used to fit a boosted tree model? Let us first write a boosted tree in a expansion series form:

Table 5.7   Forward stagewise implementation algorithm

Algorithm. Forward stagewise fitting

Initialize $f_0(\mathbf{x}) = 0$
     for $b = 1, \ldots, B$:
         Calculate:
             $(\beta_b, \phi_b) = \arg\min_{(\beta, \phi)} \sum_{i=1}^{N} L(y_i, f_{b-1}(\mathbf{x}_i) + \beta g(\mathbf{x}_i, \phi))$
         Set: $f_b(\mathbf{x}) = f_{b-1}(\mathbf{x}) + \beta_b g(\mathbf{x}, \phi_b)$
     end

$$f_B(\mathbf{x}) = \sum_{b=1}^{B} T(\mathbf{x}, \Theta_b) \tag{5.16}$$

where the tree prediction—as already seen—takes on this form:

$$T(\mathbf{x}, \Theta_b) = \sum_{j=1}^{J} \gamma_j I(\mathbf{x} \in R_j) \tag{5.17}$$

The parameters $\Theta = \{R_j, \gamma_j, j = 1, \ldots, J\}$ represent the regions $(R_j, j = 1, \ldots, J)$ obtained after the splitting variables and splitting cutoffs have been optimally found, and the prediction made within each region $(\gamma_j, j = 1, \ldots, J)$. Fitting a tree means to find $\widehat{\Theta}$ as:

$$\widehat{\Theta} = \arg\min_{\Theta} \sum_{j=1}^{J} \sum_{\mathbf{x}_i \in R_j} L(y_i, \gamma_j) \tag{5.18}$$

where $L(\cdot)$ is an appropriate loss function. As we have discussed earlier in this chapter, solving the problem entails a complex combinatorial optimization problem that we can approximate via a greedy (possibly sub-optimal) solution, as the recursive binary splitting. In theory, however, given the series expansion form of the boosted tree model, one can apply the forward stagewise algorithm to find the optimal solution, that is:

$$\widehat{\Theta}_b = \arg\min_{\Theta_b} \sum_{i=1}^{N} L(y_i, f_{b-1}(\mathbf{x}_i) + T(\mathbf{x}, \Theta_b)) \tag{5.19}$$

for each $b = 1, \ldots, B$.

Unfortunately, finding $\widehat{\Theta}_b$ entails a problem similar to solve (5.18). The issue here is not to obtain the optimal (constant) predictions γ_{jb}, but the optimal regions R_{jb} within which computing these predictions. Indeed, suppose that the optimal regions are known, then finding an optimal (constant) prediction within each region is easy as we can apply the forward stagewise procedure as follows:

$$\widehat{\gamma}_{jb} = \arg\min_{\gamma_{bj}} \sum_{\mathbf{x}_i \in R_{jb}} L(y_i, f_{b-1}(\mathbf{x}_i) + \gamma_{jb}) \tag{5.20}$$

for each $b = 1, \ldots, B$.

To compute the optimal regions, a surrogate solution of (5.19), feasible and easy to compute although potentially sub-optimal, is the so–called *gradient boosting* that uses a squared error loss function of this type:

$$\tilde{\Theta}_b = \arg\min_{\Theta} \sum_{i=1}^{N} (-G_{ib} - T(\mathbf{x}_i, \Theta))^2 \tag{5.21}$$

where G_{ib} is the gradient (i.e., the first derivative) of the original loss function $L(\cdot)$ with respect to an $f(\mathbf{x}_i)$ that in our case is equal to the boosted sum of trees provided by Eq. (5.16). Indeed, we have:

$$G_{ib} = \left[\frac{\partial L(y_i, f(\mathbf{x}_i))}{\partial f(\mathbf{x}_i)} \right]_{f(\mathbf{x}_i) = f_{b-1}(\mathbf{x}_i)} \tag{5.22}$$

We have now all the ingredients to illustrate a generic gradient tree boosting algorithm. This can be applied to regression or classification as the only difference is in the loss function considered.

This algorithm is a development of the forward stagewise procedure, that entails two fundamental steps: (i) applying (5.21) to find surrogate optimal regions, (ii) applying (5.20) to compute the optimal constant prediction in each region. Of course, this is done by sequentially adding trees as the forward stagewise procedure suggests to do. Table 5.8 outlines this procedure that we call, for the sake of simplicity, the gradient tree boosting algorithm.

It is clear now that the boosting-tree algorithm illustrated in Table 5.6 is a specialized case of the general gradient boosting-tree algorithm of Table 5.8. Let us look at this in more detail:

1. In the regression setting, the loss function is the RSS. This is a quadratic loss, as its main argument is $L(y_i, \gamma) = (y - \gamma)^2$. In this case the prediction minimizing the RSS is $\gamma = f_0 = \bar{y}$, that is the mean of the target variable y in the training dataset. If y is standardized, than the initialization becomes $f_0 = 0$, that is the one used in the boosted tree regression algorithm.
2. In boosting-tree regression, the idea to fit a tree using the residuals instead of the target variable derives directly from the fact that, with quadratic loss function, the residual coincides with the gradient. We have, in fact that $1/2 \cdot \partial(y - \gamma)^2/\partial\gamma = y - \gamma = r$. This justifies the boosting-tree regression of *learning from the error*.
3. After obtaining the optimal regions via a tree regression that uses the residuals as dependent variable, the computation of the predictions is trivial, as they are—in

Table 5.8 Gradient tree boosting algorithm for a generic loss function $L(\cdot)$

Algorithm. *Gradient tree boosting fitting*

Initialize $f_0(\mathbf{x}) = \mathrm{argmin}_\gamma \sum_{i=1}^N L(y_i, \gamma)$

---

for $b = 1, \ldots, B$:

    for $i = 1, \ldots, N$, compute the negative gradient (*pseudo–residual*):

$$r_{ib} = \left[\frac{\partial L(y_i, f(\mathbf{x}_i))}{\partial f(\mathbf{x}_i)} \right]_{f(\mathbf{x}_i) = f_{b-1}(\mathbf{x}_i)}$$

  end

    Fit a regression tree to the training dataset $(r_i, \mathbf{x}_i)$, $i = 1, \ldots, N$

    Obtain the terminal regions R_{jb}, $j = 1, \ldots, J_b$

    for $j = 1, \ldots, J_b$, compute the constant predictions:

$$\widehat{\gamma}_{jb} = \arg \min_{\gamma_{bj}} \sum_{\mathbf{x}_i \in R_{jb}} L(y_i, f_{b-1}(\mathbf{x}_i) + \gamma_{jb})$$

  end

    Update: $f_b(\mathbf{x}) = f_{b-1}(\mathbf{x}) + \sum_{j=1}^{J_b} \gamma_{jb} I(\mathbf{x} \in R_{jb})$

---

end

Output: $\widehat{f}(\mathbf{x}) = \widehat{f}_B(\mathbf{x})$

---

each new tree indexed by b—simply equal to the average of the residuals in each region indexed by (b, j).

5.7.7 Logit Boosting

We can now apply the gradient boosting-tree algorithm to the case of classification, starting from a two-class classification setting. If we consider a logistic distribution, we can easily implement a logit boosting-tree using the gradient approach. As we are in a generalized linear model setting, where we use maximum likelihood for estimating the parameters, the reference loss function is the deviance, i.e. minus two-time the log likelihood. If $y \in \{0, 1\}$, taking value one with probability p and 0 with probability $1 - p$, the likelihood function is equal to:

$$L(y, p) = \prod_{i=1}^N p^{y_i} (1 - p)^{1 - y_i} \tag{5.23}$$

and the log likelihood equal to:

$$l(y, p) = \sum_{i=1}^{N} \left[y_i \log(p) + (1 - y_i)\log(1 - p) \right] \tag{5.24}$$

We obtain the logit model when we assume:

$$p = \frac{e^{\gamma}}{1 + e^{\gamma}} \tag{5.25}$$

where the parameter γ is the *logit index* and is assumed to be a linear combination of the features, that is, $\gamma = \mathbf{x}\beta$, where β is the vector of parameters. If we substitute (5.25) into (5.24), and maximize it by taking the derivative of $l(\cdot)$ respect to γ, we obtain:

$$\frac{\partial l(\cdot)}{\partial \gamma} = \sum_{i=1}^{N} \left(y_i - \frac{e^{\gamma}}{1 - e^{\gamma}} \right) = 0 \tag{5.26}$$

which is solved when:

$$\gamma = \log\left(\frac{p}{1 - p} \right) \tag{5.27}$$

where p, estimated by $\hat{p} = 1/N \sum_{i=1}^{N} y_i$, is the proportion of the sample units having $y = 1$. We can thus initialize the logit-gradient boosting algorithm using $f_0 = \log\left(\frac{\hat{p}}{1 - \hat{p}} \right)$. Following the second part of the algorithm of Table 5.8, we need now to compute the pseudo-residuals. By putting $\gamma = f(\mathbf{x}_i)$, we have:

$$r_{ib} = \left[\frac{\partial L(y_i, f(\mathbf{x}_i))}{\partial f(\mathbf{x}_i)} \right]_{f(\mathbf{x}_i)=f_{b-1}(\mathbf{x}_i)} = \left[y_i - \frac{e^{f(\mathbf{x}_i)}}{1 + e^{f(\mathbf{x}_i)}} \right] = \left[y_i - p(\mathbf{x}_i) \right] \tag{5.28}$$

The third part of the gradient algorithm requires to find the the fitted value γ in a terminal node. This entails to find the maximum of the log likelihood based on the previously fitted values $f_{b-1}(\mathbf{x}_i)$, that is:

$$l(\cdot) = \sum_{\mathbf{x}_i \in R_{jb}} \left[y_i(f_{b-1}(\mathbf{x}_i) + \gamma) - \log(1 + e^{f_{b-1}(\mathbf{x}_i)+\gamma}) \right] \tag{5.29}$$

The maximum is obtained when the first derivative is equal to zero:

$$\frac{\partial l(\cdot)}{\partial \gamma} = \sum_{\mathbf{x}_i \in R_{jb}} \left[y_i - \frac{1}{1 + e^{-f_{b-1}(\mathbf{x}_i)-\gamma}} \right] = 0 \tag{5.30}$$

were R_{jb} represents the tree's j-th terminal node in the b-th iteration. Unfortunately, the solution of Eq. (5.30) has not a closed form, thereby requiring the use the the Newton–Raphson expansion:

$$f_b = f_{b-1} - \frac{l'(f_{b-1})}{l''(f_{b-1})} \tag{5.31}$$

where we know that:

$$l'(f_{b-1}) = \sum_{i=1}^{N} (y_i - p(\mathbf{x}_i)) \tag{5.32}$$

and where the second derivative can be showed to be equal to:

$$l''(f_{b-1}) = -\sum_{i=1}^{N} p(\mathbf{x}_i)[1 - p(\mathbf{x}_i)] \tag{5.33}$$

By plugging (5.32) and (5.33) into (5.31), we finally obtain:

$$f_b = f_{b-1} + \frac{\sum_{\mathbf{x}_i \in R_{jb}} [y_i - p(\mathbf{x}_i)]}{\sum_{\mathbf{x}_i \in R_{jb}} p(\mathbf{x}_i)[1 - p(\mathbf{x}_i)]} \tag{5.34}$$

At each of the B iterations, boosting prediction is updated using this formula, where:

$$\widehat{\gamma}_{jb} = \frac{\sum_{\mathbf{x}_i \in R_{jb}} [y_i - \widehat{p}(\mathbf{x}_i)]}{\sum_{\mathbf{x}_i \in R_{jb}} \widehat{p}(\mathbf{x}_i)[1 - \widehat{p}(\mathbf{x}_i)]} \tag{5.35}$$

that completes the algorithm set out in Table 5.8.

Table 5.9 exposes concisely the step-by-step implementation of the logit gradient tree boosting algorithm. It strictly follows the algorithm of Table 5.8, but specializes to the case of a Logit loss function.

5.7.8 Multinomial Boosting

In a multi-class setting, where the classification takes place within K different classes, the gradient boosting algorithm is similar to the logit algorithm with minor but important differences. The most relevant difference relies to the fact that the K conditional class probabilities, $p_k(\mathbf{x}_i) = \text{Prob}(y_i = k|\mathbf{x}_i)$, with $k = 1, \ldots, K$, sum up to one. Thus, we need only $K - 1$ probabilities to compute all of them.

Table 5.10 provides the pseudo-code for the implementation of the multinomial gradient tree boosting algorithm. Notice that the conditional probability to belong in class k—i.e. $\widehat{p}_k(\mathbf{x}_i)$—is modeled by the so-called *softmax* function. This is the

Table 5.9 Implementation of the Logit gradient tree boosting algorithm

---

Algorithm. *Logit gradient tree boosting fitting*

Initialize $\widehat{f}_0(\mathbf{x}) = \log(\widehat{p}/1 - \widehat{p})$, with $\widehat{p} = \bar{y}$

for $b = 1, \ldots, B$:

    for $i = 1, \ldots, N$, compute the negative gradient (*pseudo–residual*):

$$r_{ib} = y_i - \widehat{p}(\mathbf{x}_i) \text{, with } \widehat{p}(\mathbf{x}_i) = \frac{e^{f_{b-1}(\mathbf{x}_i)}}{1 + e^{1 + f_{b-1}(\mathbf{x}_i)}}$$

    end

    Fit a regression tree to the training dataset $(r_i, \mathbf{x}_i), i = 1 \ldots, N)$

    Obtain the terminal regions $R_{jb}, j = 1, \ldots, J_b$

    for $j = 1, \ldots, J_b$, compute the constant predictions:

$$\widehat{\gamma}_{jb} = \frac{\sum_{\mathbf{x}_i \in R_{jb}} [y_i - \widehat{p}(\mathbf{x}_i)]}{\sum_{\mathbf{x}_i \in R_{jb}} \widehat{p}(\mathbf{x}_i)[1 - \widehat{p}(\mathbf{x}_i)]}$$

    end

    Update: $\widehat{f}_b(\mathbf{x}) = \widehat{f}_{b-1}(\mathbf{x}) + \lambda \widehat{\gamma}_{jb}$, where λ is a learning parameter

end

Output: $\widehat{f}(\mathbf{x}) = \widehat{f}_B(\mathbf{x})$

Table 5.10 Implementation of the multinomial gradient tree boosting algorithm

---

Algorithm. *Multinomial gradient tree boosting fitting*

Initialize $\widehat{f}_{0k}(\mathbf{x}) = \log(\widehat{p}_k/1 - \widehat{p}_k)$, with $\widehat{p}_k = \bar{y}_k$

for $b = 1, \ldots, B$:

    for $k = 1, \ldots, K$:

        for $i = 1, \ldots, N$, compute the negative gradient (*pseudo–residual*):

$$r_{ibk} = y_{ik} - \widehat{p}_k(\mathbf{x}_i) \text{, with } \widehat{p}_k(\mathbf{x}_i) = \frac{e^{f_{b-1,k}(\mathbf{x}_i)}}{\sum_{k=1}^{K} e^{f_{b-1,k}(\mathbf{x}_i)}}$$

        end

    Fit a regression tree to the training dataset $(r_{ik}, \mathbf{x}_{ik}), i = 1 \ldots, N$

    Obtain the terminal regions $R_{jbk}, j = 1, \ldots, J_{bk}$

    for $j = 1, \ldots, J_{bk}$, compute the constant predictions:

$$\widehat{\gamma}_{jbk} = \frac{K-1}{K} \frac{\sum_{\mathbf{x}_i \in R_{jbk}} [y_{ik} - \widehat{p}_k(\mathbf{x}_i)]}{\sum_{\mathbf{x}_i \in R_{jbk}} \widehat{p}_k(\mathbf{x}_i)[1 - \widehat{p}_k(\mathbf{x}_i)]}$$

    end

    Update: $\widehat{f}_{b,k}(\mathbf{x}) = \widehat{f}_{b-1,k}(\mathbf{x}) + \lambda \widehat{\gamma}_{jbk}$, where λ is a learning parameter

    end

end

Output: $\widehat{f}_k(\mathbf{x}) = \widehat{f}_{Bk}(\mathbf{x})$

---

natural extension of the two-class conditional probability to the multi-class setting. The general formula of the softmax function is:

$$\sigma(\mathbf{z}_i) = \frac{e^{z_i}}{\sum_{j=1}^{K} e^{z_j}} \tag{5.36}$$

with $i = 1, \ldots, K$ and $\mathbf{z} = [z_1, \ldots, z_K] \in \mathbb{R}^K$. In Machine Learning, the softmax function is rather popular. In neural network modeling, for example, it is frequently used as the downstream normalizing transformation to predict multi-class probabilities in a classification setting. Indeed, by considering as an input the vector $\mathbf{z}$, the softmax normalizes $\mathbf{z}$ with the aim of generating a probability distribution— $p_1(\mathbf{z}), \ldots, p_K(\mathbf{z})$—producing K probabilities that are proportional to the exponential of the inputs. The aim is that of transforming a continuous signal $z_i \in (-\infty; +\infty)$ into a signal ranging between 0 and 1. More importantly, the larger the values of the signal z_i, the larger its probability, thereby producing a mapping between an increasing intensity of the signal (for example, high and positive), and its probability to be closer to 1.

5.7.9   Tuning Boosting Hyper-Parameters

Boosting predictive performance depends on fine-tuning three hyper-parameters: (i) the number of trees B, (ii) the shrinkage (or learning) parameter λ, and (iii) the number of the tree splits (d) (or, equivalently, the number of terminal nodes $d + 1$). As growing trees entails a highly nonparametric procedure, K-fold cross-validation is the standard strategy for selecting boosting best hyper-parameters.

- *Tuning B*. As boosting belongs to the larger family of additive series estimators, increasing B is likely to overfit the data even if, generally, this overfitting takes place slowly. The tuning of B, which is equivalent to find the optimal size of the series expansion, is however a common task when fitting both regression and classification trees.
- *Tuning λ*. This parameter defines at which rate the boosting algorithm learns from errors. A small value of λ entails a slower learning process, that may require a larger B to achieve an appreciable predictive performance. Likewise, higher levels of λ (closer to 1), could not exploit enough the boosting ability to lean from errors. Generally, values are between 0.01 and 0.001, depending on data characteristics. A good practice is however to cherry picking λ based on minimizing the test error over a grid of credible values of λs.
- *Tuning d*. The boosting algorithm grows trees taking the residuals (i.e. the errors) as the target variable. The tree complexity depends however on the number of the tree splits. We have already discussed as this parameter may induce a predictive bias–variance trade-off. In fact, d controls the so-called *interaction depth* of the tree, as a larger value of it means to explore a larger set of interactions among the features.

When $d = 1$, for example, the tree has the minimum interaction possible, and the tree becomes a *stamp*, as it has only two branches, with no interaction allowed. Exploring features interactions can however increase predictive accuracy, and fine-tuning d may thus be a wise choice.

Some boosting implementations adopt a re-sampling procedure. At every iteration b, the new tree is grown using only a portion of the initial set of observations (for example, the 70%). Similarly to bagging and random forests, also boosting can exploit sampling variability to increase predictive performance. Differently to bagging and random forests, however, the re-sampling is carried out *without* replacement. Generally, re-sampling proportions between 50 and 70% are valid, but one can fine-tune the proportion parameter τ to increase testing accuracy.

Computational burden is also an important issue for boosting. Implementing boosting may be in fact computationally intensive and require a careful choice of the hyper-parameters' grid. Random search, i.e., a random exploration of the grid, can be a suitable alternative to reduce computational burden compared to an exhaustive search, especially when we search the best parameters within the multivariate grid formed by all three (or four, if we consider also the re-sampling proportion) boosting hyper-parameters.

5.7.10 Feature Importance

Similarly to bagging and random forests, also in the case of boosting, we can compute feature importance measures on the basis of each feature's contribution to the reduction of the prediction error. Specifically, as boosting fits a regression tree over the pseudo-residuals even for a classification setting, one can consider the reduction in the RSS induced by splitting on a specific feature, summed or averaged over all the built trees. The future importance formula (5.12) is therefore easily adapted to boosting.

5.8 R Implementation

The software R owns various packages to estimate tree-based methods. In particular, the package `tree` allows us to estimate and plot both regression and classification trees, and carry out optimal pruning. The package `randomForest` fits bagging and random forests, whereas the package `gbm` permits to fit generalized boosted regression and classification models. For these models, optimal parameters' tuning—including cross-validation—can be carried out using the CARET package, a specialized R package for tuning a variety of learning algorithms. The syntax of the functions included in these packages is simple and flexible and will be explained along the way with practical examples.

Table 5.11 Description of the Eter dataset

Name	Description
`pub_cat`	Categorical variable for publications
`cit`	Normalized mean citation score
`core`	Core funding allocation
`third`	Third-party funding
`fee`	Students fees
`grad67`	Total graduated student at ISCED 6 and 7
`grad8`	Total graduated students at ISCED 8 (PhDs)
`er67out`	Erasmus students at ISCED 7 outcoming
`er67in`	Erasmus students at ISCED 7 incoming
`numeufp`	Number of participations to EU framework programs
`shtop10`	Share of top 10% scientists cited
`shnocit`	Share of non-cited scientists
`gdp`	Regional GDP per-capita
`age`	Age of the institution

5.8.1 Application R1: Fitting a Classification Tree Using `tree`

As illustrative example, we consider the dataset `Eter.csv`, a portion of the RISIS Eter dataset which comprises several information about higher education institutions (HEIs) in Europe. We are interest in running a classification tree on the HEIs' publication performance measured by three classes, `high`, `medium`, and `low` levels of publications (the variable `pub_cat`). We consider many features that might have a potential impact on this performance, and list in Table 5.11 the name and description of each feature (including the target variable).

The following R script (and output) shows how to use the `tree()` function for estimating the largest unpruned tree:

```
———————————————————————————————— R code and output ————————————————————————————————

# Fit an unpruned classification tree

# Clear the R environment
> rm(list = ls())

# Set the working directory
> setwd("~/Dropbox/Giovanni/Book_Stata_ML_GCerulli/tree/sjlogs")

# Load needed libraries
> library(foreign)
> library(tree)

# Read the "Eter" dataset
> Eter <- read.csv("Eter.csv")
```

```
# Form y and x matrices
> x=model.matrix(pub_cat~.,Eter)[,-1]
> y=Eter$pub_cat

# Form the complete dataset "data"
> data <- cbind(y,x)
> data <- as.data.frame(data)

# Fit a classification tree over the whole training sample
> tree.data=tree(as.factor(y)~.,data)

# Summarize the tree
> summary(tree.data)

Classification tree:
tree(formula = as.factor(y) ~ ., data = data)
Variables actually used in tree construction:
 [1] "core"    "numeufp" "third"   "er67in"  "fee"      "grad8"   "er67out" "grad67"
 [9] "shnocit" "age"     "cit"
Number of terminal nodes:  25
Residual mean deviance:  0.4949 = 111.8 / 226
Misclassification error rate: 0.09562 = 24 / 251
```

The code fits an unpruned classification tree, where, after clearing the R environment, setting the working directory and loading the needed libraries, we import the `Eter` dataset and form the target (`y`) and the feature (`x`) matrices. Then, we form the entire dataset by calling it simply `data`, and fit a classification tree over the whole training sample by calling the `tree()` function and the `summary()` function to show the results. The obtained number of terminal nodes is 25, with a training misclassification error rate of around 9%.

To validate the model, we split the initial dataset into `data.train` and `data.test` using the random indeces `train_index` generated by the `sample()` R function.

——————————————————————————— R code and output ———————————————————————————

```
# Estimate the "training error rate" over the whole dataset

# Generate the tree predictions
> yhat=predict(tree.data,type="class")

# Calculate the "training ERR" on the "overall dataset"
> ERR_all_data = mean(y!=yhat)
> ERR_all_data
[1] 0.09561753

# Validate the model.
# Split the initial dataset into "data.train" and "data.test" using
# the random indeces "train_index" generated by the "sample()" function

# Estimate the test-MSE at tree size
> attach(data)

# Set the splitting seed
> set.seed(2)
```

```
# Generate the random indeces "train_index"
> train_index = sample(1:nrow(data), nrow(data)/2)

# Generate the "data.train" dataset
> data.train=data[train_index,]

# Generate the "data.test" dataset
> data.test=data[-train_index,]

# Generate the "y_train"
> y_train=y[train_index]

# Generate the "y.test"
> y_test=y[-train_index]

# Fit the model on the train datset
> tree.data=tree(as.factor(y)~.,data,subset=train_index)
> summary(tree.data)

Classification tree:
tree(formula = as.factor(y) ~ ., data = data, subset = train_index)
Variables actually used in tree construction:
[1] "numeufp" "core"    "cit"     "grad67" "er67out" "age"
Number of terminal nodes:  16
Residual mean deviance:  0.5422 = 59.1 / 109
Misclassification error rate: 0.112 = 14 / 125

# Take prediction over the train dataset
> yhat_train=predict(tree.data,data.train,type="class")

# Take prediction over the test dataset
> yhat_test=predict(tree.data,data.test,type="class")

# Calculate the train-ERR
> Train_ERR = mean(y_train!=yhat_train)
> Train_ERR
[1] 0.112

# Calculate the test-ERR
> Test_ERR = mean(y_test!=yhat_test)
> Test_ERR
[1] 0.3333333
```

As expected, we see that the training error rate is equal to 0.112, while the testing error is larger and equal to 0.357.

We also show how to plot a tree, by fitting a classification tree of size $M = 5$ using the function prune.tree(). The following code carries out this computation, and Fig. 5.6 shows the obtained tree plot.

—————————————————————————— R code and output ——————————————————————————

```
# Fit a Regression Tree of size M

# Fit a tree over the whole dataset
> tree.data=tree(as.factor(y)~.,data)

# Summarize the results
> summary(tree.data)

Classification tree:
tree(formula = as.factor(y) ~ ., data = data)
Variables actually used in tree construction:
```

```
[1] "core"      "numeufp" "third"    "er67in"   "fee"      "grad8"    "er67out" "grad67"
[9] "shnocit" "age"      "cit"
Number of terminal nodes:   25
Residual mean deviance:   0.4949 = 111.8 / 226
Misclassification error rate: 0.09562 = 24 / 251


# Grow a pruned-tree of size M=5 using the "prune.tree()" function
> M <- 5 # tree size
> prune.data=prune.tree(tree.data,best=M)


# Summarize the results
> summary(prune.data)


Classification tree:
snip.tree(tree = tree.data, nodes = c(10L, 4L, 6L, 7L))
Variables actually used in tree construction:
[1] "core"      "numeufp" "er67in"   "fee"
Number of terminal nodes:   5
Residual mean deviance:   1.484 = 365 / 246
Misclassification error rate: 0.2988 = 75 / 251


# Plot the tree
> plot(prune.data)
> text(prune.data,pretty=0)


# Generate the tree predictions (yhat) over the whole dataset
> yhat=predict(prune.data,type="class")


# Compute the train-ERR
> Train_ERR=mean(y!=yhat)
> Train_ERR
[1] 0.2988048
```

As expected, the figure shows that the number of nodes is 5. The most important variable to explain the level of publications is the institution's core funding, that is the amount of funding received by the central governement. In the second layer, we find then the number of participations to EU framework programs, and the amount of fees payed by the students.

We now find the optimal tree size by cross-validation using the function cv.tree(). For this purpose, we first fit a classification tree over the whole training sample and generate the predictions yhat_train. Then, we set the seed and the number of folds and call the cv.tree() function.

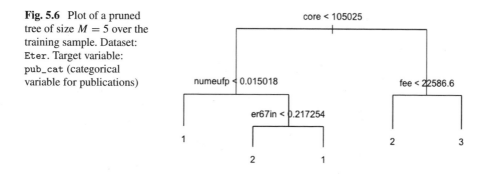

Fig. 5.6 Plot of a pruned tree of size $M = 5$ over the training sample. Dataset: Eter. Target variable: pub_cat (categorical variable for publications)

After plotting the cross-validation results, we extract the optimal tree size (best.leaves), grow the optimal pruned tree, and summarize the results. As a final step, we generate the predicted classification from the pruned optimal tree, and estimate the test error. From the optimal model, finally, we generate the predicted probabilities. The code and output follow:

———————————————————————— R code and output ————————————————————————

```
# Cross-validation with optimal tree size

# Fit a regression tree over the whole training sample
> tree.data=tree(as.factor(y)~.,data)

# Generate the predictions "yhat_train"
> yhat_train=predict(tree.data,type="class")

# Cross-validation to find the optimal tree size

# Set the seed and the number of folds
> set.seed(3) # seeds
> Nfolds=10    # N. of folds

# Cross-validate the tree model
> cv.data=cv.tree(tree.data,K=Nfolds,FUN=prune.misclass)
> attach(cv.data)

# Plot CV results
> plot(size,dev,type='b', main="Cross-validation: Error rate Vs. Number of Nodes",
 xlab="Size", ylab="ERR")

# Exstract the optimal tree size ("best.leaves")
> opt.trees = which(dev == min(dev))
> best.leaves = min(size[opt.trees])

# CV optimal tree size
> best.leaves
[1] 10

> # Grow the optimal pruned tree
prune.data.opt=prune.tree(tree.data,best=best.leaves)

# Summarize the results
> summary(prune.data.opt)

Classification tree:
snip.tree(tree = tree.data, nodes = c(15L, 26L, 12L, 14L, 9L,
20L))
Variables actually used in tree construction:
[1] "core"    "numeufp" "third"   "er67in"  "fee"      "shnocit"
Number of terminal nodes:  11
Residual mean deviance:  1.053 = 252.8 / 240
Misclassification error rate: 0.2072 = 52 / 251

# Plot the optimal tree
> plot(prune.data.opt)
> text(prune.data.opt,pretty=0)

# Generate the 'predicted classification' from the pruned optimal tree
> yhat_opt=predict(prune.data.opt,type="class")

> Test_ERR_opt=mean(yhat_train!=yhat_opt)
> Test_ERR_opt
[1] 0.2071713
```

```
# Generate the 'predicted probabilities' from the pruned optimal tree
> yprob_opt=predict(prune.data.opt,type="vector")
```

The number of optimal terminal nodes is 10, with a misclassification test error rate of 21%. Figure 5.7 plots the optimal tree.

5.8.2 Application R2: Fitting Bagging and Random–Forests Using randomForest

In this application, using again the Eter dataset, we fit the data by bagging and random forests on the whole sample using the randomForest() package. To familiarize with this function, we start by setting the main arguments of randomForest(). Specifically, we set the parameter mtry, i.e., the number of features to randomly use for splitting at each split, to 8, and the parameter ntree, i.e., the number of bootstraps, to 50. We obtain a training error around 29%.

Then, we validate the model by fitting bagging and random forests to a test sample of size 1/2 of the original dataset. First, we fit the model on the training dataset at the set parameters, and then we generate predictions in the testing dataset to calculate the test error rate. In this case, it is only slightly higher than the one we computed above (around 30%). Notably, the randomForest() function takes other two important arguments, that is, type="classification" for indicating that we are running

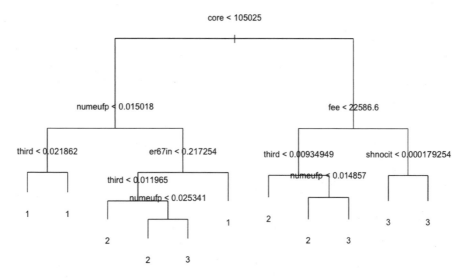

Fig. 5.7 Cross-validated optimal pruned tree. Dataset: Eter. Target variable: pub_cat (categorical variable for publications)

a classification analysis, and `importance=TRUE` allowing for computing feature importance indexes. The whole R code and output is listed below.

──────────────────────────── R code and output ────────────────────────────

```
# Fitting a classification model by bagging and random-forests

# Load the "randomForest()" package
> library(randomForest)

# Set the seed for reproducibility
> set.seed(4)

# Set parameters of the "randomForest()" function

# -> mtry = M = # of features to randomly use for splitting at each split
> M <- 8

# -> ntree = B = # of bootstraps
> B <- 50

# Fit to "data" (whole dataset) the function "randomForest()"
> bag.data=randomForest(as.factor(y)~.,
+                       data=data,
+                       mtry=M,
+                       ntree=M,
+                       importance=TRUE,
+                       type="classification")

# Obtain the predicted classification "yhat" over the entire sample
> yhat = predict(bag.data)

# Calculate the training error rate
> ERR_all_data = mean(y!=yhat,na.rm=TRUE)
> ERR_all_data
[1] 0.355102

# Validation.
# Fit the model by Bagging or Random-forests on a test sample
# of size 1/2 of the original one

# Set the seed
> set.seed(4)

# Generate the training observation index
> train_index = sample(1:nrow(data), nrow(data)/2)

# Fit the model over the training dataset
> bag.data=randomForest(as.factor(y)~.,data=data,
+                       subset=train_index,
+                       mtry=8,
+                       ntree=50,
+                       importance=TRUE,
+                       type="classification")

# Generate the prediction in the test-dataset
> yhat.test = predict(bag.data,newdata=data[-train_index,])

# Generate the "y" of the test-dataset
> y.test=data[-train_index,"y"]

# Calculate the testing error rate
> mean(yhat.test!=y.test)
[1] 0.3253968

# Estimate and plot factor-importance indeces
```

```
> importance(bag.data)
                  1          2          3 MeanDecreaseAccuracy MeanDecreaseGini
grad67   -0.0998323  3.4657208  1.2536101             2.647066         6.284954
er67out   2.6957631  0.9610986  2.0051467             3.094248         4.423066
er67in   -0.3573395  1.9204896 -0.2236835             1.055845         4.062365
grad8     1.9248262  2.1602430  0.3852291             2.378920         4.396499
numeufp   5.5572983 -0.4361489  4.4663747             6.452674        10.110603
cit       1.4782076  3.0908062  2.2151185             3.982995         2.861674
shtop10   0.6396538  1.7464493  1.3211977             2.095988         2.758744
shnocit   0.4088255  0.7837046  1.7307120             1.780055         4.302860
age       2.9367102  1.0038615  2.0866064             2.844902         3.822335
core      9.3216658  4.0163430  5.0609012            11.509791        18.596075
third     1.4152662  4.7632639  4.7095403             6.047227         9.515650
fee       3.2682384  3.8839344  4.8015654             6.095705        10.322135
> varImpPlot(bag.data)
```

We see that the test error rate is around 32%. Looking at the feature importance indexes, graphically visualized also in Fig. 5.8, we can observe that the core funding, the umber of participations to EU framework programs, and the financing from student fees are the variables that contribute more to predicting the belonging to a specific publication category. We obtain a similar ranking when using either accuracy or the Gini index as objective function.

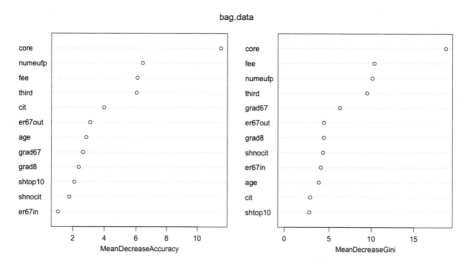

Fig. 5.8 Random forest feature importance indexes. Dataset: `Eter`. Target variable: `pub_cat` (categorical variable for publications)

5.8.3   *Application R3: Fitting Multinomial Boosting Using* gbm

In this application, we go on by using the Eter dataset and estimating a classification boosting with the R function gbm(), standing for generalized boosted regression models.

After loading the package gbm and forming the y and the x, we set the main parameters of the gbm() function, that is: n.trees (the number of sequential trees), interaction.depth (the depth of the trees), shrinkage (the learning parameter), and distribution (the distribution of the target variable y). In this application, we set distribution="multinomial".

Then, after running the gbm() function over the entire dataset, we split the data into a training and a testing dataset, fit the model on the training dataset and take the predictions (both in terms of expected probabilities ans classification) only over the testing dataset. We then compare the training and the test error rates. The following code and output show the whole procedure:

──────────────────────── R code and output ────────────────────────

```
# Fitting classification boosting

# Load the package "gbm"
> library(gbm)

# Form the "y" and the "x"
> data<-as.data.frame(Eter)
> y <- data[,1]
> x <- data[2:ncol(Eter)]
> data <-cbind(y,x)

# Set the seed for reproducibility
> set.seed(1)

# Set parameters of the "gbm()" function

# -> n.trees = B = # of bootstraps
> B <- 5000

# -> interaction.depth = ID = depth of the trees
> ID <- 4

# -> shrinkage = S = learning parameter
> S <- 0.2

# Distribution of the y
> D <- 'multinomial'

# Fit to "data" (whole dataset) the function "gbm()"
> boost.data = gbm(as.factor(y)~.,data=data,
+                n.trees=B,
+                distribution=D,
+                interaction.depth=ID,
+                shrinkage=S,
+                verbose=F)

# Summarize the boosting fit results
> summary(boost.data)
           var    rel.inf
```

```
core         core 24.015143
numeufp numeufp 15.313739
fee           fee 14.231573
grad67     grad67  7.282477
cit           cit  6.552753
third       third  6.428385
shnocit shnocit  5.886782
age           age  5.873255
er67in     er67in  4.738470
grad8       grad8  4.204735
er67out er67out  3.152709
shtop10 shtop10  2.319979


# Obtain "predicted probabilities"
> yhat_boost = predict(boost.data,data,n.trees=B,type="response")

# Obtain "predicted classification"
> pred_class_boost <- apply(yhat_boost, 1, which.max)

# Obtain the ERR over the entire dataset
> ERR_all_data = mean(y!=pred_class_boost)
> ERR_all_data
[1] 0

# Validation.
# Fit the model by boosting on a test sample of size 1/2 of the original one

# Generate the train observation index
> train_index=sample(1:nrow(data), nrow(data)/2)

# Generate the "y" of the test-dataset
> y.test=data[-train_index ,"y"]

# Fit the model on the train-dataset
> boost.data=gbm(as.factor(y)~.,data=data[train_index,],
+                 distribution=D,
+                 n.trees=B,
+                 interaction.depth=ID,
+                 shrinkage=S,
+                 verbose=F)

# Generate predicted probabilities on the test-dataset
> yhat.boost.test=predict(boost.data,
+                    newdata=data[-train_index,],
+                    n.trees=B,type="response")

# Obtain "predicted classification"
> yhat_class_boost.test <- apply(yhat.boost.test, 1, which.max)

# Calculate the test error rate
> mean(yhat_class_boost.test!=y.test , na.rm=TRUE)
[1] 0.3225806
```

While the training error rate is zero, the testing error rate is around 32%, a much larger value indeed. Also, Fig. 5.9 sets out boosting feature importance indexes. Results are comparable to those obtained in the case of the random forest algorithm. Observe that the feature importance graph is plotted upon running the line code summary(boost.data) providing also the indexes numerical results.

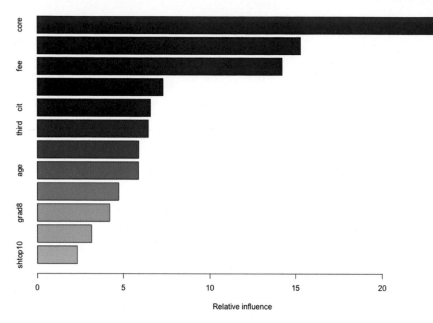

Fig. 5.9 Multinomial boosting feature importance indexes. Dataset: `Eter`. Target variable: `pub_cat` (categorical variable for publications)

5.8.4 Application R4: Tuning Random Forests and Boosting Using CARET

So far, we have not tuned yet the parameters (or hyper-parameters) of the covered tree-based methods. This is a major step, as Machine Learning is mainly focused on optimal prediction, and optimal prediction requires maximizing testing prediction accuracy (or, equivalently, minimizing testing error) over those parameters of the methods that entail a trade-off between variance and bias (the so-called hyper-parameters)[1].

In general, tuning requires exploring a multivariate grid of parameters values and search for those that maximize testing accuracy. This may be computationally burdensome, and multiple nested loops are required for carrying out this task.

The software R has a specialized package for tuning parameters of Machine Learning methods called CARET. The CARET package (standing for Classification And REgression Training), is a collection of functions allowing for streamlining the

---

[1] To compute predictions, every Machine Learning method needs to set/estimate a given set of parameters. Those parameters entailing a trade-off between prediction variance and prediction bias are sometimes referred to as *hyper-parameters*. All the others, generically, are referred to as parameters. In the literature, however, this distinction is sometimes blurred, thus I use these two terms interchangeably.

Table 5.12 Hyper-parameters tuning with cross-validation using of the CARET package. Note: one can refer to the vignette to see the other arguments of the function

Hyper–parameters' tuning via *K*–fold cross–validation using `caret`
Step 1. Set-up the `trainControl()` function
`trainControl(method = "cv", number = n, search ="grid")`
Arguments:
- `method = "cv"`: the method used to resample the dataset.
- `number = n`: number of folders to create
- `search = "grid"`: use the search grid method. For randomized search, use "random"
Step 2. Set-up the `train()` function
`train(formula, data, method, metric, trControl = trainControl(), tuneGrid)`
Arguments:
- `formula`: defines the formula of the machine learning algorithm
- `data`: defines the dataframe over which the algorithm in `formula` is applied
- `method`: defines which model to train:
- for exhaustive grid search, use: `method = "grid"`
- for randomized grid search, use: `method = "random"`
- `metric`: defines the metric used to select the optimal model:
- for regression, use: `metric = "RMSE"`
- for classification, use: `metric = "Accuracy"`, or `metric == "Kappa"`
- `trControl`: we set `trControl = trainControl()` to define the control parameters set up in Step 1
- `tuneGrid`: we set `tuneGrid = NULL` to return a data frame with all possible parameters' combination

training and testing process for Machine Learning techniques using various packages already available in R.

Tuning the parameters of a Machine Learning method using CARET is straightforward, as it requires only two main steps. In the first step we set-up the `trainControl()` function that establishes what type of tuning method one wants to carry out. Many tuning options are available, but here we focus on *K*-fold cross-validation. In the second step, we set-up the `train()` function which trains the model by calling back the `trainControl()` function set-up earlier, the `formula` we want to use, the `data`, and the type of accuracy `metric` we aim at employing. Table 5.12 illustrates the *K*-fold cross-validation procedure in a schematic simplified way.

Tuning random forests

In this section, we tune the random forests parameter `mtry`, i.e., the number of randomly variables selected at each split, using the CARET package.

After loading the CARET package and the `Eter` dataset, we carry out the two steps required by CARET to tune parameters. In step 1, we set-up the `trainControl()` function for a 10-fold cross-validation; in step 2, we set-up the `train()` function for random forests classification. Then, we set a seed for results reproducibility, form a short grid of values for `mtry` (just for illustrative purposes), and fit the `train()` function. The procedure ends up here, and we can easily visualize the results by also showing the best value of `mtry` and the maximum accuracy obtained over the grid.

What follows displays the R code and output for implementing the previous tasks:

──────────────────────────── R code and output ────────────────────────────

```
# Using the "caret" package, tune the random-forests parameter "mtry":
# number of randomly variables selected at each split

# Load the needed packages
> library(caret)

# Load the data
> data<-as.data.frame(Eter)
> y <- data[,1]
> x <- data[2:ncol(Eter)]
> data <-cbind(y,x)

# Step 1. Set-up the "trainControl()" function for a 10-fold cross-validation
> control <- trainControl(method='cv',
+                         number=10,
+                         search="grid")

# Step 2. Set-up the "train()" function for random-forests classification

# Set the seed
>set.seed(123)

# Form a grid of values for "mtry"
> tunegrid <- expand.grid(mtry=c(2,4,6))

# Fit the "train()" function
> rf_mtry <- train(as.factor(y)~.,
+                   data=data,
+                   method='rf',
+                   metric='Accuracy',
+                   tuneGrid=tunegrid,
+                   trControl=control)

# Visualize the results
> print(rf_mtry)
Random Forest

251 samples
 12 predictor
  3 classes: '1', '2', '3'

No pre-processing
Resampling: Cross-Validated (10 fold)
Summary of sample sizes: 226, 227, 226, 226, 226, 226, ...
Resampling results across tuning parameters:

  mtry  Accuracy   Kappa
  2     0.8043932  0.7048597
  4     0.7966895  0.6936989
  6     0.8165356  0.7230592

Accuracy was used to select the optimal model using the largest value.
The final value used for the model was mtry = 6.

# Get the best value of "mtry" over the grid
> rf_mtry$bestTune$mtry
[1] 6

# Show the maximum Accuracy obtained over the grid
> max(rf_mtry$results$Accuracy)
[1] 0.8165356
```

We see that, as specified, CARET used the "Accuracy" index to select the optimal model, by retaining its largest value. The final value used for the model was `mtry = 6`. By running `rf_mtry$bestTune$mtry`, we visualize the best value of `mtry` over the grid (6, as seen), and show the maximum accuracy obtained (0.81) by finally running `max(rf_mtry$results$Accuracy)`.

Tuning boosting

In this application, we tune the boosting parameters `interaction.depth` and `n.trees`, by holding the parameter `shrinkage` constant at a value of 0.1, and the parameter `n.minobsinnode` constant at a value of 20.

As done in the case of the random forests tuning, in step 1, we set-up the `trainControl()` function for a 10-fold cross-validation, while in step 2 we set-up the `train()` function for boosting classification fitting. In the second step, we first generate the parameters' grid using the `expand.grid()` function, to then use this grid as argument of the `train()` function. After fitting, we display the results of our stochastic gradient boosting, and show all the returned post-estimation objects using the `str()` function. We then go on by setting out the best accuracy value and its standard deviation over the 10-fold procedure, and by finally plotting cross-validation results.

―――――――――――――――――― R code and output ――――――――――――――――

```
# Tuning the boosting parameters "interaction.depth" and "n.trees"

# Step 1. Set-up the "trainControl()" function for a 10-fold cross-validation
> control <- trainControl(method='cv',
+                         number=10,
+                         search="grid")

# Step 2. Set-up the "train()" function for boosting classification

# Generate the parameters' grid using the "expand.grid()" function
> gbmGrid <- expand.grid(interaction.depth = c(1, 5, 9),
+                        n.trees = (1:30)*50,
+                        shrinkage = 0.1,
+                        n.minobsinnode = 20)

# Fit the model over the grid

# Set the seed
> set.seed(825)

# Fit boosting by the function "train()"
> gbmFit <- train(as.factor(y) ~ .,
+                 data = data,
+                 method = "gbm",
+                 trControl = control,
+                 verbose = FALSE,
+                 tuneGrid = gbmGrid)

# Display the results
> print(gbmFit)
Stochastic Gradient Boosting

251 samples
 12 predictor
  3 classes: '1', '2', '3'

No pre-processing
Resampling: Cross-Validated (10 fold)
```

Summary of sample sizes: 227, 226, 226, 226, 225, 227, ...
Resampling results across tuning parameters:

interaction.depth	n.trees	Accuracy	Kappa
1	50	0.6967564	0.5439260
1	100	0.7159744	0.5728395
1	150	0.7395128	0.6077654
1	200	0.7234872	0.5832328
1	250	0.7271795	0.5892072
....			
9	1300	0.7798974	0.6698782
9	1350	0.7798974	0.6692823
9	1400	0.7799103	0.6692322
9	1450	0.7839103	0.6754240
9	1500	0.7799103	0.6692985

Tuning parameter 'shrinkage' was held constant at a value of 0.1
Tuning parameter 'n.minobsinnode'
 was held constant at a value of 20
Accuracy was used to select the optimal model using the largest value.
The final values used for the model were n.trees = 1050, interaction.depth = 5, shrinkage
= 0.1 and n.minobsinnode = 20.

```
# Show the returned post-estimation objects
> str(gbmFit$results)
'data.frame':       90 obs. of  8 variables:
 $ shrinkage        : num  0.1 0.1 0.1 0.1 0.1 0.1 0.1 0.1 0.1 0.1 ...
 $ interaction.depth: num  1 5 9 1 5 9 1 5 9 1 ...
 $ n.minobsinnode   : num  20 20 20 20 20 20 20 20 20 20 ...
 $ n.trees          : num  50 50 50 100 100 100 150 150 150 200 ...
 $ Accuracy         : num  0.697 0.756 0.752 0.716 0.78 ...
 $ Kappa            : num  0.544 0.634 0.627 0.573 0.669 ...
 $ AccuracySD       : num  0.1364 0.0993 0.1108 0.135 0.1167 ...
 $ KappaSD          : num  0.202 0.146 0.165 0.201 0.174 ...

# Show the best value of "interaction.depth"
> gbmFit$bestTune$interaction.depth
[1] 5

# Show the best value of "n.trees"
> gbmFit$bestTune$n.trees
[1] 1050

# Show all results at the maximum of "Accuracy"

# Show the maximum Accuracy
> max(gbmFit$results$Accuracy)
[1] 0.7963974

# Find the index at which "Accuracy" is maximum
> best.index <- which.max(gbmFit$results$Accuracy)

# Show the all results at the best index
> gbmFit$results[best.index,]
    shrinkage interaction.depth n.minobsinnode n.trees  Accuracy     Kappa AccuracySD
KappaSD 51       0.1                 5             20    1050 0.7963974 0.6937277 0.09772454 0.1455125

# Show the standard deviation of the max accuracy
> gbmFit$results[best.index,"AccuracySD"]
[1] 0.09772454

# Plot the cross-validation results
> trellis.par.set(caretTheme())
> plot(gbmFit)
```

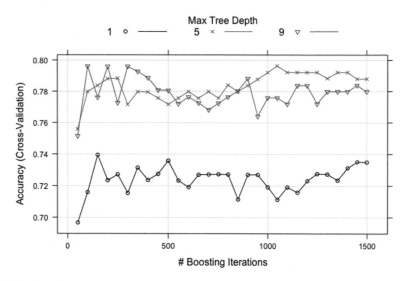

Fig. 5.10 Boosting cross-validation results over the parameters `interaction.depth` and `n.trees`. Dataset: `Eter`. Target variable: `pub_cat` (categorical variable for publications)

Results show that the best value for the `interaction.depth` is 5, and the best one for `n.trees` is 1,050. At these parameters, the accuracy is maximum and equal to 0.79 with a standard deviation of 0.097. Figure 5.10, finally, plots boosting cross-validation results showing, by different tree depth, the accuracy as a function of the number of boosting iterations.

5.9 Stata Implementation

Stata does not own a built-in command to fit tree regression/classification, as well as bagging, random forests, and boosting. Recently, however, the Stata users community has developed some commands to fit these ML models, sometimes based on R or Python functions.

The user-written commands `srtree` (Cerulli, 2022d) and `sctree` (Cerulli, 2022c), for example, implement regression and classification trees via optimal cross-validated pruning using specialized R software functions, specifically the package `tree` we used in the previous section. They basically are R wrappers estimating single trees, bagging, random forests, and boosting methods providing corresponding predictions. For random forests (bagging) and boosting, however, both `srtree` and `sctree` do not provide optimal parameters' tuning. For this purpose, we can refer to the two Stata Python-based commands `r_ml_stata_cv` (Cerulli, 2022b) and `c_ml_stata_cv` (Cerulli, 2022a), respectively for regression and classification. However, we will not present these commands, as they are wrappers of the Python

Scikit-learn platform. Here, we will focus directly on Python Scikit-learn, showing its structure and functioning.

Finally, the Stata commands `rforest` (Zou & Schonlau, 2022) and `boost` (Schonlau, 2020), are other two options to fit random forests and boosting algorithms directly in Stata. Here, we will present an application based on `rforest`, as the `boost` command works only under the Windows operative system, a limitation that we want to avoid.

5.9.1   The `srtree` and `sctree` Commands

The community-contributed Stata commands `srtree` and `sctree` are wrappers of the R software function `tree()`, fitting regression and classification trees, random forests (including bagging), and boosting. For tree fitting, these commands provide optimal pruning via K-fold cross-validation.

The `srtree` and `sctree` syntax is pretty the same. For convenience, we only outline that for `srtree`:

srtree *depvar* [*varlist*], model(*modeltype*) rversion(*R_version*)

  [prune(*integer*) cv_tree prediction(*new_data_filename*)

  in_samp_data(*filename*) out_samp_data(*filename*) ntree(*integer*)

  mtry(*integer*) inter_depth(*integer*) shrinkage(*number*) pdp(*variable*)

  seed(*integer*)]

The option `model(`*modeltype*`)` specifies the model to be estimated. It is always required to specify one model, with *modeltype* taking on these suboptions: `tree` (fitting one regression or classification tree), `randomforests` (fitting bagging and random forests), and `boosting` (fitting boosting).

The option `rversion(`*R_version*`)` specifies the R version installed in the user's operating system, with typical value as, for example, `"3.6.0"`.

The option `prune(`*integer*`)` specifies the the size M of the optimal pruned tree.

The option `cv_tree` requests to use cross-validation to find the optimal tree size (i.e., optimal pruning).

The option `prediction(`*new_data_filename*`)` requests to generate the prediction of the outcome variable. The user must specify the `new_data_filename`, i.e., the name of a dataset containing new observations having the same variables' name of the original dataset, but with the dependent variable filled in with all missing values.

The option `in_samp_data(`*filename*`)` allows to save into `filename` the in-sample predictions of the model. Within `filename`, the user can find: `y_train`, the original outcome variable in the training dataset; `yhat_train`, the prediction of the outcome variable generated by the model in the training dataset; `Train_MSE`, the training dataset's mean squared error.

The option out_samp_data(*filename*) allows to save into filename the out-of-sample predictions of the model. Within filename, the user can find: y_test, the original outcome variable in the test dataset; yhat_test, the prediction of the outcome variable generated by the model in the test dataset; Test_MSE, the test dataset's mean squared error. The option ntree(*integer*) specifies the number of bootstrapped trees to use in the bagging model.

The option mtry(*integer*) specifies the size of the subset of covariates to randomly choose when implementing the random forests model.

The option inter_depth(*integer*) specifies the complexity of the boosted ensemble. Often inter_depth = 1 works well, in which case each tree is a stump, consisting of a single split and resulting in an additive model. More generally the value of inter_depth is the interaction depth, and controls the interaction order of the boosted model, since a number of inter_depth splits can involve at most a number of inter_depth variables. This option matters only for the boosting model.

The option shrinkage(*number*) specifies the parameter controlling for the rate at which boosting learns. Typical values are 0.01 or 0.001, and the right choice can depend on the problem. Very small values can require using a very large value of the number of bootstrapped trees in order to achieve good performance. This option matters only for boosting.

The option pdp(*variable*) produces *partial dependence plot* for the specified variable. This plot illustrates the marginal effect of the selected variable on the response after integrating out the other variables. This option matters only for boosting.

The option seed(*integer*) sets the seed number allowing for reproducing the same results.

For correctly running the srtree and sctree commands in Stata, the user must install in her machine both the freeware software R and RStudio.

5.9.2 Application S1: Fitting a Regression Tree

In this application, we consider the boston_house dataset and fit a regression tree of the variable medv (median value of owner-occupied homes in $1000s) on a series of house's features. In the following code, the srtree command (run quietly to avoid showing the R software output) computes both the training and the testing MSE.

```
────────────────────────── Stata code and output ──────────────────────────
. * Load the dataset
. sysuse boston_house , clear
(Written by R.            )
. * Set outcome ("y") and features ("xvars")
. global y "medv"
. global xvars "crim zn indus chas nox rm age dis rad tax ptratio black lstat"
. * Fit an unpruned regression tree
. qui srtree $y $xvars , model(tree) rversion("3.6.0") ///
> in_samp_data("IN") out_samp_data("OUT") seed(2)
. preserve
. * Compute the train-MSE
```

```
. use IN.dta , clear
(Written by R.              )
. gen Train_DEV=(y_train-yhat_train)^2
. sum Train_DEV

    Variable |      Obs       Mean    Std. Dev.        Min        Max

   Train_DEV |      506   13.30788   29.34688    .0004675   312.5824
. di in red "Train_MSE = " r(mean)
Train_MSE = 13.307879
. * Compute the test-MSE
. use OUT.dta, clear
(Written by R.              )
. gen Test_DEV=(y_test-yhat_test)^2
. sum Test_DEV

    Variable |      Obs       Mean    Std. Dev.        Min        Max

    Test_DEV |      253     19.631   50.98504    .0006033   427.3408
. di in red "Test_MSE = " r(mean)
Test_MSE = 19.630998

. restore
```

The main Stata outputs are in the in-sample results saved in the IN.dta dataset, and the out-of-sample results saved in the OUT.dta dataset. Opening sequentially these two datasets, we can easily estimate the testing and training MSEs. As expected, the testing MSE (with a value of 19.63) is larger than the training MSE (with a value of 13.31), with the former obtained by choosing as validation sample the randomized half of the initial dataset. We are thus in the presence of some overfitting.

Figure 5.11 plots the unpruned tree obtained using the whole training dataset. The plot is grown by the *best-first* principle. This tree contains 9 terminal nodes, displaying at the end the average house value within every terminal region.

The tree growing algorithm selected the variable rm, i.e., the average number of rooms per dwelling, as the first feature to use for splitting, thus showing its relevance in driving prediction error reduction. The selected threshold is quite high, around 7 rooms per dwelling.

At the highest interaction level, we find: on the right side, variable rm again, and on the left side, variable lstat (percentage of the lower status of the population). At a lower interaction level, we can immediately notice how a higher level of criminality in the neighborhood (variable crim), above the 7 threshold, reduces sensibly the value of the house, with the terminal node showing an average of 11.98 against the largest average value of 46.82 exhibited by the 8-th terminal node (by counting nodes from the left to the right).

The srtree command also allows us to plot a tree at a given pre-fixed size M by specifying the size as an integer in the option prune(*integer*). For example, the following code plots a regression tree of size 4 (i.e., 4 leaves), and computes the training error:

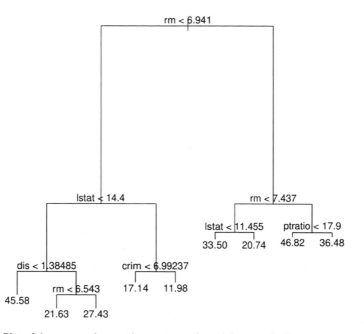

Fig. 5.11 Plot of the unpruned regression tree over the training sample. Dataset: `boston_house`. Target variable: `medv` (median value of owner-occupied homes in $1000s)

```
──────────────────────────────── Stata code and output ────────────────────────────────
. * Fit a tree of size M=4
. global M=4
. qui srtree $y $xvars , model(tree) rversion("3.6.0") ///
> prune($M) in_samp_data("IN")
. * Read results
. preserve
. use IN.dta , clear
(Written by R.            )
. gen Train_DEV=(y_train-yhat_train)^2
. sum Train_DEV

    Variable │      Obs        Mean    Std. Dev.       Min        Max
─────────────┼───────────────────────────────────────────────────────
   Train_DEV │      506    25.69947    75.16655    .001936    710.233
. di in red "Train_MSE = " r(mean)
Train_MSE = 25.699468
. restore
```

Figure 5.12 displays the plot thus obtained.

Tuning the tree means to find the optimal number of leaves. For this purpose, we re-run the previous command with option `cv_tree`:

Fig. 5.12 Plot of the 4-leave pruned regression tree over the training sample. Dataset: boston_house. Target variable: medv (median value of owner-occupied homes in $1000s)

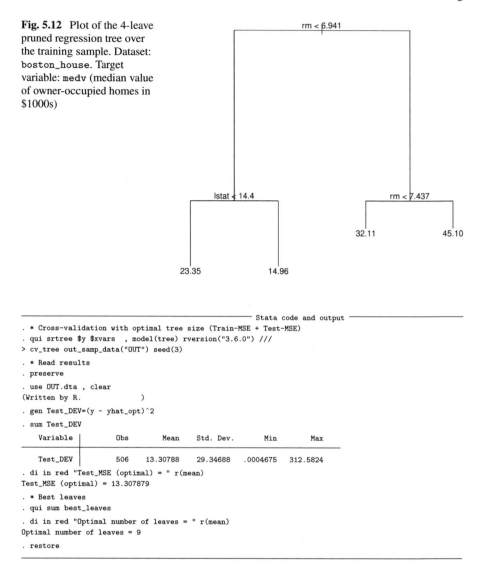

```
                                            Stata code and output
. * Cross-validation with optimal tree size (Train-MSE + Test-MSE)
. qui srtree $y $xvars  , model(tree) rversion("3.6.0") ///
> cv_tree out_samp_data("OUT") seed(3)
. * Read results
. preserve
. use OUT.dta , clear
(Written by R.              )
. gen Test_DEV=(y - yhat_opt)^2
. sum Test_DEV

    Variable │      Obs      Mean    Std. Dev.      Min       Max

    Test_DEV │      506   13.30788   29.34688   .0004675   312.5824
. di in red "Test_MSE (optimal) = " r(mean)
Test_MSE (optimal) = 13.307879
. * Best leaves
. qui sum best_leaves
. di in red "Optimal number of leaves = " r(mean)
Optimal number of leaves = 9
. restore
```

The cross-validated MSE is 13.30, which signals an improvement over the validation test seen above. However, this does not correspond to generate a smaller tree, as the optimal size is still found at 9 leaves, corresponding to the number of terminal nodes of the largest possible tree.

5.9.3 Application S2: Fitting a Classification Tree

Fitting a classification tree is similar to fit a regression tree. For this application, we consider the `carseats` dataset, collecting information on sales of child car seats at 400 different stores. Table 5.13 illustrates the name and description of the variables contained in this dataset.

The aim of this application is to predict whether a store will experience a high or a low level of sales based on a series of predictors. The target variable is `High`, taking value `Yes` if sales are larger than 8 thousands units, and `No` otherwise. Our target variable is thus binary.

The following code fits a classification tree by: loading the dataset; transforming all the factor-variables into strings (for the R software to read them correctly); setting the target outcome and the predictors; fitting the overall classification tree, finally obtaining the training and the testing MSE. By default, `sctree` provides also a graphical representation of the fitted tree.

Table 5.13 Description of `carseat` dataset containing sales of child car seats at 400 different stores

Name	Description
Sales	Unit sales (in thousands) at each location
High	Factor: Yes if Sales > 8; No otherwise
CompPrice	Price charged by competitor at each location
Income	Community income level (in thousands of dollars)
Advertising	Local advertising budget for company at each location (in thousands of dollars)
Population	Population size in region (in thousands)
Price	Price company charges for car seats at each site
ShelveLoc	Factor: quality of the shelving location for the car seats (Bad, Good, Medium)
Age	Average age of the local population
Education	Education level at each location
Urban	Factor: whether store is in an urban or rural location (No, Yes)
US	Factor: whether the store is in the US or not (No, Yes)

─────────────────────────────── Stata code and output ───────────────────────────────

```
. * Load the dataset
. sysuse carseats , clear
(Written by R.                )
. * Tranform all the factor-variables into strings for r to read them
. global y "High"
. global VARS "High US Urban ShelveLoc"
. foreach V of global VARS{
  2. decode `V´ , gen(`V´2)
  3. drop `V´
  4. rename `V´2 `V´
  5. }
. * Set the outcome and the the predictors
. global y "High"
. global xvars "CompPrice Income Advertising Population Price Age Education US Urban ShelveLoc"
. * Fit the overall classification tree
. qui sctree $y $xvars , model(tree) rversion("3.6.3") in_samp_data(IN) out_samp_data(OUT) seed(2)
. * Obtain the training error rate
. preserve
. use IN.dta , clear
(Written by R.                )
. gen Train_ERR=(y_train != yhat_train)
. sum Train_ERR
```

Variable	Obs	Mean	Std. Dev.	Min	Max
Train_ERR	400	.09	.2865402	0	1

```
. di in red "Train_ERR = " r(mean)
Train_ERR = .09
. * Obtain the testing error rate
. use OUT.dta, clear
(Written by R.                )
. gen Test_ERR=(y_test != yhat_test)
. sum Test_ERR
```

Variable	Obs	Mean	Std. Dev.	Min	Max
Test_ERR	200	.23	.4218886	0	1

```
. di in red "Test_ERR = " r(mean)
Test_ERR = .23
. restore
```

───

We see that the training error rate is equal to 9%, whereas the testing error rate is around 28%, showing that some overfitting is in place. This provides ground for finding a less complex tree by cross-validating the number of terminal nodes. The following code carries out cross-validation and then plots the optimal tree:

─────────────────────────────── Stata code and output ───────────────────────────────

```
. * Cross-validation to find the optimal tree size
. qui sctree $y $xvars , model(tree) rversion("3.6.0") cv_tree out_samp_data(OUT) seed(4)
. preserve
. use OUT.dta , clear
(Written by R.                )
. * Display the optimal tree size
. qui sum best_leaves
. di in red "Optimal tree size = " r(mean)
Optimal tree size = 7
. * Display training and testing error rates
. gen Train_ERR_optimal_nodes = (yhat!=y)
. sum Train_ERR_optimal_nodes
```

Variable	Obs	Mean	Std. Dev.	Min	Max
Train_ERR_~s	400	.1875	.3908012	0	1

```
. di in red "Training ERR rate (at optimal nodes) = " r(mean)
Training ERR rate (at optimal nodes) = .1875

. sum Test_MSE
```

Variable	Obs	Mean	Std. Dev.	Min	Max
Test_MSE	400	.2575	0	.2575	.2575

```
. di in red "Testing ERR (at optimal nodes) = " r(mean)
Testing ERR (at optimal nodes) = .2575

. restore
```

We see that the optimal number of leaves is 12 and that the cross-validated error rate is of 27%. Figures 5.13 and 5.14 sets out, respectively, the cross-validation and the tree plots.

Figure 5.13 plots the cross–validated deviance that, in this case, takes on this form:

$$\text{DEV} = -2 \sum_j \sum_k n_{jk} \log(\widehat{p}_{jk}) \tag{5.37}$$

with n_{jk} indicating the number of observations of class k belonging in the j-th terminal node. The smaller the deviance, the better the fit of the tree.

Figure 5.14 plots the optimal pruned tree. As expected, the number of leaves is 12, the optimal size.

The variable that explains most of the deviance is the shelving location of the car seat. When this is "good", however, it is the price of the seat to become optimal to

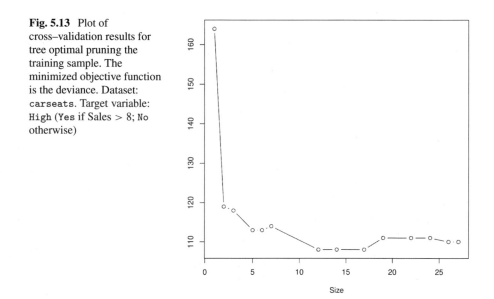

Fig. 5.13 Plot of cross–validation results for tree optimal pruning the training sample. The minimized objective function is the deviance. Dataset: `carseats`. Target variable: `High` (Yes if Sales > 8; No otherwise)

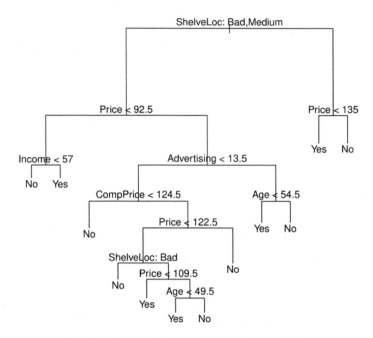

Fig. 5.14 Plot of the optimal classification tree over the training sample. Dataset: `carseats`. Target variable: `High` (`Yes` if Sales > 8; `No` otherwise)

split. In this case, a price lower than 135 dollars predicts larger sales than a price above that threshold. This is reasonable as, at the same high quality, customers should prefer items with a lower price. When the price is lower than 93.5 dollars, sales are high only if the community income level is on average larger than 57.000 dollars, i.e. only if sufficiently high. On the contrary, when the seat price is larger than 93.5 dollars, resources devoted to the advertisement, population age, and competitor prices become relevant.

5.9.4 Application S3: Random Forests with `rforest`

The `rforest` command is a Stata implementation of the random forests algorithm and provides both parameters' tuning using the `gridsearch` related command (Schonlau, 2021), and feature importance measures. The limit of this command is that it is a bit slow in its implementation, especially with large datasets. Moreover, `gridsearch` allows for tuning only two parameters at the time, although one could tune the model also without using it by suitable looping over a pre-specified grid.

For the sake of completeness, this subsection provides an application of the random forests using `rforest`. The essential `rforest` syntax is:

rforest *depvar varlist,* [type(*string*) iterations(*integer*)

    numvars(*integer*) depth(*integer*) seed(*integer*)]

where the option type(*string*) specifies the type of decision tree, and must be one of class (classification) or reg (regression). The option iterations(*integer*) sets the number of trees, with default equal to 100. The option numvars(*integer*) specifies the number of variables to randomly choose, with default equal to number of variables in *varlist*. The option depth(*integer*) defines the maximum depth of the random forest. The option seed(*integer*) set the value of the random seed.

    The essential gridsearch syntax is:

gridsearch command *depvar varlist* [*if*] [*in*], method(*str1 str2*)

    par1name(*string*) par1list(*numlist*) par2name(*string*)

    par2list(*numlist*) criterion(*str*) [options]

where, in the option method(*str1str2*), *str1* specifies the method, and *str2* the corresponding option (available options are: trainfraction, trainvar and cv. See the command help file). The option par1name(*string*) specifies the name of the tuning parameter of command. The option par1list(*numlist*) sets the values to explore for tuning the parameter. The option par2name(*string*) sets the name of the second tuning parameter of command. The option par2list(*numlist*) sets the values to explore for the second tuning parameter. The option criterion(*str*) specifies the evaluation criterion. Observe that gridsearch does not support miltinomial classification.

    In this application, we consider a regression tree, and employ the nlsw88wage dataset, the same dataset as nlsw88dwage, but with the numerical target variable wage in place of the categorical dwage. All the remaining variables of this dataset are used again as predictors.

    We run gridsearch with the rforest command as follows:

```
───────────────────────────────── Stata code and output ─────────────────
. * Load the dataset
. use nlsw88wage.dta , clear
(NLSW, 1988 extract)

. * Split the dataset into training, validation, and testing samples
. splitsample, gen(sample) split(.6 .2 .2) rseed(1010)

. preserve

. * Remover the testing units (sample=3)
. drop if sample==3
(370 observations deleted)

. replace sample=0 if sample==2
(369 real changes made)

. label define slabel 1 training 0 testing

. label values sample slabel

. * Define the target and the features
. global y "wage"

. global X age race married never_married grade    ///
>        collgrad south smsa c_city union hours ///
>            ttl_exp tenure race_1 race_2 race_3
. * Find the optimal parameters "numvars" and "depth"
. gridsearch rforest $y $X , ///
> type(reg) iter(500) seed(1) par1name(numvars) par1list(3 6 9 12) ///
```

```
> par2name(depth) par2list(0 3 6 9 12) criterion(mse) method(trainvar sample)
counter 1 1
counter 1 2
counter 1 3
counter 1 4
counter 1 5
counter 2 1
counter 2 2
counter 2 3
counter 2 4
counter 2 5
counter 3 1
counter 3 2
counter 3 3
counter 3 4
counter 3 5
counter 4 1
counter 4 2
counter 4 3
counter 4 4
counter 4 5
```

	mse	numvars	depth	seconds
1.	9.303357	3	9	1.77

```
. * Explore the grid search results
. matrix list r(gridsearch)

r(gridsearch)[20,4]
        numvars     depth       mse    seconds
r1            3         9   9.3033571       1.77
r2            3         6   9.3393002  1.1390001
r3            3        12   9.3777637  2.2809999
r4            3         0   9.4022636      3.552
r5            6         6    9.486865      1.618
r6            6         9    9.545516      2.457
r7            9         6   9.6087046  2.1559999
r8           12         6   9.6819124        2.7
r9            6        12   9.7369385  3.1210001
r10           9         9   9.7595577      3.234
r11           6         0   9.7714472      3.352
r12          12         9   9.8073072      3.951
r13           6         3   9.8547487  .90799999
r14           9        12   9.8777065  3.8959999
r15           9         3   9.9126568      1.196
r16          12        12   9.9582291  4.7360001
r17          12         3   9.9756889      1.457
r18          12         0   10.031746  5.1409998
r19           9         0   10.053437  4.2670002
r20           3         3    10.08936  .67299998

. restore
```

Let's comment of the code and its outputs. After loading the dataset, we split it into a training, a validation, and a test sample (with proportions 60%, 20%, and 20%, respectively), and define the target and the features. We remove the testing dataset, thus leaving active only the training and the validation samples. Subsequently, using gridsearch rforest, we look for the optimal values of the parameters numvars and depth, specifying [3 6 9 12] as a grid for numvars, and [0 3 6 9 12] as a grid for depth. The validation criterion is the MSE, whereas the method uses the variable sample as the train/test indicator, with sample=1 for training units and `sample=0` for validation units. Basically, this approach is equivalent to a 1-fold cross-validation. The `rforest` command allows also for a K-fold cross-validation, but its implementation is computationally intensive. See the command's help for more details.

Results shows that, across the two-dimensional grid considered, the minimum MSE is equal to 9.3, a value reached when `numvars` is equal to 3 and `depth` is equal to 9. The return matrix `r(gridsearch)` displays the overall tuning results sorted by an increasing MSE.

We can now fit the model at its best parameters, predict the outcome only on the testing dataset, and estimate the testing MSE.

```
──────────────────────────────────── Stata code and output ────
. * Fit "rforest" only on the training dataset at optimal parameters
. rforest $y $X if sample==1 , type(reg) iterations(500) numvars(3) depth(9) seed(10)

. * Obtain predictions
. predict y_hat

. * Consider only prediction for the testing dataset (sample=3)
. replace y_hat=. if sample==1 | sample==2
(1,478 real changes made, 1,478 to missing)

. * Generate the variables of squared errors
. gen sqerr=($y-y_hat)^2
(1,478 missing values generated)

. * Desplay the testing MSE
. qui sum sqerr

. di "The testing MSE is " r(mean)
The testing MSE is 11.908951
```

Results show that the testing MSE is 11.9, a value higher than any other validation MSE obtained over the grid. The proper out-of-sample prediction power of the specified model is thus weaker.

5.10 Python Implementation

Within the Machine Learning Scikit-learn platform of Python, there are specialized functions (also called *methods*) to fit and plot single trees, random forests (including bagging), and boosting. For fitting a single tree, we can use the method `DecisionTreeClassifier()` for classification trees, and `DecisionTreeRegressor()` for regression trees; for fitting random forests, we can use the methods `RandomForestClassifier()` and `RandomForest Regressor()` respectively for classification and regression; finally, for fitting boosting, we can use the method `GradientBoostingClassifier()` and `GradientBoostingRegressor()`. In what follows, we provide some application of these functions.

5.10.1 Application P1: Fitting and Plotting a Decision Tree
      in Python

Plotting a tree in Python is relatively easy and provides a larger set of information than the R implementation carried out by `srtree` and `sctree`. The scikit-learn implementation, however, adopts the depth-first approach to build the tree. For illustration purposes, we consider the popular `iris` dataset containing four features (length and width of sepals and petals) of 50 units of three species of iris (iris setosa, iris virginica, and iris versicolor). We want to classify within one of these three iris species a single instance (i.e., an iris flower) on the basis of its features. The following Python code fits and plots the resulting classification tree.

```
─────────────────────────────────── Python code and output ───────────────────────────────────
>>> # Import dependencies
... from sklearn.datasets import load_iris
>>> from sklearn import tree
>>> import matplotlib.pyplot as plt
>>> # Import the dataset and form (X,y)
... iris = load_iris()
>>> X, y = iris.data, iris.target
>>> # Initialize and fit a decision tree
... clf = tree.DecisionTreeClassifier(max_depth=3)
>>> clf = clf.fit(X, y)
>>> # Plot the tree (in 3 steps)
... # Put into two lists the names of the features and of the classes
... fn=['seplen','sepwid','petalen','petalwid']
>>> cn=['setosa', 'versicolor', 'virginica']
>>> # Prepare the frame for a plot
... fig, axes = plt.subplots(nrows = 1,ncols = 1,figsize = (10,10), dpi=500)
>>> # Generate the plot
... tree.plot_tree(clf,
...                 feature_names = fn,
...                 class_names=cn,
...                 filled = True);
... # Save the plot
[Text(1453.125, 3368.75, 'petalwid <= 0.8\ngini = 0.667\nsamples = 150\nvalue = [50, 50, 50]\nclass = seto
> sa'), Text(968.75, 2406.25, 'gini = 0.0\nsamples = 50\nvalue = [50, 0, 0]\nclass = setosa'), Text(1937.5
> , 2406.25, 'petalwid <= 1.75\ngini = 0.5\nsamples = 100\nvalue = [0, 50, 50]\nclass = versicolor'), Text
> (968.75, 1443.75, 'petalen <= 4.95\ngini = 0.168\nsamples = 54\nvalue = [0, 49, 5]\nclass = versicolor')
> , Text(484.375, 481.25, 'gini = 0.041\nsamples = 48\nvalue = [0, 47, 1]\nclass = versicolor'), Text(1453
> .125, 481.25, 'gini = 0.444\nsamples = 6\nvalue = [0, 2, 4]\nclass = virginica'), Text(2906.25, 1443.75,
> 'petalen <= 4.85\ngini = 0.043\nsamples = 46\nvalue = [0, 1, 45]\nclass = virginica'), Text(2421.875, 4
> 81.25, 'gini = 0.444\nsamples = 3\nvalue = [0, 1, 2]\nclass = virginica'), Text(3390.625, 481.25, 'gini
> = 0.0\nsamples = 43\nvalue = [0, 0, 43]\nclass = virginica')]
>>> fig.savefig('py_tree_plot1.png')
```

The `iris` dataset is part of the scikit-learn built-in datasets. Thus, we can directly load it from within this platform. After forming the array of features (X) and that of the target (y), we fit a classification tree using the function `DecisionTreeClassifier()` by requesting a tree of depth equal to 3. Figure 5.15 sets out the resulting tree obtained using the `plot_tree()` function.

Every box of the figure shows a series of information regarding the split operated at each node. The first information is about the name of the splitting variable and the value of its cutoff; the second is the value of the Gini index; the third is the number of units belonging to the region the box refers to; the fourth is the frequency of units within the region by class; the fifth is the class having the largest frequency. The Gini

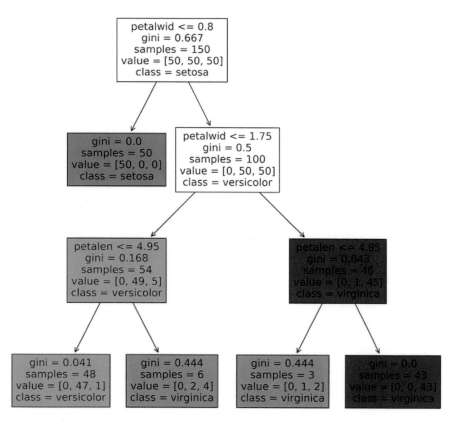

Fig. 5.15 Plot of a classification tree in Python. Dataset: `iris`. Features: length and width of sepals and petals. Target variable: three species of iris (setosa, virginica, and versicolor)

index is zero when all the units belong to the same class, thereby stopping in this case the splitting procedure.

As said, the Python implementation of the recursive binary splitting follows the depth-first principle, thus making the appearance of intermediate trees different from that produced by a best-first construction algorithm. As a consequence, when tuning a tree in Python, it is the tree depth, and not the number of leaves, that we consider as a parameter of tree complexity.

5.10.2 Application P2: Tuning a Decision Tree in Python

In this application, we show how to tune a decision tree in Python. We consider again a classification tree, and employ the `nlsw88dwage` dataset, an extract from the U.S. National Longitudinal Survey of Mature and Young Women for the year 1988. Below we show a description of this dataset, where our target categorical variable is `dwage`, taking value 1 for women earning a low salary, 2 for a medium salary, 3 for a high salary. All the remaining variables of this dataset are used as predictors.

```
──────────────────────────── Stata code and output ────────────────────────────
.* Open the dataset
. use nlsw88dwage.dta , clear
(NLSW, 1988 extract)

. describe
Contains data from nlsw88dwage.dta
  obs:         1,848                         NLSW, 1988 extract
  vars:           17                         12 Jul 2021 15:18
                                             (_dta has notes)
```

| | storage | display | value | |
variable name	type	format	label	variable label
dwage	float	%9.0g		
age	byte	%8.0g		age in current year
race	byte	%8.0g		race
married	byte	%8.0g		married
never_married	byte	%8.0g		never married
grade	byte	%8.0g		current grade completed
collgrad	byte	%16.0g		college graduate
south	byte	%8.0g		lives in south
smsa	byte	%9.0g		lives in SMSA
c_city	byte	%8.0g		lives in central city
union	byte	%8.0g		union worker
hours	byte	%8.0g		usual hours worked
ttl_exp	float	%9.0g		total work experience
tenure	float	%9.0g		job tenure (years)
race_1	byte	%8.0g		race==white
race_2	byte	%8.0g		race==black
race_3	byte	%8.0g		race==other

```
Sorted by:
```

We apply the tree classifier in Python step-by-step. We start by importing the main dependencies, i.e., all the modules needed for running the classifier correctly, and in particular the function `DecisionTreeClassifier()`. Then, we load the initial dataset in Python, we define the target and the features, and split the dataset into a training and a testing dataset using the function `train_test_split()`.

```
──────────────────────────── Python code and output ────────────────────────────
>>> # Import the needed modules
... import numpy as np
>>> import pandas as pd
>>> from sklearn.tree import DecisionTreeClassifier
>>> from sklearn.model_selection import train_test_split
>>> from sklearn.metrics import accuracy_score
>>> import os
>>> # Set the directory
... os.chdir("/Users/cerulli/Dropbox/Giovanni/Book_Stata_ML_GCerulli/tree/sjlogs")
>>> # Set the initial dataset
... dataset="nlsw88dwage.dta"
>>> # Load the stata dataset as pandas dataframe
... df = pd.read_stata(dataset)
>>> # Define ´y´ the target variable
... y=df.iloc[:,0].astype(int)
>>> # Define ´X´ as features
... X=df.iloc[:,1::]
>>> # Split the sample into "training" and "testing" datasets
... X_train, X_test, y_train, y_test = train_test_split(X, y, random_state=33)
```

Subsequently, we fit a classification tree with depth equal to 3 on the training dataset, and predict classes over the testing dataset, to then computing the testing predictive accuracy, obtaining a value equal to 0.54.

```
──────────────────────────────── Python code and output ────────────────────────
>>> # Fit a classification tree on the training dataset
... ctree=DecisionTreeClassifier(max_depth=3)
>>> ctree.fit(X_train, y_train)
DecisionTreeClassifier(ccp_alpha=0.0, class_weight=None, criterion='gini',
                       max_depth=3, max_features=None, max_leaf_nodes=None,
                       min_impurity_decrease=0.0, min_impurity_split=None,
                       min_samples_leaf=1, min_samples_split=2,
                       min_weight_fraction_leaf=0.0, presort='deprecated',
                       random_state=None, splitter='best')
>>> # Predict classes over the testing dataset
... y_pred = ctree.predict(X_test)
>>> # Calculate the predictive accuracy
... print(accuracy_score(y_test, y_pred))
0.5411255411255411
```

The following chunk of code implements in Python the tuning of the main complexity parameter, the depth of the tree, by cross-validation using the GridSearchCV function. We first define a grid of ten values (from 1 to 10) for the parameter max_depth as a Python dictionary. Then, we initialize the grid with a 10-fold cross-validation, and fit the model over the grid by putting the cross-validation results into a Pandas data frame named cv_res for visualizing the cross-validation results. Subsequently, we print the best accuracy reached (equal to 0.56) and the optimal depth of the tree (equal to 6).

```
──────────────────────────────── Python code and output ────────────────────────
>>> # Cross-validation
... # Grid search
... from sklearn.model_selection import GridSearchCV
>>> # Define the grid as a "dictionary"
... params = {"max_depth": list(range(1,11))}
>>> # Initialize the grid with a 10-fold CV
... grid = GridSearchCV(estimator=ctree,
...                     param_grid=params,
...                     cv=10,
...                     return_train_score=True)
... # Fit the model over the grid
>>> grid.fit(X_train, y_train)
GridSearchCV(cv=10, error_score=nan,
             estimator=DecisionTreeClassifier(ccp_alpha=0.0, class_weight=None,
                                              criterion='gini', max_depth=3,
                                              max_features=None,
                                              max_leaf_nodes=None,
                                              min_impurity_decrease=0.0,
                                              min_impurity_split=None,
                                              min_samples_leaf=1,
                                              min_samples_split=2,
                                              min_weight_fraction_leaf=0.0,
                                              presort='deprecated',
                                              random_state=None,
                                              splitter='best'),
             iid='deprecated', n_jobs=None,
             param_grid={'max_depth': [1, 2, 3, 4, 5, 6, 7, 8, 9, 10]},
             pre_dispatch='2*n_jobs', refit=True, return_train_score=True,
             scoring=None, verbose=0)
>>> # Put cross-validation the results into
... cv_res = pd.DataFrame(grid.cv_results_)[['mean_train_score',
```

```
...                                    'mean_test_score',
...                                    'std_test_score',
...                                    'params']]
... # Visualize the CV results
>>> print(cv_res)
   mean_train_score  mean_test_score  std_test_score           params
0          0.474748         0.474815        0.034126  {'max_depth': 1}
1          0.540164         0.515170        0.053493  {'max_depth': 2}
2          0.563013         0.514415        0.057155  {'max_depth': 3}
3          0.609105         0.547618        0.053212  {'max_depth': 4}
4          0.644139         0.556975        0.050500  {'max_depth': 5}
5          0.679572         0.562798        0.049279  {'max_depth': 6}
6          0.725026         0.547602        0.055528  {'max_depth': 7}
7          0.775529         0.531097        0.041490  {'max_depth': 8}
8          0.831567         0.534626        0.051336  {'max_depth': 9}
9          0.878704         0.515129        0.049123  {'max_depth': 10}
>>> # Print the "best accuracy" reached
... print(grid.best_score_)
0.5627984568866645
>>> # Print the optimal "tree depth"
... print(grid.best_estimator_.max_depth)
6
```

The last chunk of code fits the optimal model (i.e., a tree with a depth of 6), computes and puts the predicted probabilities and classes into a Python data frame, and export them into a Stata dataset. Once in Stata, we tabulate the predicted classes.

──────────────────────────── Python code and output ────────────────────────────

```
>>> # Fit the optimal model
... n_opt=grid.best_estimator_.max_depth
>>> ctree = DecisionTreeClassifier(max_depth=n_opt)
>>> ctree.fit(X_train, y_train)
DecisionTreeClassifier(ccp_alpha=0.0, class_weight=None, criterion='gini',
                       max_depth=6, max_features=None, max_leaf_nodes=None,
                       min_impurity_decrease=0.0, min_impurity_split=None,
                       min_samples_leaf=1, min_samples_split=2,
                       min_weight_fraction_leaf=0.0, presort='deprecated',
                       random_state=None, splitter='best')
>>> # Put the probabilites into a dataframe
... probs = pd.DataFrame(ctree.predict_proba(X))
>>> # Put the classes into a dataframe
... classes = pd.DataFrame(ctree.predict(X))
>>> # Create an array of probs and classes
... out=np.column_stack((probs,classes))
>>> # Generate a dataframe "OUT" from the array "out"
... OUT = pd.DataFrame(out)
>>> # Get to the Stata (Excel) for results
... OUT.to_stata('pred_tree.dta')
/Library/Frameworks/Python.framework/Versions/3.8/lib/python3.8/site-packages/pandas/io/stata.py:2303:
Inv
> alidColumnName:
Not all pandas column names were valid Stata variable names.
The following replacements have been made:

    0    ->    _0
    1    ->    _1
    2    ->    _2
    3    ->    _3

If this is not what you expect, please make sure you have Stata-compliant
column names in your DataFrame (strings only, max 32 characters, only
alphanumerics and underscores, no Stata reserved words)
  warnings.warn(ws, InvalidColumnName)
```

```
                                    ───── Stata code and output ─────
. * Open the dataset of results "pred_tree"
. use pred_tree.dta , clear
. rename _0 prob_low
. rename _1 prob_medium
. rename _2 prob_high
. rename _3 pred_class
. * Tabulate "pred_class"
. tab pred_class

 pred_class │     Freq.      Percent       Cum.
────────────┼───────────────────────────────────
          1 │       545        29.49       29.49
          2 │       716        38.74       68.24
          3 │       587        31.76      100.00
────────────┼───────────────────────────────────
      Total │     1,848       100.00
```

5.10.3   Application P3: Random Forests with Python

Fitting random forests and extracting feature importance measures using Python scikit-learn is rather straightforward. Specifically, in this application, we use the RandomForestRegressor() function fitting a random forests regression with the following three parameters: (i) number of splitting variables at each split (max_features) equal to 5; (ii) number of trees (n_estimators) equal to 100, and (iii) three depth (max_depth) equal to 2. Using again the boston dataset, the following code carries out this task by regressing house value on the rest of the predictors.

```
                                    ───── Python code and output ─────
>>> # Load the required dependencies
... import numpy as np
>>> import pandas as pd
>>> from sklearn.datasets import load_boston
>>> from sklearn.model_selection import train_test_split
>>> from sklearn.ensemble import RandomForestRegressor
>>> # Load the data set and split for training and testing
... boston = load_boston()
>>> X = pd.DataFrame(boston.data, columns=boston.feature_names)
>>> y = boston.target
>>> X_train, X_test, y_train, y_test = train_test_split(X, y, test_size=0.25, random_
state=12)
>>> # Fit the random forest regressor with 100 decision trees
... rf = RandomForestRegressor(max_features=5,n_estimators=100, max_depth=2)
>>> rf.fit(X_train, y_train)
RandomForestRegressor(bootstrap=True, ccp_alpha=0.0, criterion='mse',
                      max_depth=2, max_features=5, max_leaf_nodes=None,
                      max_samples=None, min_impurity_decrease=0.0,
                      min_impurity_split=None, min_samples_leaf=1,
                      min_samples_split=2, min_weight_fraction_leaf=0.0,
                      n_estimators=100, n_jobs=None, oob_score=False,
                      random_state=None, verbose=0, warm_start=False)
>>> # Obtain the random forests feature importance (FI) measures
... rf.feature_importances_
array([0.04314593, 0.00089963, 0.05348997, 0.00190417, 0.09356356,
```

```
      0.31174654, 0.007898   , 0.02775365, 0.00476056, 0.00782579,
      0.04863538, 0.0011559 , 0.397221  ])
>>> # Save the FI measures as a Pandas data frame
... FI = pd.DataFrame(rf.feature_importances_ , index=boston.feature_names)
>>> FI.columns=['importance_index']
>>> # Put FI in a Stata dataset
... FI.to_stata('rf_FI.dta')
```

```
——————————————————————————— Stata code and output ———————————————
. * Load the Stata 'rf_FI' dataset
. use rf_FI.dta , clear

. * Plot a bar plot of the FI measures
. graph hbar (mean) importance_index , ///
> over(index, sort(1) descending) ytitle(Importance measure) ///
> bar(1,color(midblue))
```

After loading the required dependencies, we open the boston dataset and split it for obtaining the training and testing datasets. Then, we fit the random forests regression with 100 decision trees, tree depth equal to 2, and splitting features equal to 5. We then obtain the random forests feature importance (FI) measures, by calling the post-estimation module feature_importances_ producing an array. For visualization purposes, we save the FI measures as a Pandas data frame, to then export this as a Stata dataset. By opening the Stata rf_FI dataset, we plot in Stata a bar plot of the FI measures, feature-by-feature. Figure 5.16 sets out the obtained bar plot, where we see that the features lstat and rm stand out.

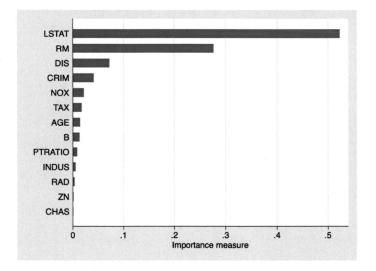

Fig. 5.16 Plot of random forests feature importance measures. Dataset: boston

5.10.4 Application P4: Boosting in Python

Fitting gradient boosting with Python scikit-learn can be easily performed using the Functions `GradientBoostingClassifier()` for classification, and `GradientBoostingRegressor()` for regression. These functions have various arguments, but the most relevant are those parameters entailing a prediction variance–bias trade-off. Indeed, as discussed earlier in this chapter, four parameters are relevant for tuning boosting:

- `n_estimators`: the number of of trees to aggregate;
- `max_depth`: the depth of the single built tree;
- `learning_rate`: the boosting learning rate;
- `subsample`: the share of training observations used to fit the model at each iterative tree building.

In this application, using again the `nlsw88dwage` dataset, we run a classification boosting for the categorical target variable `dwage` (low, medium, high salary) as a function of several women characteristics. In what follows, we break down the Python code into pieces to comment results more easily.

The first part of the code sets off by loading the dataset into memory, importing the needed Python dependencies, setting the directory, loading the Stata dataset as a Pandas dataframe, defining the target and the features, and finally splitting the sample into a "training" and a "testing" dataset.

```
───────────────────────── Stata code and output ─────────────────────────
.* Load the dataset
. use nlsw88dwage.dta , clear
(NLSW, 1988 extract)
───────────────────────── Stata code and output ─────────────────────────
```

```
───────────────────────── python (type end to exit) ─────────────────────────
>>> # Import the needed modules
... import numpy as np
>>> import pandas as pd
>>> from sklearn.ensemble import GradientBoostingClassifier
>>> from sklearn.model_selection import train_test_split
>>> from sklearn.metrics import accuracy_score
>>> import os
>>> # Set the directory
... os.chdir("/Users/cerulli/Dropbox/Giovanni/Book_Stata_ML_GCerulli/tree/sjlogs")
>>> # Set the initial dataset
... dataset="nlsw88dwage.dta"
>>> # Load the stata dataset as pandas dataframe
... df = pd.read_stata(dataset)
>>> # Define `y` the target variable
... y=df.iloc[:,0].astype(int)
>>> # Define `X` as features
... X=df.iloc[:,1::]
>>> # Split the sample into "training" and "testing" datasets
... X_train, X_test, y_train, y_test = train_test_split(X, y, random_state=33)
```

In the second part of the code, we initialize a gradient boosting classifier at specific (and possibly suboptimal) parameters using the function `GradientBoostingClassifier()`. Notice that the `random_state` argument is used to allow for results' reproducibility. Subsequently, we fit the classifier (we named gbc) on the training dataset, predict classes over the testing dataset, and calculate the predictive accuracy (i.e., one minus the testing error rate) that is equal to 0.56.

```
———————————————————————————————— Python code and output ————————
>>> # Initialize a gradient boosting classifier at suboptimal parameters
... gbc=GradientBoostingClassifier(n_estimators=100 ,
...                                 max_depth=8 ,
...                                                         learning_rate=0.1 ,
...                                                         subsample=0.8 ,
...                                                         random_state=11)
... # Fit the classifier on the training dataset
>>> gbc.fit(X_train, y_train)
GradientBoostingClassifier(ccp_alpha=0.0, criterion='friedman_mse', init=None,
                           learning_rate=0.1, loss='deviance', max_depth=8,
                           max_features=None, max_leaf_nodes=None,
                           min_impurity_decrease=0.0, min_impurity_split=None,
                           min_samples_leaf=1, min_samples_split=2,
                           min_weight_fraction_leaf=0.0, n_estimators=100,
                           n_iter_no_change=None, presort='deprecated',
                           random_state=11, subsample=0.8, tol=0.0001,
                           validation_fraction=0.1, verbose=0,
                           warm_start=False)
>>> # Predict classes over the testing dataset
... y_pred = gbc.predict(X_test)
>>> # Calculate the predictive accuracy
... print(accuracy_score(y_test, y_pred))
0.5627705627705628
```

The third part of the code is dedicated to the tuning of the boosting hyperparameters. For this purpose, we use the GridSearchCV() allowing for k-fold cross-validation. This function takes various arguments, but the most relevant are:

- estimator: the learner to tune;
- param_grid: the grid of the parameters;
- cv: the number of cross-validation folds.

The argument return_train_score=True is inserted to obtain also the training accuracy along with the training accuracy.

The parameters' grid has to be declared as a Python dictionary. Therefore, the procedure starts by defining the grid as a dictionary, and goes on by initializing the grid with a 5-fold cross-validation. The provided multivariate grid is rather small, as it is made of only $16 = 2 \times 2 \times 2 \times 2$ points. This is just for illustrative purposes, as real applications should consider larger grids. After initializing the grid, we fit the learner over it with the command grid.fit(X_train, y_train) taking as argument the train dataset. The main return object of grid.fit() is grid.cv_results_ that reports the main cross-validation results, and in particular, for each grid point: the average training accuracy (mean_train_score), the average testing accuracy (mean_test_score), the testing accuracy standard error (std_test_score), and the list of grid's parameters (params). We then put grid.cv_results_ in a Pandas dataframe we call cv_res and visualize it with the print() function. Finally, we print the "best accuracy" reached contained in the return object grid.best_score_, along with the optimal hyper-parameters contained in the return object `grid.best_estimator_`. We observe that, at the best parameters, the testing accuracy is equal to 0.58, larger than the value obtained at suboptimal parameters, but not sensibly higher (probably due to a too small grid provided).

```
─────────────────────────────────── Python code and output ───────────────────
>>> # Cross-validation
... # Grid search
... from sklearn.model_selection import GridSearchCV
>>> # Define the grid as a "dictionary"
... params = {"n_estimators": [50,100],
...           "max_depth": [5,8],
...               "learning_rate": [0.1,0.5],
...               "subsample": [0.5,0.8]
...              }
... # Initialize the grid with a 5-fold CV
>>> grid = GridSearchCV(estimator=GradientBoostingClassifier(),
...                 param_grid=params,
...                 cv=5,
...                 return_train_score=True)
... # Fit the model over the grid
>>> grid.fit(X_train, y_train)
GridSearchCV(cv=5, error_score=nan,
             estimator=GradientBoostingClassifier(ccp_alpha=0.0,
                                    criterion='friedman_mse',
                                    init=None, learning_rate=0.1,
                                    loss='deviance', max_depth=3,
                                    max_features=None,
                                    max_leaf_nodes=None,
                                    min_impurity_decrease=0.0,
                                    min_impurity_split=None,
                                    min_samples_leaf=1,
                                    min_samples_split=2,
                                    min_weight_fraction_leaf=0.0,
                                    n_estimators=100,
                                    n_iter_no_change=None,
                                    presort='deprecated',
                                    random_state=None,
                                    subsample=1.0, tol=0.0001,
                                    validation_fraction=0.1,
                                    verbose=0, warm_start=False),
             iid='deprecated', n_jobs=None,
             param_grid={'learning_rate': [0.1, 0.5], 'max_depth': [5, 8],
                         'n_estimators': [50, 100], 'subsample': [0.5, 0.8]},
             pre_dispatch='2*n_jobs', refit=True, return_train_score=True,
             scoring=None, verbose=0)
>>> # Put cross-validation the results into
... cv_res = pd.DataFrame(grid.cv_results_)[['mean_train_score',
...                             'mean_test_score',
...                             'std_test_score',
...                             'params']]
... # Visualize the CV results
>>> print(cv_res)
    mean_train_score ...                                    params
0          0.888167  ...  {'learning_rate': 0.1, 'max_depth': 5, 'n_esti...
1          0.919009  ...  {'learning_rate': 0.1, 'max_depth': 5, 'n_esti...
2          0.978174  ...  {'learning_rate': 0.1, 'max_depth': 5, 'n_esti...
3          0.990260  ...  {'learning_rate': 0.1, 'max_depth': 5, 'n_esti...
4          0.997835  ...  {'learning_rate': 0.1, 'max_depth': 8, 'n_esti...
5          1.000000  ...  {'learning_rate': 0.1, 'max_depth': 8, 'n_esti...
6          1.000000  ...  {'learning_rate': 0.1, 'max_depth': 8, 'n_esti...
7          1.000000  ...  {'learning_rate': 0.1, 'max_depth': 8, 'n_esti...
8          0.985752  ...  {'learning_rate': 0.5, 'max_depth': 5, 'n_esti...
9          1.000000  ...  {'learning_rate': 0.5, 'max_depth': 5, 'n_esti...
10         0.872121  ...  {'learning_rate': 0.5, 'max_depth': 5, 'n_esti...
```

```
11          0.999820  ...  {´learning_rate´: 0.5, ´max_depth´: 5, ´n_esti...
12          0.982145  ...  {´learning_rate´: 0.5, ´max_depth´: 8, ´n_esti...
13          1.000000  ...  {´learning_rate´: 0.5, ´max_depth´: 8, ´n_esti...
14          0.875743  ...  {´learning_rate´: 0.5, ´max_depth´: 8, ´n_esti...
15          0.999820  ...  {´learning_rate´: 0.5, ´max_depth´: 8, ´n_esti...
[16 rows x 4 columns]
>>> # Print the "best accuracy" reached
... print(grid.best_score_)
0.584401215489702
>>> # Print the optimal hyper-parameters
... print(grid.best_estimator_)
GradientBoostingClassifier(ccp_alpha=0.0, criterion=´friedman_mse´, init=None,
                           learning_rate=0.1, loss=´deviance´, max_depth=5,
                           max_features=None, max_leaf_nodes=None,
                           min_impurity_decrease=0.0, min_impurity_split=None,
                           min_samples_leaf=1, min_samples_split=2,
                           min_weight_fraction_leaf=0.0, n_estimators=50,
                           n_iter_no_change=None, presort=´deprecated´,
                           random_state=None, subsample=0.5, tol=0.0001,
                           validation_fraction=0.1, verbose=0,
                           warm_start=False)
```

The fourth part of the code extract from `grid.best_estimator_` the optimal parameters, puts them into four Python variables (opt1, opt2, opt3, opt4), initializes the model at optimal hyper-parameters, and then fits the optimal classifier. Lastly, the code predicts classes over the testing dataset, and calculates a predictive accuracy equal to 0.58, in line with the cross-validation results.

```
―――――――――――――――――――――――――― python (type end to exit) ――――――――――――――――――
>>> # Initialize the model at optimal hyper-parameters
... opt1=grid.best_estimator_.n_estimators
>>> opt2=grid.best_estimator_.max_depth
>>> opt3=grid.best_estimator_.learning_rate
>>> opt4=grid.best_estimator_.subsample
>>> gbc=GradientBoostingClassifier(n_estimators=opt1 ,
...                                max_depth = opt2 ,
...                                                     learning_rate = opt3 ,
...                                                     subsample = opt4, random_state=33)
... # Fit the optimal classifier
>>> gbc.fit(X_train, y_train)
GradientBoostingClassifier(ccp_alpha=0.0, criterion=´friedman_mse´, init=None,
                           learning_rate=0.1, loss=´deviance´, max_depth=5,
                           max_features=None, max_leaf_nodes=None,
                           min_impurity_decrease=0.0, min_impurity_split=None,
                           min_samples_leaf=1, min_samples_split=2,
                           min_weight_fraction_leaf=0.0, n_estimators=50,
                           n_iter_no_change=None, presort=´deprecated´,
                           random_state=33, subsample=0.5, tol=0.0001,
                           validation_fraction=0.1, verbose=0,
                           warm_start=False)
>>> # Predict classes over the testing dataset
... y_pred = gbc.predict(X_test)
>>> # Calculate the predictive accuracy
... print(accuracy_score(y_test, y_pred))
0.5800865800865801
```

The fifth and final part of the code exports the predicted results in Stata. Specifically, we start by putting the predicted probabilities and classes into two dataframes named respectively `pred_probs` and `pred_classes`. We then put the actual classes into another dataframe named `real_classes`, and create a Python array called out to stack by column `pred_probs`, `pred_classes`, and `real_classes`. After generating a dataframe named OUT from the out array, we export it as a Stata dataset called `pred_gbc.dta`. Off the Python shell, we open in Stata the `pred_gbc` dataset, rename the columns properly, and estimate the overall accuracy (using both the training and testing samples) obtaining an accuracy value (0.58) equivalent to previous ones.

```
─────────────────────────────── python (type end to exit) ───────────────
>>> # Export predicted results in Stata
... # Put the predicted probabilites into a dataframe
... pred_probs = pd.DataFrame(gbc.predict_proba(X))
>>> # Put the predicted classes into a dataframe
... pred_classes = pd.DataFrame(gbc.predict(X))
>>> # Put the real classes into a dataframe
... real_classes = pd.DataFrame(y)
>>> # Create an array of probs and classes
... out=np.column_stack((pred_probs,pred_classes,y))
>>> # Generate a dataframe "OUT" from the array "out"
... OUT = pd.DataFrame(out)
>>> # Obtain the results as a Stata dataset
... OUT.to_stata('pred_gbc.dta')
/Library/Frameworks/Python.framework/Versions/3.8/lib/python3.8/site-packages/pandas/
io/stata.py:2303: Inv
> alidColumnName:
Not all pandas column names were valid Stata variable names.
The following replacements have been made:
    0   ->   _0
    1   ->   _1
    2   ->   _2
    3   ->   _3
    4   ->   _4
If this is not what you expect, please make sure you have Stata-compliant
column names in your DataFrame (strings only, max 32 characters, only
alphanumerics and underscores, no Stata reserved words)
  warnings.warn(ws, InvalidColumnName)
>>> end
```

```
──────────────────────────────── Stata code and output ───────────────────
. * Open the dataset of results "pred_gbc"
. use pred_gbc.dta , clear
. * Rename the columns
. rename _0 prob_low
. rename _1 prob_medium
. rename _2 prob_high
. rename _3 pred_class
. rename _4 real_class
. * Estimate the overall accuracy (train + test)
. gen match=(real_class==pred_class)
. sum match

    Variable |        Obs        Mean    Std. Dev.       Min        Max
─────────────┼─────────────────────────────────────────────────────────
       match |      1,848   .7959957    .4030812          0          1
. di "Accuracy rate = " r(mean)
Accuracy rate = .79599567
```

We can also easily make the cross-validation results available into Stata by exporting the `cv_res` dataframe. This may be useful as `cv_res` provides us with the standard errors of the estimated testing accuracy scores. Cross-validation standard errors are essential to gauge the precision of the accuracy estimated for a learner.

5.11 Conclusions

In this chapter we provided theory and applications in R, Stata, and Python of tree-based methods. In the first part of the chapter, we addressed decision trees definition, construction, fitting, and plotting to then show how to optimally prune a tree. We then went on by presenting ensemble methods, specifically bagging, random forests and boosting, thus providing the rationale for their implementation, and showing how to fit and tune them. In the second part of the chapter, we showed how to implement tree-based algorithms on real data using R, Stata, and Python, focusing, in particular, on the functioning of the package CARET for R and on the Scikit-learn's `GridSearchCV()` method of Python for carrying out optimal tuning.

References

Blanquero, R., Carrizosa, E., Molero-Río, C., & Morales, D. R. (2021). Optimal randomized classification trees. *Computers and Operations Research, 132*, 105281.

Breiman, L. (1996). Bagging predictors. *Machine Learning, 24*(2), 123–140.

Breiman, L. (2001). Random Forests. *Machine Learning, 45*(1), 5–32.

Breiman, L., Friedman, J., Stone, C. J., & Olshen, R. A. (1984). *Classification and regression trees* (1st ed.). Boca Raton, FL: Chapman and Hall/CRC.

Bühlmann, P., & Hothorn, T. (2007). Boosting algorithms: regularization, prediction and model fitting. *Statistical Science, 22*(4), 477–505. http://projecteuclid.org/euclid.ss/1207580163.

Cerulli, G. (2022a). C_ML_STATA_CV: Stata module to implement machine learning classification in Stata. *Statistical Software Components*. Publisher: Boston College Department of Economics. https://ideas.repec.org//c/boc/bocode/s459055.html.

Cerulli, G. (2022b). R_ML_STATA_CV: Stata module to implement machine learning regression in Stata. *Statistical Software Components*. Publisher: Boston College Department of Economics. https://ideas.repec.org//c/boc/bocode/s459054.html.

Cerulli, G. (2022c). SCTREE: Stata module to implement classification trees via optimal pruning, bagging, random forests, and boosting methods. *Statistical Software Components*. Publisher: Boston College Department of Economics. https://ideas.repec.org//c/boc/bocode/s458645.html.

Cerulli, G. (2022d). SRTREE: Stata module to implement regression trees via optimal pruning, bagging, random forests, and boosting methods. *Statistical Software Components*. Publisher: Boston College Department of Economics. https://ideas.repec.org//c/boc/bocode/s458646.html.

Davison, A. C., Hinkley, D. V., & Young, G. A. (2003). Recent developments in bootstrap methodology. *Statistical Science, 18*(2), 141–157.

Efron, B. (1987). Better bootstrap confidence intervals. *Journal of the American Statistical Association, 82*(397), 171–185. http://www.jstor.org/stable/2289144.

Efron, B., & Tibshirani, R. (1993). *An introduction to the bootstrap*. Springer.

EML. 2010. Cross-Validation. In C. Sammut, & G. I. Webb (Eds.), *Encyclopedia of Machine Learning* (Vol. 249). Springer. https://doi.org/10.1007/978-0-387-30164-8_190.

Esposito, F., Malerba, D., & Semeraro, G. (1997). A comparative analysis of methods for pruning decision trees. *IEEE Transactions on Pattern Analysis and Machine Intelligence, 19*(5), 476–491.

Frank, E., Wang, Y., Inglis, S., Holmes, G., & I. H. Witten. (1998). Using model trees for classification. *Machine Learning, 32*(1), 63–76. https://doi.org/10.1023/A:1007421302149.

Freund, Y., & Schapire, R. E. (1997). A decision-theoretic generalization of on-line learning and an application to boosting. *Journal of Computer and System Sciences, 55*(1), 119–139. https://doi.org/10.1006/jcss.1997.1504.

Friedman, J., Hastie, T., & Tibshirani, R. (2000). Additive logistic regression: a statistical view of boosting (with discussion). *The Annals of Statistics, 28*(2), 337–374. http://www.jstor.org/stable/2674028.

Friedman, J. H. (2001). Greedy function approximation: A gradient boosting machine. *The Annals of Statistics, 29*(5), 1189–1232.

Geurts, P., Ernst, D., & Wehenkel, L. (2006). Extremely randomized trees. *Machine Learning, 63*(1), 3–42.

Hastie, T., Tibshirani, R., Friedman, J. (2009). *The elements of statistical learning: data mining, inference, and prediction* (2nd ed.). Springer Series in Statistics. New York: Springer. https://www.springer.com/gp/book/9780387848570.

Ho, T. K. (1995). Random decision forests. In *Proceedings of 3rd International Conference on Document Analysis and Recognition* (Vol. 1, pp. 278–282). http://ieeexplore.ieee.org/abstract/document/598994/.

James, G., Witten, D., Hastie, T., & Tibshirani, R. (2013). *An introduction to statistical learning: With applications in R* (1st ed.). New York: Springer.

Maimon, O. Z., & Rokach, L. (2014). *Data mining with decision trees: Theory and applications* (2nd ed.). World Scientific.

Quinlan, J. R. (1986). Induction of decision trees. *Machine Learning, 1*(1), 81–106. https://doi.org/10.1023/A:1022643204877.

Quinlan, J. R. (1990). Decision trees and decision-making. *IEEE Transactions on Systems, Man, and Cybernetics, 20*(2), 339–346. https://doi.org/10.1109/21.52545.

Quinlan, J. R. (1999). Simplifying decision trees. *International Journal of Human-Computer Studies, 51*(2), 497–510. https://doi.org/10.1006/ijhc.1987.0321.

Ridgeway, G. (1999). The state of boosting. *Computing Science and Statistics, 31*, 172–181.

Schonlau, M. (2020). BOOST: Stata module to perform boosted regression. *Statistical Software Components*. Publisher: Boston College Department of Economics. https://ideas.repec.org//c/boc/bocode/s458541.html.

Schonlau, M. (2021). GRIDSEARCH: Stata module to optimize tuning parameter levels with a grid search. *Statistical Software Components*. Publisher: Boston College Department of Economics. https://ideas.repec.org//c/boc/bocode/s458859.html.

Schonlau, M. (2005). Boosted regression (Boosting): An introductory tutorial and a stata plugin. *The Stata Journal, 5*(3), 330–354.

Schonlau, M., & Zou, R. Y. (2020). The random forest algorithm for statistical learning. *The Stata Journal, 20*(1), 3–29.

Sheppard, C. (2017). *Tree-based machine learning algorithms: Decision trees, random forests, and boosting*. CreateSpace Independent Publishing Platform. Google-Books-ID: TBRWtAEACAAJ.

Wang, S., & Chen, Q. (2021). The study of multiple classes boosting classification method based on local similarity. *Algorithms, 14*(2), 37. https://doi.org/10.3390/a14020037.

Zhou, X., & Yan, D. (2019). Model tree pruning. *International Journal Machine Learning and Cybernetics, 10*(12), 3431–3444.

Zou, R. Y., & Schonlau, M. (2022). RFOREST: Stata module to implement Random Forest algorithm. *Statistical Software Components*. Publisher: Boston College Department of Economics. https://ideas.repec.org//c/boc/bocode/s458614.html.

Chapter 6
Artificial Neural Networks

6.1 Introduction

An artificial neural network (ANN) is a simplified representation of a biological neural system constituting the brain of humans and animals.

In human beings and animals, the nervous system is constituted by a large number of neurons. For example, depending on the species, the brain of mammals can contain between 100 million and 100 billion neurons. A neuron is a cell able to receive information in the form of electrical and chemical signals, to then transform and re-transmit this information to other nerve cells.

Figure 6.1 shows a sketched representation of a neuron. By and large, we can see that three elements compose a neuron: cell body, axon, and dendrites. The cell body, the mind of the neuron, contains the nucleus and the cytoplasm. Dendrites propagate from the cell body and receive electrochemical messages (input signals) coming from other neurons. The axon originates from the cell body as well, and can be seen as a wire that, covered by some layers of myelin sheath, transports electrical impulses and ends with a series of nerve terminals (output signals). The myelin sheath allows for accelerating electrical transmission. A synapse, finally, is the point at which the dendrites of a neuron are connected to the nerve terminals of another neuron. Synapses are thus the neural contact points (Mai & Paxinos, 2012).

The first attempt to artificially represent in a simplified mathematical way the structure and functioning of a biological neuron was made by McCulloch and Pitts (1943). Rooted in the graph theory, the McCullock and Pitts model represents a neuron as a logic circuit gate that can be only switched-on or switched-off, thus emitting or not a specific signal. When a combined signal coming from the many dendrites' inputs overcomes a given threshold, the cell body opens up the gate and let information flow through the axon and reach the nerve terminals.

Although tracking a relevant step forward in the ability to model neurons' functioning, the McCulloch and Pitts model lacked the description of a proper neural learning mechanism, making the neurons unable to carry out complex functional

G. Cerulli, *Fundamentals of Supervised Machine Learning*, Statistics and Computing, https://doi.org/10.1007/978-3-031-41337-7_6

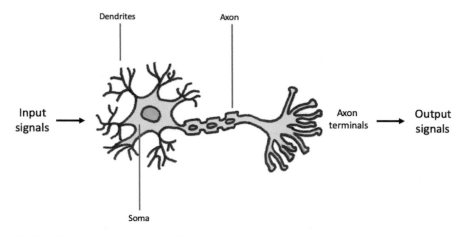

Fig. 6.1 Simplified representation of a human brain's neuron

intelligent tasks such as recognize images and understand languages. More importantly, as a pure mathematical artifact, it also lacked an empirical counterpart.

Some years later, Rosenblatt (1958) proposed for the first time the Perceptron learning rule, an algorithm able to describe how inputs aggregation via neural weights can be used to allow the neuron to decide whether to open or not the gate (*neural firing*) to let information flow. This learning rule, as we will see in more detail later on, automates the optimal updating of the weights, and can be easily fitted to the data for classification purposes (e.g., image recognition).

This model opened up the road to a large literature on neural network learning mechanisms, including deep learning architectures, currently under continuous development (Lörincz, 2011; Yamakawa, 2021). This chapter starts with a more formal definition of what an artificial neural network (ANN) is, as well as how to fit it to the data. Here, we present the popular *gradient descent* approach (based on the *back-propagation* algorithm) to fit an ANN. The chapter goes on by outlining two notable ANN architectures, that is, the Perceptron and the Adaline. The applied part of the chapter, finally, presents some ANN implementations using Python, R, and Stata.

6.2   The ANN Architecture

From a statistical learning perspective an ANN is a multi–stage regression (or classification) model which nests latent variables' construction and linkage within an input–output scheme, where the initial observable inputs are the features, and the final observable outputs are the target variable (in the regression case), or the K binary target variables in a K-class classification setting. In this sense, an ANN can handle multiple outputs, either binary, categorical multi-valued, or numerical (Abiodun et al., 2018, 2019).

An ANN can be represented either as a nested system of equations or, graphically, by a network diagram. In what follows, we will illustrate both representations. We consider the classical ANN architecture called *feedforward neural network*, as signals go unidirectionally, from the inputs to the outputs, and no feedback is allowed (Baptista & Dias, 2013).

For illustrative purposes, consider a K-class classification setting.[1] The aim of an ANN is to build a *mapping* going from the p observable features to the probability of each class $k = 1, \ldots, K$ by passing through a certain number of intermediate steps called *hidden layers*. A hidden layer is an interface that receives inputs and yields outputs that are, in turn, inputs for the subsequent layers. As final outputs, an ANN has K target outcomes $Y_k, k = 1, \ldots, K$ where each Y_k is coded as a 0/1 binary variable. In linking the p inputs to the K outputs, an ANN generates, within each hidden layer, latent features $Z_m, m = 1, \ldots, M$ (neurons) obtained as linear combinations of the p inputs, while every target Y_k is in turn modeled as a function of linear combinations of the Z_m. With only one hidden layer, the ANN mathematical system of equations takes on this form:

$$\begin{cases} Z_m = \sigma(\alpha_{0m} + \alpha_m^T X) \\ T_k = \beta_{0k} + \beta_k^T Z \\ y_k = f_k = g_k(T) \end{cases} \tag{6.1}$$

where $T = (T_1, \ldots, T_K)$, α_m^T, and β_k^T are two row vectors of p coefficients called *weights* in the ANN jargon, with coefficients α_{0m} and β_{0k} known as *biases*. The function $\sigma(\cdot)$ is the *activation* function and it typically takes the form of either the *sigmoid* function $\sigma(t) = 1/(1 + e^t)$, or of the *rectified linear unit (ReLU)* function $max[0, t]$. The function $g_k(\cdot)$ is called g-transformation and generally takes on two forms: (i) the *identity* transformation, $g_k(T) = T$, in a regression setting with only one output $y = T$; and (ii) the *softmax* transformation $g_k(T) = e^{T_k} / \sum_{j=1}^{K} e^{T_k}$, in a classification setting.

The activation function plays the role of a *signal transformer*, in that it takes as input a continuous signal—the linear combination of the features and the weights—and gives back as output a new signal generally ranging in a finite interval. Figure 6.2 shows three activation functions typically used in ANN architectures: the sigmoid, the ReLU, and the hyperbolic tangent. Every function is plotted at a different scale value s that controls for the so-called *activation rate*, where a larger value of s entails stronger activation. The sigmoid function transforms a continuous signal into a new signal ranging in the interval [0,1]. The point zero is the switching threshold where the original signal passes from negative to positive values. A larger activation rate s makes the activation induced by switching from a positive to a negative status stronger. The sigmoid is a popular function used to transform a continuous signal into a probability. The hyperbolic tangent function has a shape similar to the sigmoid,

---

[1] For convenience, notation follows that of Chap. 11 in Hastie et al. (2009).

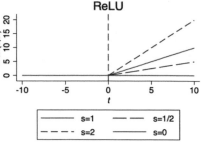

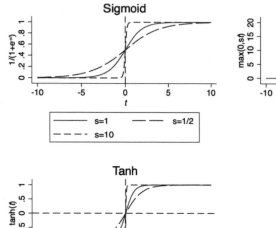

Fig. 6.2 Activation functions. Plot of the *sigmoid* $\sigma(t) = 1/(1 + e^{s \cdot t})$ and *rectified linear unit* *(ReLU)* $max[0, s \cdot t]$ functions. at different scale value s, controlling for the so-called *activation rate*. Observe that a larger value of s entails harder activation

but it allows for the signal to vary between -1 and 1. This means that the sign of the original signal is preserved, but assumes a standardized form. The ReLU, finally, puts to zero any negative value of the original signal, and induces a linear transformation only for the original positive values.

The g-transformation functions are used to convert the final ANN signals into a continuous numerical output (*identity*), or to convert them into class probabilities (*softmax*).

Figure 6.3 illustrates a graphical representation of a feedforward ANN architecture characterized by: (i) two inputs, (ii) one hidden layer, (iii) four hidden units (or neurons), and (iv) two outcomes. Let's explain this architecture. The objects included in the light grey boxes are observable components, while those included in the light blue boxes are unobservable (or latent) components. The process starts from the (observable) inputs X_1 and X_2 whose linear combination generates the latent units Z_1, Z_2, Z_3, and Z_4 through the activation function $\sigma(\cdot)$. Subsequently, linear combinations of the latent units generate the downstream units T_1, and T_2, that are the final outputs in a multi-output regression, or that are transformed into class probabilities via the g-transformation function for a K-class classification setting. The network is fully connected, in that every unit is linked to all the subsequent units through an *edge* (or *link*) by means of a specific weight (the parameters αs and βs defined above).

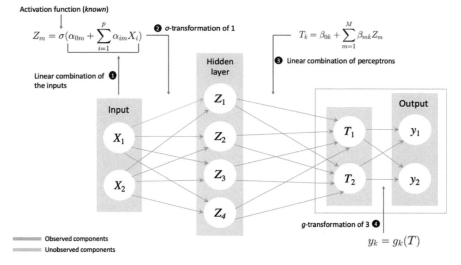

Fig. 6.3 A feedforward artificial neural network architecture characterized by: (i) two inputs, (ii) one hidden layer, (iii) four hidden units (or neurons), and (iv) two outcomes

6.2.1 ANN Imaging Recognition: An Illustrative Example

The description of a feedforward ANN devised in the previous section is useful to comprehend its basic structure. However, understanding more in-depth how an ANN processes information (in the form of *signals*) from the inputs to the outputs is not straightforward to capture, even in a simplified case as the one depicted in Fig. 6.3. In order to appreciate the actual functioning of an ANN, we consider a practical example dealing with a popular artificial intelligence task, that of recognizing an image. How can an ANN recognize an image? By and large, this process takes place in three steps:

- Each neuron specializes in recognizing a specific *portion* of the original image. This portion is contained in the original image which can be seen as a hierarchical structure.
- Each neuron emits a signal on how likely the sub-image is able to represent the original image.
- The ANN recombines these signals in a unique final signal (the output) delivering a conditional probability that the image is (or is not) the *true* original one.

To start with, consider the two simple images, A and B, represented in Fig. 6.4. We would like a machine to recognize whether the specific image supplied as an input is the actual image we have provided. The first step is to represent both images as single observations (or examples) of a dataset. This is the data transformation (or representation) of an image. In this case, the black pixels are conventionally coded by a 1, and white pixels by a 0. Of course, other types of notation can be used.

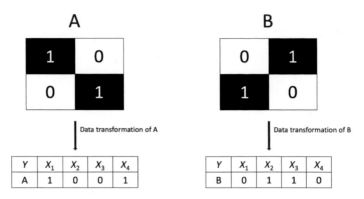

Fig. 6.4 Images A and B and their data representation

Suppose to provide our ANN with image A as an input. How can we expect the ANN to work for correctly recognizing that this is exactly image A and not image B? According to the three steps listed above, Fig. 6.5 illustrates the recognition process, split into a series of sub-tasks: inputting, firing, sub-imaging, sub-imaging likelihood, likelihood aggregation (or pooling), and final prediction.

Let's start by the inputting phase. As we can see, image A is represented by four inputs, X_1, X_2, X_3, and X_4. In the neural firing phase, the weights αs are used to form the linear combinations of the inputs, that is, C_1, C_2, C_3, and C_4. We assume here that each α_{ij} can take only two values: 1, meaning that the neuron is activated, and 0 meaning that the neuron is inactivated. Each neuron is responsible to *see* only a part of the entire image, depending on what input is accessed by the firing weights.

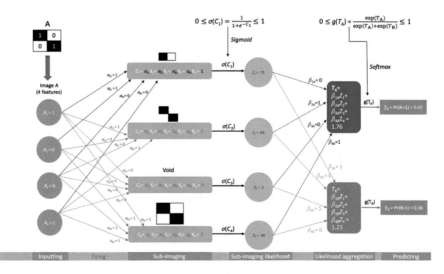

Fig. 6.5 Imaging recognition process carried out by a feedforward ANN

For example, the first neuron is specialized in recognizing only the first two pixels of the image, the second neuron the first and the fourth, and so on. This completes the sub-imaging phase. In the subsequent phase, we generate the sub-imaging likelihood by applying the sigmoid transformation to the linear combinations. In this way, we compute the latent variables Z_1, Z_2, Z_3, and Z_4 that provide a measure of how likely each sub-image represents the initial image. For example, $Z_1 = 0.73$ means that the sub-image perceived by neuron 1 has a 78% probability to represent the initial image A, while $Z_2 = 0.88$ means that the sub-image perceived by neuron 2 has a 88% probability to represent the initial image A, and so on. The likelihood aggregation phase (or pooling phase) aggregates the four likelihoods previously obtained through a new linear combination with weights βs (still taking value either 1 or 0) forming the two final output signals, one expressing how likely is that the image received as input is A, and the other expressing how likely is that the image received as input is B. By means of the softmax function, finally, these signals are transformed into probabilities. In this example, $g(T_A) = 0.64$ and $g(T_B) = 0.36$, thus this neural network correctly identifies that the inputted image is image A.

Looking at this example, one can wonder how the weights have emerged as they are. Indeed, large configurations of the weights might have generated other network configurations leading, possibly, also to mistakenly suggesting that the inputted image was image B. What assures that the weights configuration is such to provide a higher probability to recognize the supplied image?

From an evolutionary perspective, the emergence of the *right* weights (or *correct neural firing*) is pushed by natural selection. To better capture this point, let's elaborate a bit on the previous example. Suppose we are in a jungle, and that image A corresponds to a lion and image B to a sheep. Suppose that two individuals are walking in the jungle and come across a big lion. More, suppose that the first individual is, by chance, endowed with the right weights, while the other with the wrong ones. The man with the right weights is able to correctly recognize that the animal in front of him is a lion and may decide to run away quickly. On the contrary, the man with the wrong weights is unable to recognize that animal as a lion (he *sees*, instead, a sheep with higher probability) thus deciding to stay. In this way, while the first man has a high probability to survive and transmit its (right) weights to his offspring, the second man is very likely to be eaten by the lion, making him unable to transmit his wrong weights to the next generation. A reinforcement process, based on natural selection, thus takes place over the years and leads to select human beings increasingly abler to recognize lions whenever they show up in front of them. Of course, this process is not a linear one, in that any copy of the genes transmitted to the next generation is affected by error; hence, new weights configurations can emerge, compete, and establish over time based on this replication/selection process. Biologists define it as an *adaptive* system, evolving by replication through random perturbations of the weights (or genes), allowing for generating new neurons and synapses (thus, new weights) thus making it possible to improve recognition performance.

6.3   Fitting an ANN

Unlike a natural system, an artificial system as a computing machine cannot rely on a replication/selection mechanism to *calibrate* (i.e., fitting) an ANN. Therefore, how can a machine fit an ANN with the aim of increasing its predictive ability? Conceptually, an ANN is an input/output construct and, as such, it can be thought of as any other Machine Learning *mapping*, although mediated by several latent components.

Fitting (or training) an ANN thus means to find a specific configuration of its parameters, i.e., the weights, with the aim of increasing prediction. How many parameters characterize an ANN? Following the previous illustrative example made of only one hidden layer, the parameters to estimate are $\{\alpha_{0m}, \alpha_m; m = 1, \ldots, M\}$ that amount to $M(p + 1)$, and $\{\beta_{0k}, \beta_k; k = 1, \ldots, K\}$ that amount to $K(M + 1)$. For example, with $p = 3$ features, one hidden layer made of $M = 5$ neurons, and $K = 3$ outputs, an ANN requires to estimate $5(3 + 1) + 3(5 + 1) = 38$ weights.

Fitting an ANN requires first to define a fitting criterion. In a regression setting, with multiple numerical outcomes, we use the usual residual sum of squares (RSS) defined in this case as

$$R(\theta) = \sum_{k=1}^{K} \sum_{i=1}^{N} [y_{ik} - f_k(x_i)]^2 \tag{6.2}$$

where θ is the vector of all the ANN's weights. For classification, when it takes place over K classes, the objective function to minimize is the deviance (or cross-entropy):

$$R(\theta) = - \sum_{k=1}^{K} \sum_{i=1}^{N} y_{ik} \log f_k(x_i) \tag{6.3}$$

which is equal to minus the log-likelihood function of a multinomial model. In this case, $f_k(x_i) = \Pr(y_i = k|x_i)$, and the corresponding classifier is $C(x) = argmax_k f_k(x)$, that is, the Bayes optimal classifier.

As seen, a regression (or classification) ANN is characterized by a large set of parameters. Thus, a global minimizer of $R(\theta)$ may severely overfit the solution. For this reason, a penalized (or regularized) version of the previous objective functions is usually considered. In the case of regression, for example, a Ridge-type regularization can be used. In the literature, this is called *regularization by weight decay* and requires to minimize the following penalized RSS:

$$R(\theta) + \lambda P(\theta) \tag{6.4}$$

where

$$P(\theta) = \sum_{km} \beta_{km}^2 + \sum_{ml} \alpha_{ml}^2 \tag{6.5}$$

with $\lambda \geq 0$ representing the penalization parameter. Larger values of λ induce larger penalization with weights becoming increasingly smaller. The optimal tuning of λ is generally carried out by cross-validation.

6.3.1 The Gradient Descent Approach

To fit an ANN we use a specialized technique called *gradient descent* (or *back-propagation*). Gradient descent is an iterative method that changes the parameter θ in the direction indicated by the gradient.

Consider the RSS as loss function $R(\theta)$, and assume with no loss of generality that θ is a scalar. The minimization of this function over θ requires that its first derivative is zero. In this sense, the optimal estimator $\widehat{\theta}$ solve this equation:

$$g(\widehat{\theta}) = 0 \tag{6.6}$$

where $g(\theta) = \partial R(\theta)/\partial\theta$, the first derivative of the loss function over θ is the gradient. Consider a sequence of estimators $\widehat{\theta}_s, s = 1, \ldots$ indexed by s. A second-order Taylor-series expansion of the loss function around $\widehat{\theta}_s$ is

$$R(\theta) = R(\widehat{\theta}_s) + g(\widehat{\theta}_s)(\theta - \widehat{\theta}_s) + \frac{1}{2}g'(\widehat{\theta}_s)(\theta - \widehat{\theta}_s)^2 \tag{6.7}$$

The first-order condition implies that

$$\partial R(\theta)/\partial\theta = g(\widehat{\theta}_s) + g'(\widehat{\theta}_s)(\theta - \widehat{\theta}_s) = 0 \tag{6.8}$$

By solving for $\widehat{\theta} = \widehat{\theta}_{s+1}$, we obtain

$$\widehat{\theta}_{s+1} = \widehat{\theta}_s - \frac{1}{g'(\widehat{\theta}_s)}g(\widehat{\theta}_s) \tag{6.9}$$

This optimization procedure is known as the Newton-Raphson method. The gradient descent algorithm is obtained from Eq. (6.9) when we put $1/g'(\widehat{\theta}_s) = \gamma_s$, with $\gamma_s > 0$ called *learning rate* often constant at a specific value over each iteration. The iterative gradient descent algorithm thus provides the following updating rule of the parameter:

$$\widehat{\theta}_{s+1} = \widehat{\theta}_s - \gamma_s g(\widehat{\theta}_s) \tag{6.10}$$

Equation (6.10) only requires the computation of the gradient at each iteration. As an example, consider an exponential regression model with only the intercept. For this model, the loss function is $(1/2N)\sum_{i=1}^{N}(y_i - e^{\theta})^2$, and the corresponding gradient is $1/N\sum_{i=1}^{N}(y_i - e^{\theta})e^{\theta} = (\bar{y} - e^{\theta})e^{\theta}$, where $\bar{y} = 1/N\sum_{i=1}^{N}y_i$. In this case, the gradient descent iterative process requires that $\widehat{\theta}_{s+1} = \widehat{\theta}_s + \gamma_s(\bar{y} - e^{\widehat{\theta}_s})e^{\widehat{\theta}_s}$. An

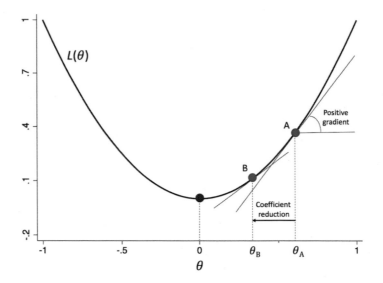

Fig. 6.6 Representation of the gradient descent algorithm

even simpler example is the linear regression with one feature and without the constant term. In this case, the loss function is $(1/2N) \sum_{i=1}^{N}(y_i - \theta x)^2$ with gradient $g(\theta) = -\bar{x} + \bar{x}_2 \theta$, where $\bar{x} = 1/N \sum_{i=1}^{N} x_i$, and $\bar{x}_2 = 1/N \sum_{i=1}^{N} x_i^2$. In this case, the gradient is a linear function of the parameter θ and the parameter updating rule becomes $\widehat{\theta}_{s+1} = \widehat{\theta}_s - \gamma_s (\bar{x}_2 \widehat{\theta}_s - \bar{x})$.

Iteration by iteration, the gradient descent pushes the parameter to move in the opposite direction of the gradient. Figure 6.6 provides a simplified representation of the gradient descent algorithm. Here, the loss function $L(\theta)$ is clearly minimized at $\theta = 0$. Suppose, however, to start from point A. In this point, the tangent to the loss function has a positive slope, and thus the sign of the gradient (which is, in this special case, the slope itself) is positive. This entails a *reduction* of the current θ, passing from θ_A to θ_B. A steeper tangent at point A indicates that the gradient is sizable and that the adjustment will be stronger there than in points like B where the tangent is less steep. This lends support to the intuition that the gradient descent forces a reduced adjustment strength of the parameter in the neighborhood of the minimum.

6.3.2 The Back-Propagation Algorithm

To see how the gradient descent is used to fit an ANN, consider the loss function of equation (6.2) rewritten in this way:

$$R(\theta) = \sum_{i=1}^{N} \sum_{k=1}^{K} [y_{ik} - f_k(x_i)]^2 = \sum_{i=1}^{N} R_i \tag{6.11}$$

The derivatives of this function with respect to the parameters β_{km} and α_{ml} are

$$\frac{\partial R(\theta)}{\partial \beta_{km}} = -2[y_{ik} - f_k(x_i)]g_k'(\beta_k^T z_i)z_{mi} \tag{6.12}$$

and

$$\frac{\partial R(\theta)}{\partial \alpha_{ml}} = -\sum_{k=1}^{K} 2[y_{ik} - f_k(x_i)]g_k'(\beta_k^T z_i)\beta_{km}\sigma'(\alpha_m^T x_i)x_{il} \tag{6.13}$$

Once these derivatives are known, we can implement a gradient descent update that, at the $(r+1)$-st iteration, takes on this form:

$$\beta_{km}^{(r+1)} = \beta_{km}^{(r)} - \gamma_r \sum_{i=1}^{N} \frac{\partial R_i(\theta)}{\partial \beta_{km}^r} \tag{6.14}$$

and

$$\alpha_{ml}^{(r+1)} = \alpha_{ml}^{(r)} - \gamma_r \sum_{i=1}^{N} \frac{\partial R_i(\theta)}{\partial \alpha_{ml}^{(r)}} \tag{6.15}$$

where $\gamma_r > 0$ is the so-called *learning rate*, setting the speed of learning.

Finding by iteration the optimal parameters using Eqs. (6.14) and (6.15) entails a cyclical problem, as computing Eq. (6.14) requires to know the parameters included in Eq. (6.15) and vice versa. Fortunately, it can be proved that these two derivatives are linked one another by an equation, known as *back-propagation equation*. Indeed, one can rewrite:

$$\frac{\partial R_i}{\partial \beta_{km}} = \delta_{ki} z_{mi} \tag{6.16}$$

and

$$\frac{\partial R_i}{\partial \alpha_{ml}} = s_{mi} x_{il} \tag{6.17}$$

where δ_{ki} and s_{mi} are known as *back-propagation errors*. According to their definition, these errors are linked by the following equation:

$$s_{mi} = \sigma'(\alpha_m^T) \sum_{k=1}^{K} \beta_{km} \delta_{ki} \tag{6.18}$$

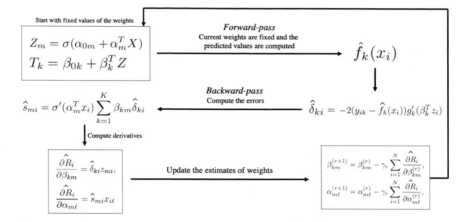

Fig. 6.7 Workflow of the back–propagation algorithm based on gradient descent to fit an ANN

that states a direct relationship between δ_{ki} and s_{mi}. This equation is of primary importance for the back-propagation algorithm to work.

Figure 6.7 illustrates the workflow of the back-propagation algorithm when the gradient descent algorithm is used to fit an ANN. This algorithm is straightforward to describe: (1) we start with initial fixed values of the weights; (2) given the knowledge of the weights, we compute the predicted values of the target variable (*forward-pass*); (3) given weights and predictions, we estimate the back-propagation errors (*backward-pass*), and thus (4) the derivatives; finally, (5) we update the estimates of the weights and re-run the same procedure until convergence.

We can fit an ANN both in an *online* and in an *offline* mode. In the online mode, the learning process takes place sequentially, i.e., observation-by-observation. It means that each update of the weights is obtained using the information coming from a single piece of information (a new instance) until convergence is achieved. In the ANN context, a popular online learning algorithm is the *stochastic gradient descent* (SGD). The SGD follows the same workflow outlined in Fig. 6.7, but it performs it observation-by-observation. This requires data to be put in random order before starting the algorithm, with a reshuffle when the entire dataset has been used.

In offline learning, on the contrary, one produces the best predictor by learning over the entire training dataset at once. It means that each update takes place using, in each iteration, the whole dataset. ANNs are however usually fitted by setting a compromise between pure online and pure offline learning, a method that is called *batch learning*. In batch learning, the initial dataset is randomly split into different batches and trained sequentially offline in every single batch. One complete sweep over the entire dataset is then called *epoch*. For example, with 200 observations and a batch size of 5, one epoch is completed after 40 updates. Hence, if we request the algorithm to have an epoch size of, let's say, 1,000, the entire training process will entail 40,000 batches (or updates).

Online learning (as the one performed by the SGD) is generally faster and is particularly appealing with huge datasets (millions of observations), although it may be more unstable than offline learning. Offline learning (as the one performed by the batch gradient descent) is generally more stable than pure online learning but it is slower, and should be the best choice with datasets of medium size.

When fitting an ANN by batch learning, a too large value of the learning rate γ_r can overshoot the solution, while a too small value can increase the computation time to reach convergence. Usually, setting a value in between 0.01 and 0.001 is an acceptable compromise for this parameter, although it may also be possible to optimize it as a tuning parameter using, for example, cross–validation.

Some variants of the gradient descent algorithms allow for adjusting the learning rate during the training process by decreasing its value at a certain pace over consecutive iterations. This procedure is sometimes called *adaptive learning*, as it generally allows for a better behavior of the algorithm in the proximity of the minimum of the loss function.

In theory, as an ANN is a complex but still parametric model, one might perform a more efficient fitting procedure using second-order learning that exploits the second derivative matrix of $R(\theta)$ (i.e., the Hessian matrix). The calculus of the Hessian may be however prohibitive as the dimension of this matrix may be conceivably very large in many applications.

6.3.3 Specific Issues Arising when Fitting an ANN

Some specific issues generally come up when fitting an ANN to the data. In what follows, we discuss the most relevant ones among those one can come across with in standard applications.

Choosing weights initial values. Typically, the initial value of the weights is set to be random values close to zero, but not exactly zero as this would produce zero derivatives, thus preventing gradient descent to work properly. At the same time, if one starts with too large weights the algorithm can produce poor solutions.

Dealing with overfitting. As we have seen, due to the large number of parameters to fit, ANNs tend to be plagued by high overfitting. A Ridge-type regularization can reduce the overfitting phenomenon, along with a tuning of the number of hidden layers and neurons (see below).

Scaling of the inputs. The type of scaling adopted for the inputs can affect the scaling of the weights, and this can have a remarkable impact on the final ANN predictions. Generally, two types of scaling are used by Machine Learning scholars: (1) *data normalization*, and (2) *data standardization*. Data normalization entails a rescaling of the input x over the original range of the input values, thereby producing a new input ranging from 0 to 1:

$$x_{norm} = \frac{x - \min(x)}{\max(x) - \min(x)} \tag{6.19}$$

Data standardization implies a rescaling of the distribution of the input x, thus producing a new input having mean equal to zero and unit variance:

$$x_{std} = \frac{x - \text{mean}(x)}{\sigma_x} \tag{6.20}$$

Typically, when inputs are standardized, we draw initial weights from random uniform weights within the range [0.7, +0.7].

Choosing the number of hidden units and hidden layers. When computation burden is an issue, ANNs are estimated with a choice of the number of hidden units (i.e., the neurons) generally between 5 and 100. Indeed, the larger the number the inputs, the larger the number of neurons. At present computational power, it is however recommended to tune this parameter via cross-validation, thus finding the *optimal* number of neurons. Similarly, the choice of the number of hidden layers can be carried out by rule-of-thumbs or guided by theoretical insights. Constructing hierarchical structures of latent features, however, poses severe interpretative issues especially in medical and socio-economic applications, where hierarchical learning architectures have a weak correspondence with the initial input variables. On the contrary, in other fields, as in imaging recognition or in natural language processing (NLP), one can reach a clearer interpretation of what adding a new layer entails, as we have outlined in Sect. 6.2.1 with regard to images' decomposition through latent variables' generation. Also the number of hidden layers can be tuned via cross-validation. It is however unclear the benefit brought by cross-validating layers and neurons when one uses cross-validation to estimate the regularization parameter of the ANN-Ridge regression/classification.

Coping with multiple minima. The ANN loss function, both for regression and classification, is *nonconvex* and, as such, it can have more than one local minimum. In cases like this, the solution of the back-propagation algorithm can depend on the initial values of the weights. There are three procedures to cope with this issue: (i) *lowest (penalized) error*: one can try various random starting values for the weights, carry out the fit, and then select the solution that has provided the smallest penalized error; (ii) *prediction averaging*: in this case, we first consider the predictions obtained by fitting the ANN over many configurations of the initial weights, to then averaging them obtaining one final aggregated prediction: (iii) *boostrap aggregation*: in this case, we first generate several ANN predictions by a bootstrap resampling of the original dataset, to then take an average of these predictions to keep as the final prediction.

Addressing computational cost. Fitting an ANN is computationally onerous. The complexity of the fit depends on several parameters, including the number of observations, the number of features, the total number of neurons (included in the various hidden layers considered), and the number of training epochs. A longstanding research has tried over the year to reduce ANN computational cost. More recently, as we will see, the emergence of *convolutional neural networks* (CNNs) is partly justified also by the need to impede the activation of those neurons whose contribution to the ANN goodness-of-fit is poor, thus reducing computational burden.

6.4 Two Notable ANN Architectures

6.4.1 The Perceptron

Proposed by Rosenblatt (1957), the Perceptron is a simple neural network characterized by only one hidden layer, and only one neuron. It was proposed as a way to model the firing process of a single neuron in the brain. Because a single neuron can only be in one of the two states (either *switched-off* or *switched-on*), the outcome can be modeled as a binary variable y taking either value "+1" (switched-on) or "−1" (switched-off).

In a supervised learning binary classification setting, the Perceptron algorithm can be used to classify a new instance into one of two classes, indicated respectively by "−1" and "+1".

Formally, defining a Percepton is rather straightforward. Consider a training instance (i.e., an observation) i over which we observe a (column) vector of p known features $\mathbf{x}_i$, and an outcome $y_i \in [-1; +1]$. Suppose also to have a vector of unknown weights θ. A decision rule to establish whether unit i belongs to class "+1" or to class "−1" can be

$$\text{if } \theta^T \mathbf{x}_i \geq \tau \text{ then } y_i = +1 \tag{6.21}$$

$$\text{if } \theta^T \mathbf{x}_i < \tau \text{ then } y_i = -1 \tag{6.22}$$

where τ is an unknown threshold. We can reparametrize this decision rule, by rescaling the threshold to be equal to zero:

$$\text{if } \theta^T \mathbf{x}_i \geq 0 \text{ then } y_i = +1 \tag{6.23}$$

$$\text{if } \theta^T \mathbf{x}_i < 0 \text{ then } y_i = -1 \tag{6.24}$$

where we redefine $\mathbf{x} = [1, x_1, \cdots, x_p]^T$ and $\theta = [\theta_0 = -\tau, \theta_1, \ldots, \theta_p]^T$, and the model is thus characterized by $p + 1$ features and $p + 1$ weights.

The Perceptron classification rule is represented in Fig. 6.8, where we can easily observe that it is characterized by a sharp jump at threshold zero.

How can a Perceptron learn from data the way to correctly classify observations? This has to do with the Perceptron learning rule, that follows the gradient descent algorithm. Interestingly, the gradient descent provides in this case a straightforward interpretation of the weights' updating mechanism, thus making it intuitive the way in which gradient descent actually works.

In what follows, we first provide a more formal derivation of the Perceptron learning algorithm, to then leave space to its meaning and interpretation.

To begin with, we consider the Perceptron loss function. In an online learning, where the learning process takes place sequentially, i.e., observation-by-observation, the observation index "i" can be replaced by the sequential index s. Therefore, the loss function associated with the s-th observation/iteration takes the following form:

Fig. 6.8 The Perceptron
classification rule

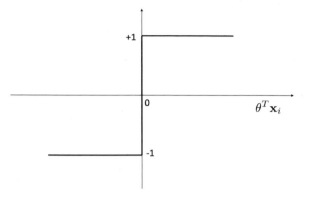

$$L_s(\theta) = \max[0, -y_s \theta^T \mathbf{x}_s] \qquad (6.25)$$

which is a convex function. We can see how Eq. (6.25) quantifies the losses. Consider first the case of correct classification. This occurs when $y_s = +1$ and $\theta^T \mathbf{x}_s \geq 0$, or when $y_s = -1$ and $\theta^T \mathbf{x}_s < 0$, implying that $-y_s \theta^T \mathbf{x}_s < 0$ and a value of the loss equal to zero. Consider now the case of incorrect classification. This occurs when $y_s = +1$ and $\theta^T \mathbf{x}_s < 0$, or $y_s = -1$ and $\theta^T \mathbf{x}_s \geq 0$. In this case, the loss will be positive with a magnitude depending on the size of $\theta^T \mathbf{x}_s$. Hence, in the incorrect classification case, the larger the distance from zero of $\theta^T \mathbf{x}_s$, the larger the value of the loss. This makes the Perceptron's loss increasingly sensitive to increasingly larger misclassifications.

The gradient associated with the loss function (6.25) is equal to

$$\nabla L_s(\theta) = \frac{\partial L_s(\theta)}{\partial \theta} = -y_s \mathbf{x}_s \mathbf{1}[y_s \theta^T \mathbf{x}_s \leq 0] \qquad (6.26)$$

where ∇ indicates the gradient more concisely. An equivalent way to write Eq. (6.26) is

$$\nabla L_s(\theta) = -\frac{1}{2}(y_s - \widehat{y}_s)\mathbf{x}_s \qquad (6.27)$$

with $\widehat{y}_s = \mathbf{1}[\theta^T \mathbf{x}_s \geq 0]$, and the gradient descent updating rule becomes

$$\theta_{s+1} = \theta_s + \frac{1}{2}\gamma(y_s - \widehat{y}_s)\mathbf{x}_s \qquad (6.28)$$

with the learning rate $\gamma > 0$. Equation (6.28) makes the interpretation of the weights' correction induced by the Perceptron easier. Indeed, when there is not misclassification, that is when $y_s = \widehat{y}_s$, then the weights are not updated. On the contrary, when $y_s \neq \widehat{y}_s$, then an update takes place.

Table 6.1 A numerical example of the weights' updating rule for the Perceptron, when we consider only one feature $x = 1$

s	θ_s	x_s	$\widehat{y}_s$	y_s	$(y_s - \widehat{y}_s)x_s$	θ_{s+1}	$\Delta\theta_s$
Misclassification type A							
0	8	1	1	−1	−2	6	−2
1	6	1	1	−1	−2	4	−2
2	4	1	1	−1	−2	2	−2
3	2	1	1	−1	−2	0	
Misclassification type B							
0	−8	1	−1	1	2	−6	2
1	−6	1	−1	1	2	−4	2
2	−4	1	−1	1	2	−2	2
3	−2	1	−1	1	2	0	
Correct classification type A							
0	−8	1	−1	−1	0	−8	0
1	−8	1	−1	−1	0	−8	0
2	−8	1	−1	−1	0	−8	0
3	−8	1	−1	−1	0	−8	
Correct classification type B							
0	8	1	1	1	0	8	0
1	8	1	1	1	0	8	0
2	8	1	1	1	0	8	0
3	8	1	1	1	0	8	

To appreciate how the Perceptron's gradient descent works, Table 6.1 illustrates a simple numerical example. We consider one single feature $x = 1$, implying that $\theta x = \theta$, $\gamma = 1/2$, and an initial weight either equal to $\theta_0 = 8$ or $\theta_0 = -8$. The table considers all the four possibilities one can have when classifying a new instance, that is, $[y, \widehat{y}] \in \{(1, -1), (-1, 1), (-1, -1), (1, 1)\}$. Consider the case $[y, \widehat{y}] = (-1, 1)$ (misclassification type A). The table shows how the weight scales, according to the changes of size $\Delta\theta_s = \theta_{s+1} - \theta_s$. This decrement of the weight is proportional to x, but in this case it is constant and equal to -2, as $x = 1$. If we let the updating procedure to run, we see that the weight is pushed in the right direction, that is, toward zero. This means that the updates are corrections, trying to make prediction increasingly closer to the true y. Figure 6.9 sets out the direction (indicated by the arrows) of the scaling of the weight in the case of a positive starting weight equal to 8 (implying that $\widehat{y} = 1$), and a true outcome $y = -1$. We call this case misspecification type A. We clearly see that the weight is pushed toward zero (the decision boundary),

Fig. 6.9 Weight's updating rule for the Perceptron as illustrated in Table 6.1. *Note* we consider only one feature x. For the sake of simplicity, we consider all instances with $x = 1$, so that $\theta x = \theta$. The initial weight is equal to 8

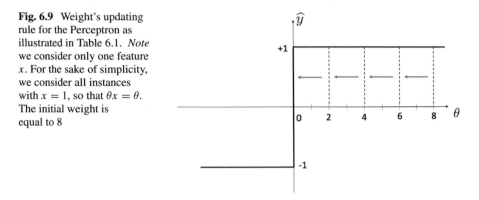

that is, in the direction of becoming incrementally more correct. This is the right direction. A similar phenomenon but with opposite behavior (not reported in the figure) takes place when one starts with negative weight, and a true outcome $y = +1$ (see misclassification type B in Table 6.1). Finally, when classification is correct, there is no changes of the weight (as $\Delta\theta_s = 0$ at every iteration s).

It can be proved that the convergence of the Perceptron's weights to their optimal values takes place only when the two classes are separable by a linear decision boundary, and the learning rate γ is enough small (Raschka & Mirjalili, 2019). In the more general case of non-linearly separable classes, the Perceptron updating process of the weights never goes to an end. In this case, one can fix a maximum number of epochs (i.e., passes through the entire training dataset), or a cut-off over the number of misclassifications that we can tolerate.

Figure 6.10 shows the Perceptron learning architecture. We can immediately observe how the learning process via gradient descent takes place when the error is different from zero. This generates a continuous update of the weights taking place according to the formula (6.28). This architecture can be easily translated into a programming code fitting the Perceptron to the data.

In this direction, Table 6.2 illustrated the pseudo-code for programming the Perceptron fit. In Sect. 6.5.1, we will show how to implement it in Python.

6.4.2   The Adaline

Proposed by Widrow and Hoff (1960), the Adaline (ADAptive LInear NEuron) is another single-layer neural network that can be seen as a Perceptron's companion algorithm.

Indeed, the two network architectures, Perceptron and Adaline, are very similar, except for defining a different loss function. The difference in the loss function produces a different gradient formula, and thus a different updating rule of the weights.

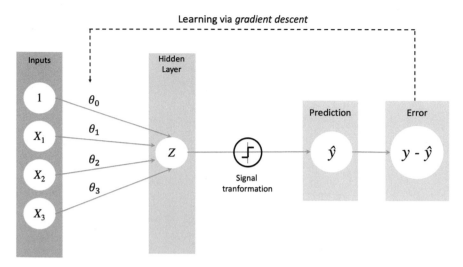

Fig. 6.10 The Perceptron learning architecture. *Note* the learning process via gradient descent takes place when the error is different from zero

Table 6.2 The Perceptron algorithm

Perceptron algorithm	
Initialize $\theta = 0$	$\triangleright$ Initialize the weights
for $h = 1$ to H:	$\triangleright$ Loop over H epochs
for $i = 1$ to N:	$\triangleright$ Loop over N observations
$\widehat{y} = \text{sign}(\theta^T \mathbf{x}_i)$	$\triangleright$ Based on the current weights, make a prediction +1 or -1
$\theta \leftarrow \theta + \frac{1}{2}\gamma(y_i - \widehat{y})\mathbf{x}_i$	$\triangleright$ Update weights if an error occurs
end for	
end for	
return θ	$\triangleright$ Return the final weights

Furthermore, the two algorithms have different activation functions: a *linear* function in the case of the Adaline, and a *unit-step* function in the case of the Perceptron.

The Adaline's loss function is the residual sum of squares (RSS):

$$L(\theta) = \frac{1}{2} \sum_{i=1}^{N} (y_i - \theta^T \mathbf{x}_i)^2 \qquad (6.29)$$

Compared to the Perceptron's loss function, the Adaline loss function (6.29) has two upsides: (i) the use of a continuous linear activation (instead of a unit-step function) makes the loss globally differentiable; (ii) the function is convex, so optimization via gradient descent works smoothly.

For each weight θ_j, the gradient associated with this loss function is the partial derivative of the loss function with respect to θ_j, that is

$$\frac{\partial L(\theta)}{\partial \theta_j} = -\sum_{i=1}^{N}(y_i - \theta^T \mathbf{x}_i)x_{ij} \tag{6.30}$$

and the gradient descent updating learning rule for the weight θ_j is

$$\theta_j := \theta_j + \gamma \sum_{i=1}^{N}(y_i - \theta^T \mathbf{x}_i)x_{ij} \tag{6.31}$$

The Adaline learning updating rule seems identical to the Perceptron one. However, two important differences emerge: (i) the prediction of y, that is $\widehat{y}$, is in the Adaline case a real number, not a integer class label as in the Perceptron case; (ii) when updating the weights, the Adaline uses all observations in the training set, that is, it carries out what is called an *offline learning*. In contract, as we discussed earlier, the Perceptron uses *online learning*, which updates the weights incrementally, observation after observation.

Figure 6.11 sets out the Adaline learning architecture. It is similar to the Perceptron one, except for the elements stressed above. Finally, Table 6.3 illustrates the pseudo-code to implement the Adaline algorithm. Also in this case the similarity with the Perceptron's pseudo-code is evident.

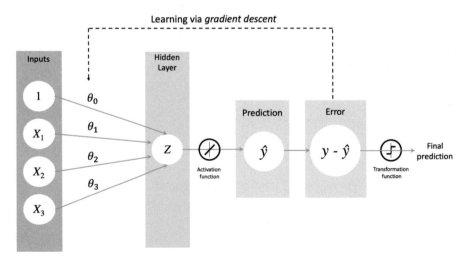

Fig. 6.11 The Adeline learning architecture. *Note* the learning process via gradient descent takes place using a linear prediction. The final continuous predicted outcome is then transformed into an integer label

Table 6.3 The Adaline learning algorithm

Adaline algorithm	
Initialize $\theta = 0$	$\triangleright$ Initialize the weights
for $h = 1$ to H:	$\triangleright$ Loop over H epochs
$\quad \widehat{y} = \mathbf{x}\theta$	$\triangleright$ Based on the current weights, make a continuous prediction
$\quad \theta \leftarrow \theta + \gamma x^T (y - \widehat{y})$	$\triangleright$ Update weights if an error occurs
end for	
return θ	$\triangleright$ Return the final weights
return $\widehat{y}_{\text{adaline}} = \mathbf{1}_+[\widehat{y} \geq 0]$	$\triangleright$ Return the predicted label

6.5 Python Implementation

Python has various dedicated packages to fit ANNs. In what follows, we will run some ANN applications with Scikit-learn and Keras. The latter, in particular, is a wrapper of the powerful Tensorflow platform for deep learning. Before presenting these applications, however, it seems instructive to program the Perceptron algorithm from scratch in Python. This is what the first application carries out.

6.5.1 Application P1: Implementing the Perceptron in Python

Python is an object-oriented programming (OOP) language. As such, one can define in Python specific *classes* characterized by a number of *parameters* and *attributes* as blueprints emanating single *instances*. An instance is an unique object built from a class. A class is initialized by typing at the very beginning of the program `class.<class_name>(object)`.

In what follows, we define the class `Perceptron`.[2] This can be defined by a special function `__init__()` taking a series of parameters (or arguments), where the first argument must be the `self` parameter. Whatever attribute of `self` can be produced by `self.<attribute_name>=value` that assigns a specific value to the attribute.

Our Perceptron classifier is characterized by three parameters:

1. `gamma`: the learning rate (taking values between 0.0 and 1.0). It is a float.
2. `n_pass`: the number of passes through the training dataset (i.e., the number of epochs). It is an integer.
3. `random_state`: a randomly generated seed for initializing the weights and guaranteeing results replicability. It is an integer.

---

[2] This implementation of the Perceptron in Python is based on that of Raschka and Mirjalili (2019), pp. 26–30.

We also consider the following attributes of the Perceptron class:

1. `theta`: the weights. It is a 1d-array.
2. `err`: the number of misclassifications (or updates) occurring in each epoch. It is a list.

Finally, we define a fitting function—`fit(X,y)`—taking as parameters the features and the target, that is

1. `X`: the training feature matrix, with $shape = [n\_obs, n\_features]$, where n_obs is the number of observations and $n\_features$ is the number of features (or predictors). It is an array.
2. `y`: the target variable, with $shape = [n\_obs]$. It is an array.

The core of the Perceptron program is in the body of the function `fit()`, where we basically apply the algorithm illustrated in Table 6.2. After initializing the weights from normal distribution with mean 0 and variance 0.01, we run two nested loops, one over the chosen number of epochs and, within it, one over the observations (*online learning*). The variable `update` stores a value different from zero only when the prediction is incorrect, that is, when an error occurs; therefore, `theta` changes only when `update` is different from zero (observe that `self.theta[0]` accounts for the bias term). Then, we accumulate the errors committed in each epoch into the variable `errors`, to finally store in the attribute `err` the errors made in each epoch.

The `fit()` function calls the `predict()` function, which in turn calls the `index()` function. The latter provides as return the index value $\theta^T x_s$ used to predict either $+1$ or -1 whenever the value of index is respectively larger than zero or smaller than zero (this is, in fact, what is returned by the `predict()`'s return value through `np.where()`).

```
────────────────────────── python (type end to exit) ──────────────────────────
>>> # Import the needed dependencies
... import os
>>> import numpy as np
>>> import pandas as pd
>>> import matplotlib.pyplot as plt
>>>
>>> #-------------------- Perceptron implementation --------------------
... class Perceptron(object):
...
...     def __init__(self, gamma=0.01, n_pass=50, random_state=1):
...         self.gamma = gamma
...         self.n_pass = n_pass
...         self.random_state = random_state
...
...     def fit(self, X, y):
...
...         R = np.random.RandomState(self.random_state)
...         self.theta = R.normal(loc=0.0, scale=0.01, size=1 + X.shape[1])
...         self.err = []
...
...         for _ in range(self.n_pass):
...             errors = 0
...             for x_s, y_s in zip(X, y):
...                 update = self.gamma * (y_s - self.predict(x_s))
...                 self.theta[1:] += 0.5 * update * x_s
...                 self.theta[0] += 0.5 * update
```

```
...            errors += int(update != 0.0)
...        self.err.append(errors)
...     return self
...
...     def index(self, X):
...        """Calculate net input"""
...        return np.dot(X, self.theta[1:]) + self.theta[0]
...
...     def predict(self, X):
...        """Return class label at each iteration s"""
...        return np.where(self.index(X) >= 0.0, 1, -1)
... #---------------------------------------------------------------------
>>>
>>> # Load the Iris dataset
... df = pd.read_csv("iris.data",
...                  header=None,
...                  encoding='utf-8')
...
... # Run the perceptron classifier
>>> ppn = Perceptron(gamma=0.1, n_pass=10 , random_state=1)
>>>
>>> # Select only Iris-setosa ("-1") and Iris-versicolor ("+1")
... y = df.iloc[0:100, 4].values
>>> y = np.where(y == 'Iris-setosa', -1, +1)
>>>
>>> # Extract as features only Sepal-length and Petal Length
... X = df.iloc[0:100, [0, 2]].values
>>>
>>> # Fit the perceptron model
... ppn.fit(X, y)
<__main__.Perceptron object at 0x147807820>
>>>
>>> # Plot the 'Number of updates' as a function of 'Epochs'
... plt.plot(range(1, len(ppn.err) + 1), ppn.err, marker='8', color='red')
[<matplotlib.lines.Line2D object at 0x147173970>]
>>> plt.xlabel('Epochs')
Text(0.5, 0, 'Epochs')
>>> plt.ylabel('Number of updates')
Text(0, 0.5, 'Number of updates')
>>> plt.show()
>>>
>>> # Plot data
... plt.scatter(X[:50, 0], X[:50, 1],
...             color='red', marker='+', label='Iris-setosa')
<matplotlib.collections.PathCollection object at 0x14710aee0>
>>> plt.scatter(X[50:100, 0], X[50:100, 1],
...             color='green', marker='o', label='Iris-versicolor')
<matplotlib.collections.PathCollection object at 0x14732cd00>
>>> plt.xlabel('Sepal-length')
Text(0.5, 47.04444444444444, 'Sepal-length')
>>> plt.ylabel('Petal-length')
Text(85.06944444444443, 0.5, 'Petal-length')
>>> plt.legend(loc='upper left')
<matplotlib.legend.Legend object at 0x14713cf70>
>>> plt.show()
```

The code goes on by fitting the Perceptron program to the Iris dataset, by selecting only two target's categories—Iris-setosa ("-1") and Iris-versicolor ("+1")—and two features—sepal length and petal length. Then, it plots the number of updates (i.e., the errors made) by every epoch, with the number of epochs fixed to 10 beforehand. Observe that the learning parameter gamma is set to 0.1. The plot is visible in Fig. 6.12, where the algorithm reaches perfect prediction (i.e., zero updates) at epoch number

Fig. 6.12 Perceptron
updates as a function of the
epochs

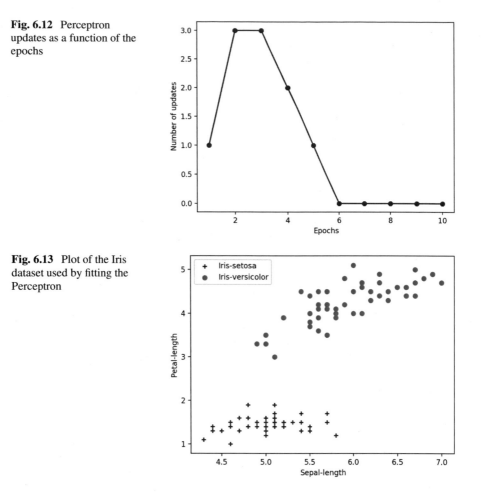

Fig. 6.13 Plot of the Iris
dataset used by fitting the
Perceptron

6. This happens because the two classes are linearly separable as we can clearly see
by plotting the scatter of the data as in Fig. 6.13.

6.5.2   Application P2: Fitting ANNs with Scikit-Learn

In   this   application,   we   use   the   Python   Scikit-learn   function
`MLPClassifier()` to fit a 2-layer ANN, by optimally tuning over the number of
neurons included in each hidden layer. In what follows, we first display the code to
then provide its explanation.

```
──────────────────────────────── python (type end to exit) ────────────────
>>> # IMPORT THE NEEDED PYTHON PACKAGES
... from sklearn.neural_network import MLPClassifier
>>> from sklearn.model_selection import GridSearchCV
>>> from sklearn import datasets, preprocessing
>>> from sklearn.model_selection import train_test_split
>>> import pandas as pd
>>> import numpy as np
>>>
>>> # LOAD THE "iris" DATASET
... iris = datasets.load_iris()
>>>
>>> # DEFINE THE FEATURES (X) AND THE TARGET (y)
... # ("iris.data" and "iris.target" are "arrays")
... X, y = iris.data[:, :4] , iris.target
>>>
>>> # SPLIT THE SAMPLE INTO "TRAINING" AND "TESTING" DATASETS
... X_train, X_test, y_train, y_test = train_test_split(X, y, random_state=33)
>>>
>>> # STANDARDIZE THE FEATURES
... # For the training set
... scaler = preprocessing.StandardScaler().fit(X_train)
>>> X_train = scaler.transform(X_train)
>>> # For the testing set
... scaler = preprocessing.StandardScaler().fit(X_test)
>>> X_test = scaler.transform(X_test)
>>>
>>> # INITIALIZE THE NEURAL NETWORK AT GIVEN PARAMETERS
>>> # NOTE: "hidden_layer_sizes = (m1,m2,m3,...,mK)"
... # defines a K-layer network where:
... # - the first layer contains m1 neurons
... # - the second layer contains m2 neurons
... # - the third layer contains m2 neurons
... # - the last layer contains mK neurons
>>> # SET THE SEED
... R=1010
>>>
>>> # INITIALIZE THE NETWORK
... nn_clf = MLPClassifier(solver='lbfgs',
...                        alpha=1e-5,
...                        hidden_layer_sizes=(5,2,5,6),
...                        random_state=R)
...
... # DEFINE THE PARAMETER VALUES THAT SHOULD BE SEARCHED FOR CROSS-VALIDATION
>>> # WE ONLY CONSIDER A 2-LAYER NETWORK
>>> # CREATE A PARAMETER GRID AS A LIST OF 2D TUPLES FOR "hidden_layer_sizes"
... grid=[]
>>> for i in range(2,10,2):
...   for j in range(2,6,2):
...       g=(i,j)
...       grid.append(g)
>>> print(grid)
[(2, 2), (2, 4), (4, 2), (4, 4), (6, 2), (6, 4), (8, 2), (8, 4)]
>>>
>>> # GENERATE THE GRID (JUST ONE NUMBER) FOR "random_state"
... gridR=[1]
>>>
>>> # PUT THE GENERATED GRIDS INTO A PYTHON DICTIONARY
... param_grid={'hidden_layer_sizes': grid , 'random_state':gridR }
>>> print(param_grid)
{'hidden_layer_sizes': [(2, 2), (2, 4), (4, 2), (4, 4), (6, 2), (6, 4),
 (8, 2), (8, 4)], 'random_stat
> e': [1]}
>>>
>>> # INSTANTIATE THE GRID SEARCH
... grid = GridSearchCV(MLPClassifier(),
...                     param_grid,
```

```
...                          cv=10,
...                          scoring='accuracy',
...                          return_train_score=True)
...
... # FIT THE GRID TO THE TRAINING SET
>>> grid.fit(X_train, y_train)
GridSearchCV(cv=10, error_score=nan,
             estimator=MLPClassifier(activation='relu', alpha=0.0001,
                                     batch_size='auto', beta_1=0.9,
                                     beta_2=0.999, early_stopping=False,
                                     epsilon=1e-08, hidden_layer_sizes=(100,),
                                     learning_rate='constant',
                                     learning_rate_init=0.001, max_fun=15000,
                                     max_iter=200, momentum=0.9,
                                     n_iter_no_change=10,
                                     nesterovs_momentum=True, power_t=0.5,
                                     random_state=None, shuffle=True,
                                     solver='adam', tol=0.0001,
                                     validation_fraction=0.1, verbose=False,
                                     warm_start=False),
             iid='deprecated', n_jobs=None,
             param_grid={'hidden_layer_sizes': [(2, 2), (2, 4), (4, 2), (4, 4),
                                                 (6, 2), (6, 4), (8, 2),
                                                 (8, 4)],
                         'random_state': [1]},
             pre_dispatch='2*n_jobs', refit=True, return_train_score=True,
             scoring='accuracy', verbose=0)
>>>
>>> # VIEW THE RESULTS: ONLY TRAINING VS. TEST ERROR
... pd.DataFrame(grid.cv_results_)[['mean_train_score','mean_test_score']]
   mean_train_score  mean_test_score
0          0.276782         0.276515
1          0.651792         0.652273
2          0.348218         0.348485
3          0.381941         0.383333
4          0.927574         0.927273
5          0.689495         0.687879
6          0.589396         0.587121
7          0.714297         0.715152
>>>
>>> # EXAMINE THE BEST MODEL
... print(grid.best_score_)
0.9272727272727271
>>> print(grid.best_params_)
{'hidden_layer_sizes': (6, 2), 'random_state': 1}
>>> print(grid.best_estimator_)
MLPClassifier(activation='relu', alpha=0.0001, batch_size='auto', beta_1=0.9,
              beta_2=0.999, early_stopping=False, epsilon=1e-08,
              hidden_layer_sizes=(6, 2), learning_rate='constant',
              learning_rate_init=0.001, max_fun=15000, max_iter=200,
              momentum=0.9, n_iter_no_change=10, nesterovs_momentum=True,
              power_t=0.5, random_state=1, shuffle=True, solver='adam',
              tol=0.0001, validation_fraction=0.1, verbose=False,
              warm_start=False)
>>> print(grid.best_index_)
4
>>>
>>> # STORE THE TWO BEST PARAMETERS INTO TWO VARIABLES
... opt_nn=grid.best_params_['hidden_layer_sizes']
>>>
>>> # USING THE BEST PARAMETERS TO MAKE PREDICTIONS
>>> # TRAIN YOUR MODEL USING ALL DATA AND THE BEST KNOWN PARAMETERS
... nn_clf = MLPClassifier(solver='lbfgs',
...                        alpha=1e-5,
...                        hidden_layer_sizes=opt_nn,
...                        random_state=R)
...
```

```
... # FIT THE MODEL TO THE TRAINING SET
>>> nn_clf.fit(X_train, y_train)
MLPClassifier(activation='relu', alpha=1e-05, batch_size='auto', beta_1=0.9,
              beta_2=0.999, early_stopping=False, epsilon=1e-08,
              hidden_layer_sizes=(6, 2), learning_rate='constant',
              learning_rate_init=0.001, max_fun=15000, max_iter=200,
              momentum=0.9, n_iter_no_change=10, nesterovs_momentum=True,
              power_t=0.5, random_state=1010, shuffle=True, solver='lbfgs',
              tol=0.0001, validation_fraction=0.1, verbose=False,
              warm_start=False)
>>>
>>> # MAKE TRAINING OUTCOME PREDICTION
... y_hat_train = nn_clf.predict(X_train)
>>>
>>> # MAKE TRAINING PROBABILITY PREDICTION
... prob_train = nn_clf.predict_proba(X_train)
>>>
>>> # MAKE TESTING OUTCOME PREDICTION
... y_hat_test = nn_clf.predict(X_test)
>>>
>>> # MAKE TESTING "PROBABILITY" PREDICTION
... prob_test = nn_clf.predict_proba(X_test)
>>>
>>> # ESTIMATE TESTING ACCURACY
... test_accuracy = 1- np.mean((y_hat_test != y_test))
>>> print(test_accuracy)
0.631578947368421
```

After importing the needed Python dependencies, we load the `iris` dataset, and define the features (`X`) and the target (`y`) (observe that `iris.data` and `iris.target` are Python arrays).

Then, we split the sample into training and testing datasets, and standardize the features in both datasets, going on by initializing the neural network at given parameters. We have to notice that the argument `hidden_layer_sizes` = (`m1,m2,m3,...,mk`) defines a k-layer network where the first layer contains `m1` neurons, the second layer contains `m2` neurons, the third layer contains `m3` neurons, and so forth. Of course, the last layer contains `mk` neurons.

After setting the seed to allow for results' reproducibility, we initialize the network by defining the parameter values that should be searched for cross-validation. In this application, we only optimally tune the number of neurons contained in each layer for a 2-layer network. In theory, however, one could also cross-validate the number of hidden layers itself. For the sake of conciseness, we do not go through this further step.

For fine-tuning the number of neurons, we create a parameter grid as a list of 2-d tuples to be inserted as arguments into `hidden_layer_sizes()`. In each tuple, the first value refers to the number of neurons in the first layer, and the second value to the number of neurons in the second layer. For this application, we consider a grid made of these eight elements: (2, 2), (2, 4), (4, 2), (4, 4), (6, 2), (6, 4), (8, 2), and (8, 4).

We then proceed by putting the generated grid into a Python dictionary, and instantiate the grid search using the Python function `GridSearchCV()`.

After fitting the grid to the training set, we access the results only by contrasting the training and testing classification error rate. We then store the two best parameters thus obtained into two Python variables, and use them to make (probability and class) training and testing predictions.

Using the testing dataset actual target variable and its optimal prediction, we estimate the testing accuracy that, in this case, is equal to 0.63.

6.5.3   Application P3: Fitting ANNs with `Keras`

Keras is a popular Python deep learning API that runs on top of the Theano and TensorFlow platforms. As a wrapper of these two platforms, Keras is a higher level platform enabling the user to develop, fit, and evaluate neural networks deep learning models in a easy and fast way. Keras is suited for building complex ANN architectures. Building a model in Keras entails the following sequential steps:

1. load the data and split it between features (X) and outcome (y);
2. define a specific Keras model;
3. compile the specified Keras model;
4. fit the model;
5. evaluate model performance;
6. perform predictions.

In Keras, we define a model as a sequence of layers. For this purpose, we first generate a *sequential* model, and then add as many layers as our network architecture requires. Table 6.4 illustrates a template setting out how to develop and fit a model using Keras.

We set off by importing the needed Keras dependencies and in particular the `Sequential` and `Dense` classes. Then, we generate the feature matrix (X), and the output (y) variable. For 1,000 cases, we draw $p = 100$ features from a uniform distribution in the interval [0.0; 1.0). The outcome variable is generated as a one-dimensional array of random integers of values either 0 or 1.

To initialize a model, we use the `sequential()` function. This tells Python that we are setting up a sequential model, that is, an ANN model made of various layers sequentially ordered.

The `add(Dense())` function adds the first fully connected layer and sets up its properties. Here, we indicate that this first hidden layer must contain 32 neurons (or hidden units), and use the ReLu as an activation function. Notice that with the option `input_dim=100`, we define an input layer made of 100 observable units (corresponding to the 100 feature variables). Thus, the input layer is not set up beforehand as a stand-alone layer but it is defined within the command that sets up the first hidden layer.

We finally add a terminal layer made of one single node using the sigmoid as an activation function, since we are working in a (binary) classification setting.

Once the model has been defined, we need to *compile* it. Compiling a model in Keras means specifying some additional properties required for the proper training of the network. For this purpose, we use the function `compile()` where we insert three arguments: (i) `optimizer`: the type of gradient descent algorithm to use for

Table 6.4 Example of a Keras modeling workflow

```
# Import needed dependencies
import numpy as np
from keras.models import Sequential
from keras.layers import Dense

# Form input (X) and output (y) variables
X = np.random.random((1000,100))
y = np.random.randint(2,size=(1000,1))

# Define the network architecture using Keras model
model = Sequential()
model.add(Dense(32, activation='relu', input_dim=100))
model.add(Dense(1, activation='sigmoid'))

# Compile the Keras model
model.compile(optimizer='rmsprop', loss='binary_crossentropy', metrics=['accuracy'])

# Fit the keras model on the data
model.fit(X,y,epochs=10,batch_size=32)

# Evaluate the goodness-of-fit of the model
_, accuracy = model.evaluate(X,y)
print('Accuracy: %.2f' % (accuracy*100))

# Make probability predictions
predictions_prob = model.predict(X)

# Make class predictions
predictions_class = (model.predict(X) ≥ 0.5).astype(int)
```

estimating the weights (in this example, we employ the Root Mean Squared Propagation, or RMSProp); (ii) `loss`: the loss function to evaluate the gradient descent performance (in our example, we use the binary cross-entropy loss as it is well suited for a binary classification problem); (iii) `metrics`: the evaluation metrics we can use to measure the goodness-of-fit (in this case we consider only the classification accuracy, namely, the rate of correct classification matches).

After compiling the model, we can fit it using the function `fit()`, taking as main arguments the arrays of the features and that of the outcome, the number of epochs, and the batch size.

After fitting the model, we can assess its accuracy in the training dataset by using the `evaluate()` function, which returns a list of two values: (i) that of the loss function, and (ii) that of the accuracy. As we are interested only in retrieving the accuracy, when calling this function we type in the left-hand-side the symbol "`_,`", telling Python to ignore the first returned value, namely, that of the loss.

Finally, in the last two steps, we generate both probability and class predictions using the `predict()` function.

```
───────────────────────── python (type end to exit) ─────────────────────────
>>> # Import needed dependencies
... import numpy as np
>>> from keras.models import Sequential
>>> from keras.layers import Dense
>>>
>>> # Form input (X) and output (y) variables
... X = np.random.random((1000,100))
>>> y = np.random.randint(2,size=(1000,1))
>>>
>>> # Define the network architecture using keras model
... model = Sequential()
>>> model.add(Dense(32, activation='relu', input_dim=100))
>>> model.add(Dense(1, activation='sigmoid'))
>>>
>>> # Compile the keras model
... model.compile(optimizer='rmsprop',
...               loss='binary_crossentropy',
...               metrics=['accuracy'])
...
... # Fit the keras model on the data
>>> model.fit(X,y,epochs=10,batch_size=32)
Epoch 1/10
 1/32 [..............................] - ETA: 17s - loss: 0.6903 - accuracy: 0.5938
>
32/32 [==============================] - 1s 738us/step - loss: 0.7027 - accuracy: 0.5150
Epoch 2/10
 1/32 [..............................] - ETA: 0s - loss: 0.7142 - accuracy: 0.4688
>
32/32 [==============================] - 0s 672us/step - loss: 0.6931 - accuracy: 0.5260
Epoch 3/10
 1/32 [..............................] - ETA: 0s - loss: 0.7051 - accuracy: 0.5312
>
32/32 [==============================] - 0s 651us/step - loss: 0.6862 - accuracy: 0.5610
Epoch 4/10
 1/32 [..............................] - ETA: 0s - loss: 0.6639 - accuracy: 0.6250
>
32/32 [==============================] - 0s 633us/step - loss: 0.6808 - accuracy: 0.5690
Epoch 5/10
 1/32 [..............................] - ETA: 0s - loss: 0.7618 - accuracy: 0.4688
>
32/32 [==============================] - 0s 542us/step - loss: 0.6764 - accuracy: 0.5740
Epoch 6/10
 1/32 [..............................] - ETA: 0s - loss: 0.6642 - accuracy: 0.6250
>
32/32 [==============================] - 0s 546us/step - loss: 0.6722 - accuracy: 0.5910
Epoch 7/10
 1/32 [..............................] - ETA: 0s - loss: 0.7166 - accuracy: 0.4688
>
32/32 [==============================] - 0s 537us/step - loss: 0.6674 - accuracy: 0.6000
Epoch 8/10
 1/32 [..............................] - ETA: 0s - loss: 0.7300 - accuracy: 0.4688
>
32/32 [==============================] - 0s 540us/step - loss: 0.6641 - accuracy: 0.5930
Epoch 9/10
 1/32 [..............................] - ETA: 0s - loss: 0.6328 - accuracy: 0.5625
>
32/32 [==============================] - 0s 537us/step - loss: 0.6616 - accuracy: 0.5920
Epoch 10/10
 1/32 [..............................] - ETA: 0s - loss: 0.6286 - accuracy: 0.6562
`
32/32 [==============================] - 0s 538us/step - loss: 0.6579   accuracy: 0.6100
<keras.callbacks.History object at 0x1aaa255b0>
>>>
```

```
>>> # Evaluate the goodness-of-fit of the model
... _, accuracy = model.evaluate(X,y)
 1/32 [............................] - ETA: 2s - loss: 0.6337 - accuracy: 0.6562
>
32/32 [==============================] - 0s 578us/step - loss: 0.6493 - accuracy: 0.6380
>>> print('Accuracy: %.2f' % (accuracy*100))
Accuracy: 63.80
>>>
>>> # Make probability predictions
... predictions_prob = model.predict(X)
>>>
>>> # Make class predictions
... predictions_class = (model.predict(X) > 0.5).astype(int)
```

6.6 R Implementation

There are several packages for fitting ANNs in R. Among the most popular, we have functions as mlp(), neuralnet(), and nnet(). Also, other packages (and related functions), as for example NeuralNetTools, supply complementary analyses, as for example plotting networks' and computing feature-importance measures with the aim of easing ANNs interpretability.

In this section, we consider two applications using respectively the mlp() and neuralnet() functions, as well as some functions of the NeuralNetTools package. In the first application, mainly aimed at familiarizing with fitting an ANN in R, we will fit an ANN to the classical iris dataset, but we will also show how to plot the network and obtain feature-importance measures. In the second application, we will move a little farther by showing also how to optimally tuning an ANN using K-fold cross-validation. This latter is a more general application, partly similar to what we did using the Python function MLPClassifier() although, in this specific case, we will consider a regression, not a classification, setting.

6.6.1 Application R1: Fitting an ANN in R Using the mlp() Function

In this application we fit an ANN in R using the mlp() function from the package RSNNS. The following R code implements this task, providing also two graphical outputs, one concerning a plot of the network, and one related to the feature-importance measures.

––––––––––––––––––––––––––––––– R code and output –––––––––––––––––––––––––––––––

```
# Fitting a neural network in R using the function "mlp()"

# Clean the R environment
> rm(list = ls())
```

```
# Load the needed libraries
> library("NeuralNetTools")
> library("RSNNS")

# Set a random seed
> set.seed(130)

# Load the "iris" dataset
> data(iris)

# Shuffle the observations
> iris <- iris[sample(1:nrow(iris),length(1:nrow(iris))),1:ncol(iris)]

# Set the features (X)
> X <- iris[,1:4]

# Set the target (y)
> y <- decodeClassLabels(iris[,5])

# Split the data into training (85%) and testing (15%) sets
> iris <- splitForTrainingAndTest(X, y, ratio=0.15)

# Standardize training and testing sets
> iris <- normTrainingAndTestSet(iris)

# Fit the data using "mlp()"
> model <- mlp(iris$inputsTrain,
+              iris$targetsTrain,
+              size=c(2,5),
+              learnFuncParams=0.1,
+              maxit=50,
+              inputsTest=iris$inputsTest,
+              targetsTest=iris$targetsTest)

# Plot the network
> x_names <- c("Sepal length","Sepal width","Petal length","Petal width")
> y_names <- c("Setosa","Versicolor","Virginica")

> plotnet(model,
+         x_names=x_names,
+         y_names = y_names,
+         circle_cex=5,
+         rel_rsc = c(2, 2),
+         pos_col="black",
+         neg_col="gray")

# Produce training and testing probability predictions
> test_pred <- predict(model,iris$inputsTest)
> train_pred <- fitted.values(model)

# Generate the training confusion matrix
> A <-confusionMatrix(iris$targetsTrain,train_pred)

# Compute the training accuracy
> train_acc = (A[1,1]+A[2,2]+A[3,3])/sum(A)*100
> print(train_acc)
[1] 78.74016

# Generate the testing confusion matrix
> confusionMatrix(iris$targetsTest,test_pred)
      predictions
targets 1 2 3
      1 4 0 0
      2 0 4 6
      3 0 0 9

# Compute testing accuracy
```

```
> B <- confusionMatrix(iris$targetsTest,test_pred)
> test_acc = (B[1,1]+B[2,2]+B[3,3])/sum(B)*100
> print(test_acc)
[1] 73.91304

# Compute feature-importance measures for output "setosa"
> olden(model, out_var="Output_setosa", x_lab=x_names, bar_plot=FALSE)

                 importance
Sepal length   -13.19704
Sepal width     17.16205
Petal length   -30.83266
Petal width    -29.62323

# Graph feature-importance measures for output "setosa"
> olden(model, out_var="Output_setosa", x_lab=x_names, bar_plot=TRUE)

# END
```

The code starts by cleaning the R environment, loading the needed libraries and the `iris` dataset. We then shuffle the observations, and set a random seed to allow us to reproduce the same results. After forming the matrices of the features (X), and of the output (y), we split the data into a training set (made of the 85% of the total number of observations), and a testing set (made of the remaining 15%). Before carrying out the fit, however, an important step is that of normalizing the data. For this purpose, we use the built-in function `normTrainingAndTestSet()` doing this both for the training and testing sets. We are now ready to run `mlp()`.

As arguments, it is easy to see that `mlp()` firstly takes both the training and testing data (as features and target matrices). The additional arguments are however those that really matter for the fit. Specifically, the argument `size` takes as many values as the number of layers that we want to add on the network. In this case, we set a network structure made of two hidden layers containing respectively 2 and 5 hidden units (or neurons). The option `learnFuncParams` sets the gradient descent learning parameter to 0.1, while we request a maximum number of iterations of the optimization algorithm equal to 50 by properly setting the option `maxit`.

After fitting, we plot the network using the function `plotnet()` from the package `NeuralNetTools`. It is a quite flexible function, as it allows, for example, to give a different color to positive and negative weights. The plot is visible in Fig. 6.14, where the links for positive weights are colored in black, and those for negative weights in gray.

The code goes on by producing training and testing probability predictions, and then by generating the training and testing confusion matrices. From them, we can finally compute the training and testing accuracy, respectively equal to 78.74% and 73.91%.

The very last part of the code provides feature-importance measures based on the approach of Olden et al. (2004). For each outcome and for each input, this approach computes feature-importance by summing the product of the connection weights taking place from the single input node to the single output node. This is somewhat similar to estimating indirect effects in structural equation models. Compared to

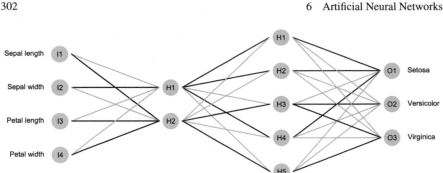

Fig. 6.14 Plot of a 2-layer ANN using the R function `plotnet()` on the Iris dataset

Fig. 6.15 Graph of
feature-importance measures
for the output "Irsis-setosa"
using the Iris dataset.
Method: Olden et al. (2004)

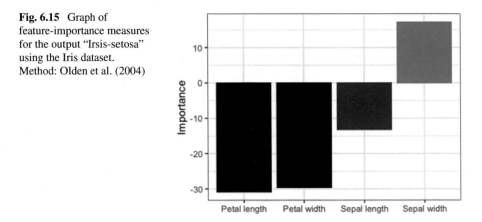

traditional feature-importance measures, generally based on the contribution of each
feature to the reduction of the RSS (regression) or of a measure of misclassification
(classification), this approach has the advantage to provide both a relative size of the
importance and its sign. From an interpretative point of view, this may help to give
results a deeper understanding by singling out what inputs have an overall positive
or negative effect over the probability that a certain output occurs.

In the last part of the code, we obtain the feature-importance measures for the
output "Iris-setosa" using the function `olden()`. We can observe that all the features
have a negative impact on the probability to classify an instance as "Iris-setosa",
except for the sepal width having a positive effect. In terms of magnitude, the largest
negative effect is obtained by the petal length. Figure 6.15 displays these feature-
importance measures using a bar plot.

6.6.2   Application R2: Fitting and Cross-Validating an ANN in R Using the `neuralnet()` Function

In this application, we fit an ANN in R with the function `neuralnet()` and perform optimal tuning via K-fold cross-validation. Also, we compare the testing predicting performance of a linear model with that of an ANN. Below, we display the entire code.

──────────────────────── R code and output ────────────────────────

```
# Fitting an ANN in R with neuralnet() and optimal tuning via K-fold cross-validation

# Clean the environment
> rm(list = ls())

# Load the "Boston" dataset
> set.seed(500)
> library(MASS)
> data <- Boston

# Form the training and test datasets
> index <- sample(1:nrow(data),round(0.75*nrow(data)))
> train <- data[index,]
> test <- data[-index,]

# Fit a linear model (LM) on the training and estimate train-MSE
> lm.fit <- glm(medv~., data=train)

# Obtain LM predictions and compute the test-MSE
> pr.lm <- predict(lm.fit,test)
> MSE.lm <- sum((pr.lm - test$medv)^2)/nrow(test)

# Standardize the data before fitting -> (x-min)/(max-min)
> maxs <- apply(data, 2, max)
> mins <- apply(data, 2, min)
> scaled <- as.data.frame(scale(data, center = mins, scale = maxs - mins))
> train_ <- scaled[index,]
> test_ <- scaled[-index,]

# Load the package "neuralnet"
> library(neuralnet)

# Note that "scale()" returned a matrix to be coerced into a dataframe.
# Form the formula
> n <- names(train_)
> f <- as.formula(paste("medv ~", paste(n[!n %in% "medv"], collapse = " + ")))

# Fit the ANN using the "neuralnet()" function
> nn <- neuralnet(f,
+                 data=train_,
+                 hidden=c(5,3),    # 2-layer ANN with 5 and 3 neurons
+                 linear.output=T) # linear-output = regression
>
# Plot the network
> plot(nn)

# Predicting "medv" using the neural network
> pr.nn <- compute(nn,test_[,1:13])

# Scale back the prediction (to allow for a meaningful comparison with the linear model)
> pr.nn_ <- pr.nn$net.result*(max(data$medv)-min(data$medv))+min(data$medv)
> test.r <- (test_$medv)*(max(data$medv)-min(data$medv))+min(data$medv)
```

```
# Compute the test-MSE of the ANN
> MSE.nn <- sum((test.r - pr.nn_)^2)/nrow(test_)

# Compare "MSE.lm" and "MSE.nn"
> print(paste(MSE.lm,MSE.nn))
[1] "31.2630222372615 16.4595537665717"

# Graph the performance of the ANN and LM on the test set
> par(mfrow=c(1,2))

> plot(test$medv,pr.nn_,col='red',main='Real vs predicted NN',pch=18,cex=0.7)
> abline(0,1,lwd=2)
> legend('bottomright',legend='NN',pch=18,col='red', bty='n')

> plot(test$medv,pr.lm,col='blue',main='Real vs predicted lm',pch=18, cex=0.7)
> abline(0,1,lwd=2)
> legend('bottomright',legend='LM',pch=18,col='blue', bty='n', cex=.95)

# K-fold cross-validation (CV) for the number of neurons in each layer

# Perform CV for the linear model
> library(boot)
> set.seed(200)
> lm.fit <- glm(medv~.,data=data)
> cv.glm(data,lm.fit,K=10)$delta[1]
[1] 23.17094

# Perform CV for the ANN

# Initialize the number of folds and a void scalar for the CV error
> set.seed(450)
> K <- 10
> cv.error <- NULL

# Generate a progress-bar to check the CV running speed and completion
> library(plyr)
> pbar <- create_progress_bar('text')
> pbar$init(K)

# Form the "grid" for optimal grid-search:

# Form the grid for the first hidden-layer
> grid1=c(1,2,3)
> n.elements.grid1 <- length(grid1)

# Form the grid for the second hidden-layer
> grid2=c(1,2,3)
> n.elements.grid2 <- length(grid2)

# Compute the number of elements of the whole grid
> n.elements.grid <- n.elements.grid1*n.elements.grid2

# Initialize an empty matrix "CV" that will contain
# the CV error at each grid point
> CV <- matrix(NA, nrow = 3, ncol = n.elements.grid)

# Initialize a counter to 1
> s <- 1

# Run a 3-level loop to carry out CV:

# For each element of the grid:
# -> loop over the K folds. At each iteration:
#      - generate randomly a testing and a training set
#      - fit "neuralnet()" to the training set to then predict over the testing set
#      - save into "mse" the average of the obtained K errors
# -> Fill the CV matrix with the grid values and the corresponding test-MSE
```

```
> for(n1 in c(1,2,3)){
+ for(n2 in c(1,2,3)){
+       for(i in 1:K){
+               index <- sample(1:nrow(data),round(0.9*nrow(data)))
+               train.cv <- scaled[index,]
+               test.cv <- scaled[-index,]
+               nn <- neuralnet(f,data=train.cv,hidden=c(n1,n2),linear.output=T)
+               pr.nn <- compute(nn,test.cv[,1:13])
+               pr.nn <- pr.nn$net.result*(max(data$medv)-min(data$medv))+min(data$medv)
+               test.cv.r <- (test.cv$medv)*(max(data$medv)-min(data$medv))+min(data$medv)
+               cv.error[i] <- sum((test.cv.r - pr.nn)^2)/nrow(test.cv)
+               pbar$step()
+       }
+ mse <- mean(cv.error)
+ CV[1,s] <- n1
+ CV[2,s] <- n2
+ CV[3,s] <- mse
+ s <- s+1
+ }
+ }

|==============================================================================| 100%

# Extract from CV the minimum test-MSE and corresponding optimal hyper-parameters
> CV[,which.min(CV[3,])]
[1] 3.00000 2.00000 9.38863
>
# Form and visualize a table for the results
> M <- as.matrix(CV[,which.min(CV[3,])])
> colnames(M) <- c("Optimal_CV_reults")
> rownames(M) <- c("Neurons_layer_1:", "Neurons_layer_2:", "CV_test_error:")
> M

                 Optimal_CV_reults
Neurons_layer_1:           3.00000
Neurons_layer_2:           2.00000
CV_test_error:             9.38863
```

After cleaning the R environment, we load the `Boston` dataset and form the training and test sets. As target variable, we consider the variable `medv`, that is, the median value of a house.

To start with, we fit to the training set a linear model (LM) that we will lately compare to our ANN fit. Once predictions of the linear model are computed, we can estimate the training MSE of this model.

Manually, we first standardize the data before fitting, to then load the package `neuralnet` including the `neuralnet()` function. Then, we move on by generating an R formula providing the target and the features, as required by the `neuralnet()` function. We fit a 2-layer ANN with respectively 5 and 3 neurons, where the argument `linear-output = regression` indicates that we are in a regression setting.

After fitting, we can plot the network using the built-in `plot()` function. The plot is visible in Fig. 6.16.

The code goes on by predicting `medv` using the fitted neural network and by scaling back the prediction values to allow for a meaningful comparison with the predictions of the linear model. We then compare the testing MSE of the LM (31.26) and ANN (16.45), clearly showing a better performance of the ANN. In order to visualize the

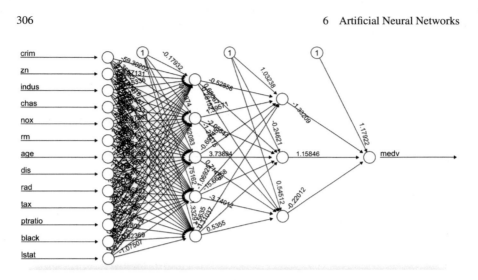

Fig. 6.16 Plot of a 2-layer ANN using the `plot()` function with respectively 5 and 3 neurons. Dataset: `Boston`; target variable: `medv`: the median value of a house. Features 13 house characteristics

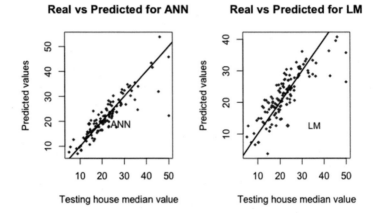

Fig. 6.17 Plot of actual versus predicted target values in the ANN and LM. Dataset `Boston`; target variable: `medv`: the median value of a house. Features: 13 house characteristics

quality of the fit in the two models, we graph the performance of the ANN and LM on the test set. Figure 6.17 provides the result, where we can observe how the actual and predicted values are more aligned for the ANN than for the LM, thus confirming an overall better testing performance of the former compared to the latter.

The code proceeds by providing a K-fold cross-validation (CV) for both models. For the ANN, we use this procedure also for optimal tuning, that is to determine the optimal number of neurons in each layer. To carry-out CV for the LM, we use the built-in function `cv.glm()`, whereas for the ANN we do it manually. In this latter case, we implement the following procedure:

1. Initialize the number of folds (K), a random seed, and a void scalar for the CV error.
2. Generate a progress-bar to check the CV running speed and completion.
3. Form the *grid* for optimal grid-search:

   a. form the grid for the first hidden-layer
   b. form the grid for the second hidden-layer
   c. compute the number of elements of the whole grid

4. Initialize an empty matrix, CV, that will contain the CV error at each grid point.
5. Run a 3-level loop that, for each element of the grid:

   a. loop over the K folds. At each iteration:

   - generate randomly a testing and a training set
   - fit neuralnet() to the training set to then predict over the testing set
   - save into mse the average of the obtained K errors

   b. Fill the matrix CV with the grid values and the corresponding test-MSE.

After this 3-level loop, we extract from CV the minimum test-MSE and corresponding optimal hyper-parameters using the function which.min(), and form and visualize a table for the results. Cross-validation results show that the optimal ANN is characterized by 3 neurons in the first layer and 2 neurons in the second layer. This ANN reaches a CV test error of 9.39, less than half of 23.17 which is the CV test error achieved by the LM. Cross-validation results thus confirm the better out-of-sample prediction performance of an (optimal) neural network compared to a linear regression model.

6.6.3 Application R3: Tuning ANN Parameters Using CARET

Tuning ANN parameters can be easily carried out using the CARET package, as we did in the case of regression and classification trees. In this section, we present two applications fitting and cross-validating an ANN with CARET, the first one using the nnet() function (for a one-layer ANN), and the second one using the mlpML() function (for a multi-layer ANN).

For illustrative purposes, the first application uses the Boston dataset, and aims at predicting again the median value of the houses' price as a function of 13 predictors. The code is listed below.

————————————————————————— R code and output —————————————————————————

```
# Fitting and cross-validating and ANN with "caret" using "nnet"

# Clear all
> rm(list = ls())

# Set the directory
> WD <- "~/Dropbox/Giovanni/Corsi_Stata_by_G.Cerulli/Course_R_ML_Ircres_WS_2021"
> setwd(WD)

# Load "caret"
> library(caret)

# Loading the data
> my_data <- read.csv("boston_tree.csv", sep=",")
> data<-as.data.frame(my_data)
> y <- data[,1]
> x <- data[2:ncol(my_data)]
> data <-cbind(y,x)

# Step 1. Define the "trainControl()" function

# trainControl(method = "cv", number = n, search ="grid"):
#  - method = "cv": the method used to resample the dataset
#  - number = "n": number of folders to create
#  - search = "grid": use the search grid method. For randomized method, use "random"

# Set-up 10 folds to repeat 3 times
> control <- trainControl(method='cv',
+                         number=10,
+                         search="grid")

# Step 2. Define the "train()" function

# train(formula, df, method = "rf", metric= "Accuracy", trControl = trainControl(), tuneGrid
 = NULL):
#  - formula: defines the formula of the algorithm
#  - method: defines which model to train.
#  - metric = "accuracy" -> defines how to select the optimal model
#  - trControl = "trainControl()": defines the control parameters
#  - tuneGrid = NULL": returns a data frame with all the possible parameters' combination

# Fit an ANN with "caret" using "nnet"

# Set the seed
> set.seed(123)

# Tune "size" and "decay" by forming a grid of their values
> tunegrid <- expand.grid(size=c(2,3,5),
+                         decay=c(0.1,0.3,0.5,0.7,0.9))

# Use "train()"
nn <- train(y~.,
            data=data,
            method='nnet',
            metric='RMSE',
            tuneGrid=tunegrid,
            trControl=control)

# Visualize the results
> print(nn)
Neural Network

506 samples
 13 predictor
```

```
No pre-processing
Resampling: Cross-Validated (10 fold)
Summary of sample sizes: 454, 457, 456, 455, 455, 455, ...
Resampling results across tuning parameters:

  size  decay  RMSE       Rsquared    MAE
  2     0.1    23.40227   0.06037253  21.53071
  2     0.3    23.40232   0.04895328  21.53077
  2     0.5    23.40237   0.07183289  21.53082
  2     0.7    23.40243   0.06132116  21.53088
  2     0.9    23.40246   0.10100391  21.53092
  3     0.1    23.40226   0.07536680  21.53070
  3     0.3    23.40230   0.04286069  21.53074
  3     0.5    23.40234   0.09174852  21.53079
  3     0.7    23.40238   0.05313978  21.53083
  3     0.9    23.40241   0.06768669  21.53087
  5     0.1    23.40225   0.03396521  21.53069
  5     0.3    23.40228   0.07247778  21.53072
  5     0.5    23.40231   0.03273151  21.53075
  5     0.7    23.40234   0.07371509  21.53078
  5     0.9    23.40236   0.03150439  21.53081

RMSE was used to select the optimal model using the smallest value.
The final values used for the model were size = 5 and decay = 0.1.
```

After clearing the R environment, setting the current directory, and loading the data, we proceed in two steps:

- Step 1. We define the CARET's `trainControl()` with the following arguments:

    - `method = "cv"`, indicating the method used to resample the dataset;
    - `number = "10"`, setting the number of cross-validation folds to 10;
    - `search = "grid"`: setting an exhaustive grid-search. For randomized grid-search, one can use `search = "random"`

- Step 2. We define the CARET's `train()` function with the following arguments:

    - `formula`, defining the model's specification;
    - `data = data`, setting the dataset to use;
    - `method = "nnet"`, defining which R model/function to train;
    - `metric = "accuracy"`, setting how to select the optimal model;
    - `trControl=control`, defining the control parameters, based on step 1;
    - `tuneGrid=tunegrid`: returning a data frame with all the possible parameters' combination.

We choose to tune two parameters: `size`, the number of neurons in the single layer, and `decay`, the parameter for weight decay, by forming a grid of their values via the function `expand.grid()`. Then, we train the model with these settings, and visualize the results using `print(nn)`. The main result is a table, listing each element of the size/decay grid with the related RMSE (root mean squared error), R-squared, and MAE (*mean absolute error*, representing the average of the absolute difference between the actual and predicted values). The algorithm uses the RMSE to select the

optimal model using the smallest value available in the grid. We see that the final
values used for the model were size = 5 and decay = 0.1.

In our second application, we use the auto dataset, and our aim is that of predicting
the price of a car based on a series of characteristics of the car, including whether
the maker is foreign or domestic, the number of repairs, the weight, etc. We thus
fit and cross-validate an ANN with CARET using the mlpML() (multi-layer ANN)
function. The whole code is listed below.

──────────────────────────────── R code and output ────────────────────────────────

```
# Fitting and cross-validating and ANN with "caret" using "mlpML"

# Clear all
> rm(list = ls())

# Set the directory
> WD <- "~/Dropbox/Giovanni/Book_Stata_ML_GCerulli/neuralnet/sjlogs"
> setwd(WD)

# Load "caret"
> library(caret)
> library(RSNNS)

# Loading the data
> my_data <- read.csv("auto.csv", sep=",")
> data<-as.data.frame(my_data)

# Form the training and test datasets
> set.seed(1)
> index <- sample(1:nrow(data),round(0.75*nrow(data)))
> data.train <- data[index,]
> data.test  <- data[-index,]

# Form "y_train" and "x_train"
> y_train <- data.train[,1]
> x_train <- data.train[2:ncol(my_data)]

# Form "y_test" and "x_test"
> y_test <- data.test[,1]
> x_test <- data.test[2:ncol(my_data)]

# Cross-validation using "caret"

# Generate a grid by setting the neurons per layer (3-layer assumed)
> mlp_grid = expand.grid(layer1 = 1:3,
+                        layer2 = 1:3,
+                        layer3 = 1:3)

# Define the "trainControl()" function
> ctrl <- trainControl(method = "cv",
+                      number = 5,
+                      verboseIter = TRUE,
+                      returnData = FALSE)

# Fit "mlpML" over the grid
> mlpCV = caret::train(x_train,
+                      y_train,
+                      method='mlpML',
+                      metric='RMSE',
+                      preProc = c('center', 'scale'),
+                      trControl = ctrl,
+                      tuneGrid = mlp_grid)
```

```
# Visualize the results
> print(mlpCV)

Multi-Layer Perceptron, with multiple layers

Pre-processing: centered (10), scaled (10)
Resampling: Cross-Validated (5 fold)
Summary of sample sizes: 42, 41, 44, 41, 40
Resampling results across tuning parameters:

  layer1  layer2  layer3  RMSE      Rsquared        MAE
  1       1       1       4515.184  1.336170e-02  3746.948
  1       1       2       4469.778  4.732603e-03  3653.499
  1       1       3       3014.296           NaN  2118.805
  1       2       1       4513.617  3.089852e-01  3760.659
  1       2       2       2967.120           NaN  2062.168
  1       2       3       3344.963           NaN  2441.369
  1       3       1       5279.063  2.447471e-01  4620.403
  1       3       2       5320.203  3.850175e-01  4514.113
  1       3       3       3072.491           NaN  2151.437
  2       1       1       4866.706  5.003747e-02  4419.658
  2       1       2       3068.257           NaN  2168.174
  2       1       3       4612.550  4.304445e-05  3936.388
  2       2       1       6128.277  2.553142e-01  5367.533
  2       2       2       3033.382           NaN  2101.801
  2       2       3       3872.858  3.449612e-01  2960.160
  2       3       1       3844.116  8.759764e-02  3059.643
  2       3       2       3730.556  5.316001e-02  2830.132
  2       3       3       3096.829           NaN  2218.817
  3       1       1       3771.944  2.917156e-01  2889.239
  3       1       2       3091.263           NaN  2082.456
  3       1       3       2984.989           NaN  2405.394
  3       2       1       4859.372  6.843609e-03  4373.763
  3       2       2       3982.751  2.701806e-01  3417.230
  3       2       3       2905.630           NaN  2088.706
  3       3       1       4687.152  2.811013e-01  4045.408
  3       3       2       4665.685           NaN  3861.416
  3       3       3       3115.733           NaN  2221.033

RMSE was used to select the optimal model using the smallest value.
The final values used for the model were layer1 = 3, layer2 = 2
 and layer3 = 3.

# Obtain the values of the best parameters
> mlpCV$bestTune
   layer1 layer2 layer3
24      3      2      3

# Obtain the minimum RMSE
> min(mlpCV$results$RMSE)
[1] 2905.63

# Obtain the minimum MAE
> min(mlpCV$results$MAE)
[1] 2062.168
```

We set out by loading the `auto` dataset and forming the training and testing datasets. We then generate a grid by setting the neurons for each layer in a 3-layer ANN, using again the `expand.grid()` function. For the sake of simplicity, we form the grid by assuming at most three neurons per layer. As a consequence, our grid is made of $3 \times 3 \times 3 = 27$ elements. Then, as required by CARET, we define the `trainControl()` function, and fit `mlpML()` over the grid to the training data. Finally, we can visualize the cross-validation results by typing `print(mlpCV)`. Using the RMSE to select the optimal model, we find that the optimal number of neurons by layer is *layer 1* = 3, *layer 2* = 2, and *layer 3* = 3. At these values of the parameters, the corresponding minimum RMSE is equal to 2905.63, and the minimum MAE is equal to 2062.17.

6.7 Stata Implementation

Stata has no built-in commands to fit ANNs. However, in his GitHub page (`https://nbalov.github.io`), Nikolay H. Balov has provided a non-official Stata command, `mlp2`, to fit a multi-layer Perceptron with 2 hidden layers. In what follows, we present three examples of the use of `mlp2` as provided by the author in his webpage. The first application aims at predicting the health insurance type chosen by some individuals; the second application predicts the Titanic disaster survival probability; and the third, finally, deals with handwritten numerals recognition.

6.7.1 Application S1: Fitting an ANN in Stata Using the `mlp2` Command

In this application, we consider the dataset `sysdsn1` containing health insurance information about 616 psychologically depressed individuals in the United States. We aim at predicting the type of insurance taken out by these subjects based on a series of demographic characteristics. The insurance is categorized as follows:

- *indemnity plan*: a regular fee-for-service insurance;
- *prepaid plan*: a fixed up-front payment allowing unlimited use;
- *no insurance*: no presence of subscribed insurance.

We consider a series of demographic factors that are likely to predict individual insurance choice, and in particular, `age` (the age on the subject), `male` (whether or not male), `nonwhite` (whether he/she is not white), and `site` (site of study). The code and related outputs are listed below.

```
───────────────────── Stata code and output ─────────────────────
. * Insurance type prediction using "mlp2"
.
. * Load the dataset
. webuse sysdsn1 , clear
(Health insurance data)
.
. * Describe the dataset
. describe
Contains data from https://www.stata-press.com/data/r17/sysdsn1.dta
 Observations:            644                    Health insurance data
```

```
      Variables:            13              28 Mar 2020 13:10
```

Variable name	Storage type	Display format	Value label	Variable label
patid	float	%9.0g		ID
noinsur0	byte	%8.0g		No insurance at baseline
noinsur1	byte	%8.0g		No insurance at year 1
noinsur2	byte	%8.0g		No insurance at year 2
age	float	%10.0g		NEMC (ISCNRD-IBIRTHD)/365.25
male	byte	%8.0g		NEMC patient male
ppd0	byte	%8.0g		Prepaid at baseline
ppd1	byte	%8.0g		Prepaid at year 1
ppd2	byte	%8.0g		Prepaid at year 2
nonwhite	byte	%9.0g		Nonwhite
ppd	byte	%8.0g		
insure	byte	%9.0g	insure	Type of insurance
site	byte	%9.0g		Site of study

```
Sorted by: patid

.
. * Set the seed
. set seed 1010

.
. * Tabulate the outcome-variable "insure" (made of 3 categories)
. tab insure , mis
```

Type of insurance	Freq.	Percent	Cum.
Indemnity	294	45.65	45.65
Prepaid	277	43.01	88.66
Uninsure	45	6.99	95.65
.	28	4.35	100.00
Total	644	100.00	

```
.
. * Fit the neural network by "mlp2 fit"
. mlp2 fit insure age male nonwhite i.site, layer1(100) layer2(100) lrate(0.05)
(28 missing values generated)
```

Multilayer perceptron		input variables =	4
		layer1 neurons =	100
		layer2 neurons =	100
Loss: softmax		output levels =	3
Optimizer: sgd		batch size =	50
		max epochs =	100
		loss tolerance =	.0001
		learning rate =	.05
Training ended:		epochs =	26
		start loss =	1.05443
		end loss =	.84056

```
..
. * Display the return objects
. ereturn list

matrices:
             e(gamma0) :  1 x 3
              e(beta0) :  1 x 100
             e(alpha0) :  1 x 100
              e(gamma) :  100 x 3
               e(beta) :  100 x 100
              e(alpha) :  6 x 100

.
. * Predict the "insurance probabilities" by "mlp2 predict"
```

```
. mlp2 predict, genvar(ypred) truelabel(insure)
(28 missing values generated)
Prediction accuracy:  .5967479674796748

.
. * Generate the "predicted values" ("ysim") by "mlp2 simulate"
. mlp2 simulate, genvar(ysim)

.
. * Tabulate true-y-values vspredicted y-values
. tab ysim insure , nolabel mis
```

		Type of insurance			
ysim	1	2	3	.	Total
1	144	106	20	12	282
2	122	150	20	12	304
3	27	21	5	4	57
.	1	0	0	0	1
Total	294	277	45	28	644

```
.
. * Generate the matched prediction indicator
. generate match = (ysim == insure) if insure!=.
(28 missing values generated)
. tab match
```

match	Freq.	Percent	Cum.
0	317	51.46	51.46
1	299	48.54	100.00
Total	616	100.00	

```
.
. * Calculate the "prediction accuracy"
. summarize match
```

Variable	Obs	Mean	Std. dev.	Min	Max
match	616	.4853896	.5001927	0	1

After loading the dataset and describing it, we use `mlp2 fit` to fit a 2-layer ANN containing 100 neurons in each layer, with a learning rate equal to 0.05. The results' table shows the main information about the output we obtained. The estimated multi-layer Perceptron contains 4 input variables, the hidden layers 1 and 2 both contain 100 neurons. By default, the batch size is set to 50, and the number of epochs carried out for the stochastic gradient descent algorithm to converge is 49. We started with a loss of 1.08287, and we ended up with loss of .835013. By typing `ereturn list`, we can access the main command's outputs, that is, the matrices of weights.

Subsequently, using the `mlp2 predict` command, we obtain the predicted insurance-type probabilities (`ypred`), and then generate the *simulated* predicted classes `ysim` (i.e., the predicted types of insurance) using `mlp2 simulate`. Finally, given the previous computations, we generate the matched prediction indicator variable, `match`, and calculate, by summarizing it, a training prediction accuracy equal to .485.

6.7.2 Application S2: Comparing an ANN and a Logistic Model

The task of this application is fitting and comparing two models for predicting Titanic's passenger survival: a logistic regression model and a two-layer neural network. We use the dataset `titanic_train.dta`, where our binary target variable is `survived` (taking value 1 if surviving the disaster and 0 otherwise), and the predictors are `sex`, `age`, `pclass` (ticket class), `sibsp` (number of siblings/spouses), `parch` (number of parents/children), and `fare` (passenger fare). The training dataset includes a total of 891 passengers, but the number of complete records is 741 due to missing values. The code and related outputs are listed below.

─────────────────────────────── Stata code and output ───────────────────────────────

```
. * Comparing surviving titanic disaster probability

. * Clear the environment
. clear all

. * Load the dataset "titanic_train.csv"
. cd "~/Dropbox/Giovanni/Corsi_Stata_by_G.Cerulli/Course_Stata_Machine_Learning/My_ML_Course/Neural_Net
> works/Labs_Neural_Network"
/Users/giocer/Dropbox/Giovanni/Corsi_Stata_by_G.Cerulli/Course_Stata_Machine_Learning/My_ML_Course/Neur
> al_Networks/Labs_Neural_Network

. import delimited titanic_train.csv
(encoding automatically selected: ISO-8859-2)
(12 vars, 891 obs)

. * Encode the string variable sex into a binary variable, "bsex"
. egen bsex = group(sex)

. * Validation setting

. * - Two-thirds of observations for training and one-third for validation.
. * - Compare the prediction performance of the two models.

. * Fit a logistic regression model on the training dataset with the "logit" command
. logit survived bsex age pclass sibsp parch fare in 1/600, or

Iteration 0:    log likelihood = -321.42313
Iteration 1:    log likelihood = -221.62832
Iteration 2:    log likelihood = -220.06392
Iteration 3:    log likelihood = -220.05786
Iteration 4:    log likelihood = -220.05786

Logistic regression                             Number of obs  =      474
                                                LR chi2(6)     =   202.73
                                                Prob > chi2    =   0.0000
Log likelihood = -220.05786                     Pseudo R2      =   0.3154
```

survived	Odds ratio	Std. err.	z	P>|z|	[95% conf. interval]	
bsex	.0738699	.0188864	-10.19	0.000	.0447549	.1219256
age	.9633228	.0094899	-3.79	0.000	.9449014	.9821034
pclass	.3360143	.0675789	-5.42	0.000	.2265505	.4983685
sibsp	.6957472	.1035633	-2.44	0.015	.519695	.9314389
parch	1.029911	.1696705	0.18	0.858	.7457102	1.422424
fare	.9984295	.0032647	-0.48	0.631	.9920513	1.004849
_cons	1872.28	1634.887	8.63	0.000	338.1398	10366.82

```
Note: _cons estimates baseline odds.

.
```

```
. * Estimate logistic predictive accuracy

. * - make predictions based on the fitted model
. * - predict performance based on observations from 601 to 891 (testing dataset)
. * - use prediction accuracy as a the performance criterion

. predict logitpred
(option pr assumed; Pr(survived))
(177 missing values generated)
. replace logitpred = floor(0.5+logitpred)
(714 real changes made)
. generate match = (logitpred == survived)
. summarize match in 601/891
```

Variable	Obs	Mean	Std. dev.	Min	Max
match	291	.6597938	.4745945	0	1

```
. * Neural network regression model

. * - specify the model using the "mlp2" command: two hidden-layers with 10 neurons each
. * - put option "lrate(0.5)" to increase the default learning rate of the optimizer

. . set seed 12345
. . mlp2 fit survived bsex age pclass sibsp parch fare in 1/600, layer1(10) layer2(10) lrate(0.5)
```

Multilayer perceptron	input variables	=	6
	layer1 neurons	=	10
	layer2 neurons	=	10
Loss: softmax	output levels	=	2
Optimizer: sgd	batch size	=	50
	max epochs	=	100
	loss tolerance	=	.0001
	learning rate	=	.5
Training ended:	epochs	=	100
	start loss	=	.579942
	end loss	=	.355872

```
. * Validation: use the last third of the dataset to validate the prediction performance of the model

. mlp2 predict in 601/891, genvar(ypred)
Prediction accuracy: .8666666666666667

. * Calculate "by hand" the prediction accuracy
. gen pred_label=(ypred_1>ypred_0) in 601/891 if ypred_1!=. & ypred_0!=.
(651 missing values generated)
. generate match2 = (pred_label == survived) in 601/891 if ypred_1!=. & ypred_0!=. & survived!=.
(651 missing values generated)
. summarize match2 in 601/891
```

Variable	Obs	Mean	Std. dev.	Min	Max
match2	240	.8666667	.3406451	0	1

After loading the entire dataset, we choose to use about two-thirds (i.e., the first 600 units) of the dataset for training and the remaining one-third for validation. Then, we compare the prediction performance of the two models, the logistic regression and the 2-layer ANN. We start by fitting our logistic regression model using the logit command. Results on the odd ratios show that age, having a sibling or a spouse, traveling a lower class, and being a man, all negatively affect survival chances. On the other hand, having a child or a parent increases survival chances. Sex and ticket class seem to have the highest impact.

We go on by computing logistic predictive accuracy based on out-of-sample observations from 601 to 891. The result shows a validation accuracy of the logistic regression model of about 66%. When it comes to the neural network regression model, we decide to apply the `mlp2` command by letting the 2 hidden layers have 10 neurons each. Also, we insert the option `lrate(0.5)` to increase the default learning rate of the stochastic gradient descent (SGD) optimizer. We can observe that after 100 steps, the optimization loss has decreased from about 0.58 to 0.36.

For validation, we still use the same out-of-sample observations employed for the logistic regression, and find a prediction accuracy of around 87%, largely higher than that of the logistic model. This result stresses the higher predictive power of an ANN over a simple logistic model, a performance that is 32% larger. However, the two models are not identical in terms of complexity, as the considered neural network has 202 parameters, while the logistic regression has only 7 parameters. This induces a larger computational burden in the case of the ANN. Notice, finally, that the neural network interpretative power is weaker compared to the logistic regression, as an ANN tells us very little in terms of predictor importance and statistical significance of the parameters.

6.7.3   Application S3: Recognizing Handwritten Numerals Using `mlp2`

In this application, that draws on Balov's examples of the use of `mlp2` (see, again, the GitHub website of the author), we use `mlp2` for fitting a neural network for carrying out handwritten numerals pattern recognition. We use again the `msist` ("modified national institute of standards and technology") database. Below is the code we run.

```
────────────────────────────────────── Stata code and output ──────────
. * Fitting an ANN for handwritten numerals recognition using "mlp2"
.
. *  Load the training dataset
. sysuse mnist-train.dta , clear
(Training MNIST: 28x28 images with stacked pixel values v* in [0,1] range)
.
. * Visualize individual images
.
. * Write the program "draw_test_digit"
. cap prog drop draw_test_digit
. program draw_test_digit
  1. args ndigit
  2. sysuse mnist-train, clear
  3. quietly drop y
  4. quietly keep if _n==`ndigit´
  5. cap quietly gen y = 1
  6. quietly reshape long v, i(y) j(x)
  7. quietly replace y = 28-floor((x-1)/28)
  8. quietly replace x = x - 28*(28-y)
  9. quietly twoway (contour v y x, crule(intensity) ecolor(black) ///
> clegend(off) heatmap), ///
```

```
> yscale(off) xscale(off) legend(off) ///
> graphregion(margin(zero)) plotregion(margin(zero)) ///
> graphregion(color(white)) aspectratio(1) nodraw ///
> name(d`ndigit´, replace)
 10. end
.
. * Display the "first" observation
. draw_test_digit 1
(Training MNIST: 28x28 images with stacked pixel values v* in [0,1] range)
. graph combine d1
.
. * Select a group of digits and display them in one graph
. local list 2 22 4 15 17 29 31 50 10 54 12 48 33 84 30 80 56 98 34 88
. local disp_list
. foreach ndigit of local list {
  2. draw_test_digit `ndigit´
  3. local disp_list `disp_list´ d`ndigit´
  4. }
(Training MNIST: 28x28 images with stacked pixel values v* in [0,1] range)
(Training MNIST: 28x28 images with stacked pixel values v* in [0,1] range)
(Training MNIST: 28x28 images with stacked pixel values v* in [0,1] range)
(Training MNIST: 28x28 images with stacked pixel values v* in [0,1] range)
(Training MNIST: 28x28 images with stacked pixel values v* in [0,1] range)
(Training MNIST: 28x28 images with stacked pixel values v* in [0,1] range)
(Training MNIST: 28x28 images with stacked pixel values v* in [0,1] range)
(Training MNIST: 28x28 images with stacked pixel values v* in [0,1] range)
(Training MNIST: 28x28 images with stacked pixel values v* in [0,1] range)
(Training MNIST: 28x28 images with stacked pixel values v* in [0,1] range)
(Training MNIST: 28x28 images with stacked pixel values v* in [0,1] range)
(Training MNIST: 28x28 images with stacked pixel values v* in [0,1] range)
(Training MNIST: 28x28 images with stacked pixel values v* in [0,1] range)
(Training MNIST: 28x28 images with stacked pixel values v* in [0,1] range)
(Training MNIST: 28x28 images with stacked pixel values v* in [0,1] range)
(Training MNIST: 28x28 images with stacked pixel values v* in [0,1] range)
(Training MNIST: 28x28 images with stacked pixel values v* in [0,1] range)
(Training MNIST: 28x28 images with stacked pixel values v* in [0,1] range)
(Training MNIST: 28x28 images with stacked pixel values v* in [0,1] range)
(Training MNIST: 28x28 images with stacked pixel values v* in [0,1] range)
. graph combine `disp_list´
.
. * Classify handwritten digits using "mlp2"
.
. * Load the dataset and set the seed
. sysuse mnist-train, clear
(Training MNIST: 28x28 images with stacked pixel values v* in [0,1] range)
. set seed 12345
.
. * Run "mlp2 fit"
. mlp2 fit y v*, layer1(100) layer2(100)
```

Multilayer perceptron	input variables =	784
	layer1 neurons =	100
	layer2 neurons =	100
Loss: softmax	output levels =	10
Optimizer: sgd	batch size =	50
	max epochs =	100
	loss tolerance =	.0001
	learning rate =	.1
Training ended:	epochs =	18
	start loss =	2.59628
	end loss =	.001498

```
. * Perform in-sample prediction using "mlp2 predict"
. mlp2 predict , genvar(ypred)
Prediction accuracy: .9999166666666667

.
. * Validate the results by out-of-sample (test) prediction accuracy
.
. * Load the test datset
. use mnist-test , clear
(Testing MNIST: 28x28 images with stacked pixel values v* in [0,1] range)
. set seed 12345
.
. * Run "mlp2 predict"
. mlp2 predict, genvar(ypred)
Prediction accuracy: .9739
.
. * Look at some of the test digits that are missclassified
. mat list e(pred_err_ind)
e(pred_err_ind)[1,261]
```

	c1	c2	c3	c4	c5	c6	c7	c8	c9	c10	c11	c12	c13	c14	c15	c16
r1	116	150	246	248	322	341	382	446	484	496	579	583	614	620	629	692

	c17	c18	c19	c20	c21	c22	c23	c24	c25	c26	c27	c28	c29	c30	c31	c32
r1	811	845	901	939	957	966	1004	1015	1040	1113	1129	1157	1182	1183	1192	1195

	c33	c34	c35	c36	c37	c38	c39	c40	c41	c42	c43	c44	c45	c46	c47	c48
r1	1227	1243	1248	1320	1365	1394	1396	1501	1521	1523	1528	1531	1550	1610	1622	1679

	c49	c50	c51	c52	c53	c54	c55	c56	c57	c58	c59	c60	c61	c62	c63	c64
r1	1682	1710	1722	1733	1755	1791	1801	1814	1902	1942	1985	1988	2005	2017	2025	2036

	c65	c66	c67	c68	c69	c70	c71	c72	c73	c74	c75	c76	c77	c78	c79	c80
r1	2044	2054	2071	2110	2119	2131	2136	2149	2186	2187	2225	2294	2326	2330	2388	2415

	c81	c82	c83	c84	c85	c86	c87	c88	c89	c90	c91	c92	c93	c94	c95	c96
r1	2427	2448	2489	2598	2608	2649	2655	2743	2811	2837	2851	2883	2922	2940	2946	2954

	c97	c98	c99	c100	c101	c102	c103	c104	c105	c106	c107	c108	c109	c110	c111	c112
r1	3061	3074	3118	3131	3285	3331	3334	3374	3406	3423	3442	3491	3504	3521	3550	3559

	c113	c114	c115	c116	c117	c118	c119	c120	c121	c122	c123	c124	c125	c126	c127	c128
r1	3560	3568	3575	3598	3627	3682	3728	3752	3777	3781	3797	3809	3812	3819	3837	3854

	c129	c130	c131	c132	c133	c134	c135	c136	c137	c138	c139	c140	c141	c142	c143	c144
r1	3894	3903	3907	3942	3944	3947	3977	4001	4066	4079	4124	4141	4149	4155	4177	4200

	c145	c146	c147	c148	c149	c150	c151	c152	c153	c154	c155	c156	c157	c158	c159	c160
r1	4202	4225	4249	4270	4272	4290	4420	4426	4434	4438	4498	4537	4548	4553	4568	4594

	c161	c162	c163	c164	c165	c166	c167	c168	c169	c170	c171	c172	c173	c174	c175	c176
r1	4602	4636	4741	4752	4762	4808	4815	4824	4834	4861	4881	4967	5069	5079	5139	5141

	c177	c178	c179	c180	c181	c182	c183	c184	c185	c186	c187	c188	c189	c190	c191	c192
r1	5332	5451	5601	5635	5643	5677	5735	5750	5888	5937	5938	5956	5973	5974	5983	5993

	c193	c194	c195	c196	c197	c198	c199	c200	c201	c202	c203	c204	c205	c206	c207	c208
r1	5998	6046	6060	6072	6167	6169	6422	6512	6548	6556	6560	6561	6565	6575	6577	6598

	c209	c210	c211	c212	c213	c214	c215	c216	c217	c218	c219	c220	c221	c222	c223	c224
r1	6599	6622	6626	6633	6652	6663	6756	6766	6784	6848	6957	7435	7452	7473	7824	7900

	c225	c226	c227	c228	c229	c230	c231	c232	c233	c234	c235	c236	c237	c238	c239	c240
r1	7916	7922	8095	8247	8278	8294	8312	8326	8409	8441	8521	8523	8528	8585	8850	9007

	c241	c242	c243	c244	c245	c246	c247	c248	c249	c250	c251	c252	c253	c254	c255	c256
r1	9010	9016	9025	9317	9588	9635	9665	9680	9699	9739	9746	9750	9756	9771	9780	9784

	c257	c258	c259	c260	c261
r1	9812	9840	9859	9945	9981

```
. list y ypred* in 116
```

116.	y	ypred_0	ypred_1	ypred_2	ypred_3	ypred_4	ypred_5	ypred_6	ypred_7
	4	2.62e-08	7.62e-08	2.71e-06	5.71e-07	.0739827	.0000209	.0052555	3.23e-07

ypred_8	ypred_9
.0004435	.9202936

Fig. 6.18 Graphical
representation of a numeral

In this context, we are working in a classification setting. Our training dataset, `mnist-train.dta`, contains 60,000 images, whereas our testing dataset, `mnist-test.dta`, 10,000 images. Every image has dimension 28×28, with total of 784 pixels. The pixels, stacked horizontally, are represented by the variables from `v1` to `v784`. Each single pixel contains as information a float number in the interval [0,1]. Our target variable y, finally, is a label variable taking values from 0 to 9 corresponding to the digit represented in the image.

We start by first visualizing individual images. For this purpose: (i) we first extract an individual record and expand its pixel values in a long-format, thus creating a new dataset; (ii) use the `twoway contour` command with the `heatmap` option to draw the single image. For implementing this task in Stata, Balov provides a short but effective program for this code, `draw_test_digit`, a program to create a graph displaying the "ndigit". Using this program, we are able to display the first record from the training dataset, that represents a "5", by simply typing `draw_test_digit 1`. The graphical visualization is represented in Fig. 6.18. Then, we select a series of digits and display them in a single graph. This produces the image in Fig. 6.19.

The code goes on by classifying handwritten digits by a 2-layer neural network with 100 neurons per layer. We load the dataset again and set the seed, and then we run `mlp2 fit`. After 27 epochs, the algorithm converges with a considerable reduction of the loss, starting from 2.51268 and ending at .000419. We perform in-sample

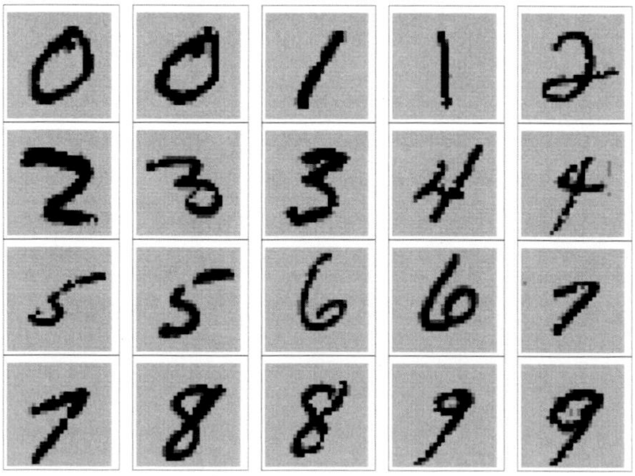

Fig. 6.19 Graphical
representation of a series of
numerals

prediction using `mlp2 predict`, thus obtaining an in-sample prediction accuracy of 0.999, signaling strong overfitting. We note that option `genvar(ypred)` provides a "stub" for new variables holding the predicted class probabilities, generating variables from `ypred_0` to `ypred_9` of probability values that sum to one. As usual, the numeral with maximum probability will be the predicted class used for calculating the prediction accuracy, where the accuracy is given by the proportion of correctly predicted numerals in the sample.

The final step concerns the validation of the results using out-of-sample (or testing) prediction accuracy. For this purpose, we load the testing dataset `mnist-test` and run again `mlp2 predict`, thus obtaining an out-of-sample prediction accuracy rather high, equal to .9742. This means that our test error is smaller than 3%. We can list the test dataset's numerals that are missclassified as the indices of the missclassified units are saved in the matrix `e(pred_err_ind)`: we simply type `mat list e(pred_err_ind)`. As an example, we consider the numeral 116, a test image representing a "4" but that was erroneously classified as "9". This is also reasonable, as the way people generally write a "4" can be often similar to the way they write a "9".

6.8 Conclusion

Artificial neural networks (ANNs) are nowadays considered as a leading approach in Machine Learning. In this chapter, we have provided a thorough description of basic *feedforward* neural networks, that is, network architectures where the connections between nodes are complete (full nodes' communication), and the nodes themselves do not form connection cycles. We have showed how one can train and tune an ANN of this kind in Python, R, and Stata. In the next chapter, we will relax the assumption that an ANN has fully connected nodes, and that feedback between nodes is absent. We thus present more sophisticated network architectures able to increase prediction performance in many application contexts, as imaging recognition and sequences predictions. We refer to the so-called *deep learning* literature and specifically to convolutional and recurrent neural networks.

References

Abiodun, O. I., A. Jantan, A. E. Omolara, K. V. Dada, N. A. Mohamed, & H. Arshad. (2018). State-of-the-art in artificial neural network applications: A survey. *Heliyon, 4*(11), e00938. https://www.sciencedirect.com/science/article/pii/S2405844018332067.

Abiodun, O. I., M. U. Kiru, A. Jantan, A. E. Omolara, K. V. Dada, A. M. Umar, O. U. Linus, H. Arshad, A. A. Kazaure, & U. Gana. (2019). Comprehensive Review of Artificial Neural Network Applications to Pattern Recognition. *IEEE Access, 7*, 158820–158846. https://doi.org/10.1109/ACCESS.2019.2945545.

Baptista, D., & Dias, F. M. (2013). A survey of artificial neural network training tools. *Neural Computer Application, 23*(3–4), 609–615. http://dblp.uni-trier.de/db/journals/nca/nca23.html#BaptistaD13a.

Hastie, T., Tibshirani, R., & Friedman, J. (2009). *The Elements of Statistical Learning: Data Mining, Inference, and Prediction* (2nd ed.). Springer Series in Statistics, New York: Springer. https://www.springer.com/gp/book/9780387848570.

Lőrincz, A. (2011). Learning the states: A brain inspired neural model. In J. Schmidhuber, K. R. Thórisson & M. Looks (Eds.) *AGI* vol. 6830 of *Lecture Notes in Computer Science* (pp. 315–320). Springer.

Mai, J., & Paxinos, G. (2012). *The human nervous system* (3rd ed.). Academic Press.

McCulloch, W. S., & Pitts, W. (1943). A logical calculus of the ideas immanent in nervous activity. *The Bulletin of Mathematical Biophysics, 5*(4), 115–133.

Raschka, S., & Mirjalili, V. (2019). *Python machine learning* (3rd ed.). Packt Publishing.

Rosenblatt, F. (1958). The perceptron: A probabilistic model for information storage and organization in the brain. *Psychological Review, 65*(6), 386–408.

Yamakawa, H. (2021). The whole brain architecture approach: Accelerating the development of artificial general intelligence by referring to the brain. *Neural Networks, 144*, 478–495. https://doi.org/10.1016/j.neunet.2021.09.004.

Chapter 7
Deep Learning

7.1 Introduction

Deep learning algorithms represent a subset of Machine Learning algorithms built on sophisticated multi-layer artificial neural network (ANN) architectures. The *deep* labeling, to be contrasted with the *shallow* labeling, has mainly to do with the construction of ANNs characterized by a hierarchical stratification of layers that stand between the observed inputs and the observed outputs.

Deep learning has proved to be highly effective in carrying out many intelligent tasks, including image recognition, automatic speech recognition, natural language processing (including text generation), high-frequency financial data forecasting, image reconstruction, anomaly detection, and so forth.

In deep learning, this multiple-layer structure has the role to extract higher-level features from the initial input-features, thus leading to identify only the salient elements needed, for example, for pattern prediction/recognition. Such feature decomposition and transformation progressively identify sub-structures within the original data, some having a major role in allowing for pattern recognition, some having a less relevant role. For example, an image can be decomposed and transformed into many sub-components, that can be in turn decomposed and transformed into other sub-components. However, only a few of these components are those that really matter for image successful recognition.

To give an intuition of why this feature-extraction mechanism works, consider Fig. 7.1. In this figure, I have represented two objects, a *real* cat and a *stylized* cat. From a human perception standpoint, both figures can be easily recognized and thus classified as a cat, but the one on the left panel, the stylized one, contains a much smaller amount of information compared to the one placed on the right panel (the real cat). We may conclude that raw data generally contains too much information compared to what one really needs for pattern recognition. The raw signal is somewhat too rich in details, many of them only adding up larger noise than useful and exploitable information for effective pattern recognition.

G. Cerulli, *Fundamentals of Supervised Machine Learning*, Statistics and Computing, https://doi.org/10.1007/978-3-031-41337-7_7

Fig. 7.1 A stylized versus a real cat. Stylized cat drawing: courtesy of Giovanni Cerulli

Stylized cat   Real cat

In the literature, two classes of (supervised) deep learning models have become popular in applications: (i) convolutional neural networks (CNNs), and (ii) recurrent neural networks (RNNs). As it will become clearer later on, CNNs have proved to be particularly successful for imaging recognition, while RNNs for sequences prediction (including both quantitative and qualitative sequences). This chapter presents these two deep learning architectures and provides Python applications using the Keras API. However, before starting with the description and then application of these methods, it is useful to understand why deep learning is particularly successful in certain predictive domains. We make the point that this is particularly true when data present some natural *ordering*.

7.2 Deep Learning and Data Ordering

The feature-extraction mechanism is not a prerogative of deep learning algorithms. As we saw in the previous chapter, this is also the basis of the functioning of shallow ANNs, although deep ANNs can enlarge the formation of increasingly salient higher-level futures. What is thus the real value-added of deep learning? In this section, we make the point that deep learning is particularly effective when raw data present some form of inherent *ordering*.

Data ordering is very central in statistical learning as this has a remarkable impact on models' predictive ability. The classical assumption in statistical learning is that data inputs are independent and identically distributed (IID). The IID assumption entails that the ordering of the observations is random, that is, it does not matter whether let's say, unit 5 comes after or before unit 3. This information is in fact irrelevant if data are assumed to be IID.

Data, however, may have an inner ordering taking in general two different forms: (i) *sequential ordering*, and (ii) *spatial ordering*. In what follows, I provide a brief description of these two types of data ordering, suggesting the link they can have with the two deep learning approaches we present in this chapter, that is, CNNs and RNNs.

Sequential ordering

Sequential ordering entails that data instances appear one after another according to some logical pattern. In this case, whether an instance comes before or after another instance does matter. A typical example of data characterized by a sequential ordering is times-series data. In time-series, time dictates the ordering, as it establishes clearly what comes first and what comes second. This ordering entails a *cumulative information process* in that the next instance brings with itself information coming from the previous instances thus yielding a chain of cumulative information storing. This implies that observations (i.e., instances) are no longer independent, as they become dependent along the time sequence. This dependence matters and can be rigorously defined using conditional probability notation as follows:

$$P(Y_t | Y_{t-1}, Y_{t-2}, Y_{t-3}, \ldots) \neq P(Y_t) \tag{7.1}$$

This equation simply states that the probability of Y to take specific values at time t, by knowing that the times series has taken values $Y_{t-1}, Y_{t-2}, Y_{t-3}, \ldots$ before, is different from the marginal (or standalone) probability that Y takes a specific value in t by ignoring what happened before. For example, suppose one wants to predict tomorrow's temperature (and thus, possibly, weather conditions). It is clear that information about today's and yesterday's temperature is relevant to predict tomorrow temperature, as temperature data have strong time cumulative ordering based on the physics of the atmosphere. We may say that data in t incorporate information coming from previous data thus engendering a form of *path dependence* characterized by either a short- or long-term memory.

When we restrict the definition of this dependence only to its linear component, we speak of *serial correlation*, a mechanism typically modeled in time-series econometrics. It is important however to bear in mind that serial dependence is more general than serial correlation as it includes whatever probabilistic dependence than just the linear one, as entailed by Eq. (7.1).

Recurrent Neural Networks (RNNs) are suitable ANNs to model nonlinear serial correlation and have proved to be particularly able to predict sequences, including text sequences in natural language processing (NLP), sound and video sequences, as well as time-series.

Spatial ordering

Spatial ordering entails another type of violation of the IID assumption. In this regard, *spatial dependence* is an inner property of spatial data originating from the the general *contiguity* of space (Nikparvar & Thill, 2021). Spatial dependence relies on the intuitive insight that "close objects are more related than distant objects".

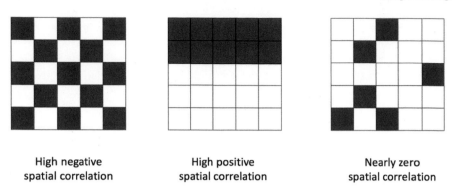

High negative   High positive   Nearly zero

spatial correlation   spatial correlation   spatial correlation

Fig. 7.2 Example of negative, positive, and zero spatial autocorrelation

There are various indicators to measure spatial dependence, but the most popular is Moran's I, a measure of spatial *autocorrelation*:

$$I = \frac{N}{\sum_i \sum_j W_{ij}} \frac{\sum_i \sum_j W_{ij}(x_i - \bar{x})(x_j - \bar{x})}{\sum_i (x_i - \bar{x})^2} \qquad (7.2)$$

where N is the number of spatial units indexed by i and j; x is the variable of interest; $\bar{x}$ is the mean of x; W_{ij} is a generic element of a matrix of spatial weights with zeroes on the diagonal.

The matrix of spatial weights has a sizable role in shaping the degree of spatial autocorrelation. Generally, the weights are measured as the inverse of the distance between pairs, but other contiguity rules can be used. The I index can take either positive, negative or zero values, and it ranges between -1 and $+1$. A positive (negative) spatial autocorrelation indicates that, on average, adjacent spatial units tend to have similar (dissimilar) values. In this sense, the Moran index is a type of dissimilarity index.

Figure 7.2 illustrates three spatial patterns referring to a negative, positive, and zero correlation. In the first pattern, a spatial unit (let's say, a black square) tends to be surrounded by different units (i.e., white squares), thus implying a negative spatial correlation; in the second pattern, a spatial unit (let's say, a black square) tends to be surrounded by similar units (i.e., black squares), thus implying a positive spatial correlation; the third panel, finally, shows a situation similar to a random distribution of black and white squares, thus indicating a zero spatial correlation.

In imaging recognition, spatial dependence is at the basis of good prediction ability. Images are in fact characterized by a high degree of spatial ordering. For example, for facial recognition, the fact that patches representing, let's say, the nose of a person are always located in the middle between patches corresponding to the right eye and left eye provides a natural ordering that in some way helps improve prediction, as regular patterns of this sort exist in pretty all the facial images one can collect. Also, regular imaging patterns lend support to the possibility of building data hierarchical architectures that can be exploited to detect salient compositions of original data inputs. This is the philosophy adopted by Convolution Neural Networks (CNNs), whose main task is to generate surrogate features in a hierarchical manner to detect and then retain only salient signals from the original signals with the aim of reducing the computational burden compared to traditional fully connected ANNs. Indeed, as images (and not only images) often contain more information than that needed for recognition purposes, a fitting procedure processing the whole information pool in one shot might be highly inefficient both as a learning architecture and computationally.

7.3 Convolutional Neural Networks

Convolutional neural networks (or CNNs) are a class of artificial neural network architectures inspired by the functioning of the human brain's visual cortex. Pioneered by LeCun et al. (1989), CNNs have been proved to reach outstanding ability in carrying out many artificial intelligence tasks, such as images recognition and classification, segmentation of images, natural language processing, construction of interfaces between human brain and computer, as well as high-frequency time-series forecasting.

What is the logic behind the functioning of a CNN? Let's consider an image. Generally, for recognizing a specific object contained in an image, let's say a cat, not all the elements (i.e., a certain set of adjacent pixels) are of the same importance. Some patches are in fact more relevant than others. We may say that, in general, an image contains more information than needed for recognition purposes.

This over-information content provides ground for *salient feature extraction*, i.e., a data reduction process that allows to hierarchically extract low-dimensional information from raw data. In a multi-layer neural network, this process takes place sequentially and hierarchically with earlier layers extracting lower level features to then form higher level features.

At the heart of the CNN functioning, there is the construction of the so-called *feature maps*. They are synthetic projections outputs of a *local patch* of pixels belonging to the input image. Figure 7.3 provides a visual representation of a feature map projection for a cat image.

The good ability of CNNs to recognize structured patterns (as the cat in Fig. 7.3) lays on two salient characteristics of a CNN: (i) *sparse connection*, and (ii) *parameter sharing*.

Sparse connection refers to the fact that one single element in the feature map is linked only to a small patch of pixels. Indeed, in high-dimensional inputs settings, as in the case of images, it is useless to connect all neurons of subsequent layers to form the feature maps, as a fully connected ANN does not account for the spatial ordering owned by the image. In contrast, CNNs recognize this spatial local correlation of neurons belonging to adjacent layers, thus reducing the number of active links compared to a fully connected ANN, and thus connecting only a small area

Fig. 7.3 Example of a
feature map projection in a
convolutional neural network

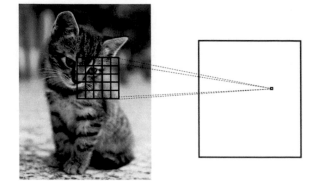

of input cells. *Parameter sharing*, in turn, refers to the fact that, in building feature maps, the same set of weights are employed for different image patches.

Thanks to these two CNN properties, the fitting size of the network—i.e., the number of free parameters to estimate (the weights), drops remarkably down compared to a fully connected ANN. This makes CNNs more appealing than ANNs in spatially-correlated settings, as irrelevant information is disregarded, while the reduced parametric size of the model attenuates computational complexity.

7.3.1 The CNN Architecture

A CNN architecture is composed of the following sub-structures: (i) a given number of *convolutional layers*, (ii) a given number of *pooling layers*, and (iii) a given number of final *fully connected layers* ending with a final outcome layer.

A *convolutional layer* is a shared-weight structure made of the so-called *convolution kernels* (or *filters*) sliding along the input features and generating feature maps through translation equivariant responses.

A *pooling layer* is used to reduce data dimension by transforming a cluster of neurons at one layer into a single neuron in the subsequent layer. This procedure does not involve weights estimation, but only arithmetic operations (maximum and average pooling).

Fully connected layers form a standard ANN, similar to those illustrated in the previous chapter, and terminate with the final outcome nodes.

In what follows, we describe these three CNN building-blocks starting from the description of the convolutional operation needed to build the convolutional layer.

7.3.2 The Convolutional Operation

The convolutional operation is at the core of the CNN architecture. In order to proceed gradually, we first consider the one-dimensional (1D) convolution, to then illustrate the two-dimensional (2D) convolution. Before going ahead, however, it is worth presenting the notion of *padding* size and that of *stride* for a generic vector or matrix.

As for the padding, Fig. 7.4 illustrates two examples, one referring to a padding of size 3 for a vector x, and one referring to a padding of size 1 for a matrix M. Padding simply means to join frames of zeros around vectors or matrices. The number of frames is the padding size.

As for the stride, it refers to the number of pixels shifts over an input matrix M, given a filter of size m. For example, if the stride is set equal to 1, the filter is moved along the matrix by 1 pixel at a time, whereas if the stride is set equal to 2, the filter is moved along the matrix by 2 pixels at a time, and so forth. Figure 7.5 illustrates the movement of the filter along the matrix M when the filter is 2×2, and the stride is either equal to 1 or 2.

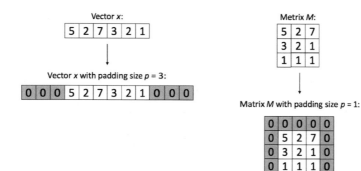

Fig. 7.4 Examples of a padding of size 3 (for a vector x), and size 1 (for a matrix M)

Stride equal to 2 with a 2 x 2 filter

Stride equal to 1 with a 2 x 2 filter

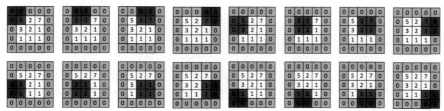

Fig. 7.5 Example of a stride set either to 1 and to 2 using a 2×2 filter along the matrix M

We can observe that the choice of the stride implies a trade-off between the quality of the image representation and the computational workload. Indeed, a larger stride implies a rougher representation of the image in the feature map, but it entails a smaller number of convolutions, thus saving computational time. On the contrary, a smaller stride provides a finer representation of the image in the feature map, but implies a larger set of convolutions, thus increasing the computational burden.

The 1D convolution

Consider two row vectors, a $1 \times n$ vector x (the input), and $1 \times m$ vector ω (the filter), with $m \leq n$. The convolutional operator produces a new $o \times m$ vector y, so that

$$y = \mathbf{X}\omega_r^T \tag{7.3}$$

where $\mathbf{X}$ is a $o \times m$ matrix formed by stacking sliding window sub-vectors of x of size m, and ω_r is the rotation of ω. The size of the convolutional output is

$$o = \left\lfloor \frac{n + 2p - m}{s} \right\rfloor + 1 \tag{7.4}$$

where the symbol $\lfloor$ indicates the floor operator (returning the largest integer that is equal or smaller than the input), p is the padding size, and s is the stride size.

To provide an example, consider a row vector $x = (10, 50, 60, 10, 20, 40, 30)$, where $n = 7$, and a row vector $\omega = (1/9, 1/3, 1/6)$, where $m = 3$. By assuming zero padding for simplicity, and a stride $s = 2$, we have that

$$o = \left\lfloor \frac{7 - 3}{2} \right\rfloor + 1 = 3 \tag{7.5}$$

The matrix $\mathbf{X}$ takes on this form:

$$\mathbf{X}_{o \times m} = \begin{bmatrix} 10 & 50 & 60 \\ 60 & 10 & 20 \\ 20 & 40 & 30 \end{bmatrix}$$

while the rotated ω is $\omega_r = (1/6, 1/3, 1/9)$. This produces the following convolutional output:

$$\underset{o \times 1}{y} = \underset{o \times m}{\begin{bmatrix} 10 & 50 & 60 \\ 60 & 10 & 20 \\ 20 & 40 & 30 \end{bmatrix}} \underset{m \times 1}{\begin{bmatrix} 1/6 \\ 1/3 \\ 1/9 \end{bmatrix}} = \underset{o \times 1}{\begin{bmatrix} 25 \\ 21 \\ 48 \end{bmatrix}}$$

Figures 7.6 and 7.7 show the way in which we formed the matrix $\mathbf{X}$ respectively for a 1D sliding window and a 2D sliding window. Let's focus on the 1D sliding window. Here, we consider a sliding window of size $m = 3$ and stride $s = 2$. In this setting, we can see that we move the rotated ω (in light blue color) by two positions,

	1	2	3	4	5	6	7
	10	50	60	10	20	40	30
1	1/6	1/3	1/9				
2			1/6	1/3	1/9		
3					1/6	1/3	1/9

Fig. 7.6 1D sliding window approach with size $m = 3$ and stride $s = 2$ to build the matrix $\mathbf{X}$

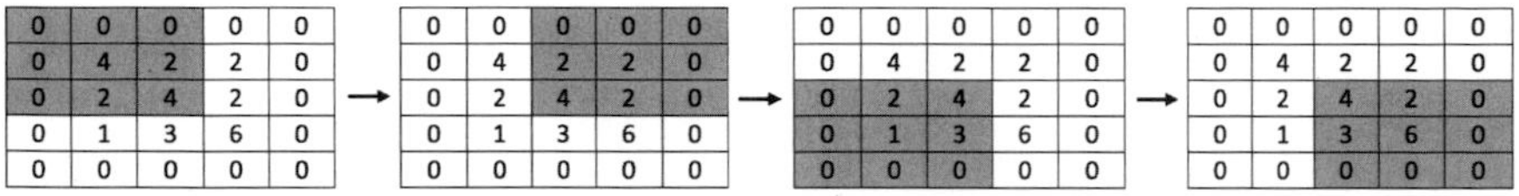

Fig. 7.7 2D sliding window approach with size $m = 3$ and stride $s = 2$ to build the matrix $\mathbf{X}$

starting from the left to the right. The first element of the output y, for example, is thus computed as

$$y_1 = 10 \cdot \tfrac{1}{6} + 50 \cdot \tfrac{1}{3} + 60 \cdot \tfrac{1}{9} = 25$$

The matrix $\mathbf{X}$ is thus formed by stacking the sub-vectors of x corresponding to the current window.

In the previous example, we chose a zero padding. In general, the choice of a padding of size p affects the way in which the boundary cells of vector x are treated compared to the cells occupying more central positions within this vector. Indeed, without padding, central cells tend to be used more times than boundary cells to produce the outcome y, thus making the convolutional outcome more unbalanced toward central cells. A positive padding can balance output computation by giving cells similar relevance.

In applications, there are three types of padding choices: *full*, *same*, and *valid*. Full padding sets $p = m - 1$, implying a relevant increase in the output dimension. For this reason, this mode is not widely used in real applications. Same pudding chooses p with the aim of giving the output y the same size of vector x. Valid padding, finally, entails to set $p = 0$, which means that there is no padding at all, as in the example illustrated above. Figure 7.8 sets out a comparison of full, same, and valid pudding in a 2D convolution (see next section). We can see that full padding increases the size of the output compared to the size of the initial input, whereas valid pudding reduces the size of the output compared to the input. Same pudding, in contrast, keeps the size of the output equal to that of the input. In applications, same pudding is generally preferred to the other two pudding modes.

The 2D convolution

The 2D convolution operates in a similar way as the 1D convolution except that x now indicates a $n_1 \times n_2$ matrix, and ω a $m_1 \times m_2$ matrix. In order to compute the

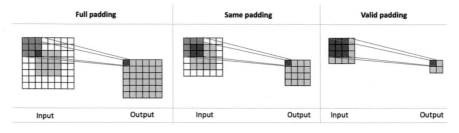

Fig. 7.8 A comparison of full, same, and valid pudding in convolutional operations

convolutional outcome matrix y, we have a different formula than (7.3), taking on this form:

$$y = \varphi(\mathbf{X}[\odot]w_r) \tag{7.6}$$

where $\varphi(\cdot)$ is a function returning the sums of the elements of each block matrix forming the matrix $\mathbf{X}$, and $[\odot]$ indicates the matrix *penetrating face* product, an extension of the Hadamard product to block matrices. As this formula may be tricky to be understood, in what follows we provide a simple practical example of how it works.

Suppose to start with a 4×4 matrix x (i.e., $n_1 = n_2 = 4$), and let's pad it by setting $p = 1$, and allowing a stride $s = 2$. The resulting 5×5 padded matrix is

$$x_{padded} \atop 5 \times 5 = \begin{bmatrix} 0\,0\,0\,0\,0 \\ 0\,4\,2\,2\,0 \\ 0\,2\,4\,2\,0 \\ 0\,1\,3\,6\,0 \\ 0\,0\,0\,0\,0 \end{bmatrix}$$

Suppose that the rotated matrix w_r is

$$w_r \atop 3 \times 3 = \begin{bmatrix} 0.5 & 0.2 & 1 \\ 0.5 & 1 & 1 \\ 0.3 & 1 & 0.5 \end{bmatrix}$$

By applying formula (7.6), we thus obtain that

$$y = \varphi\left(\left[\begin{array}{ccc|ccc} 0&0&0&0&0&0 \\ 0&4&2&2&2&0 \\ 0&2&4&4&2&0 \\ \hline 0&2&4&4&2&0 \\ 0&1&3&3&6&0 \\ 0&0&0&0&0&0 \end{array} \right] [\odot] \begin{bmatrix} 0.5 & 0.2 & 1 \\ 0.5 & 1 & 1 \\ 0.3 & 1 & 0.5 \end{bmatrix} \right) = \varphi\left(\left[\begin{array}{ccc|ccc} 0&0&0&0&0&0 \\ 0&4&2&1&2&0 \\ 0&2&2&1.2&2&0 \\ \hline 0.5&2.2&5&2&0.4&0 \\ 0.5&2&4&1.5&6&0 \\ 0.3&1&0.5&0&0&0 \end{array} \right] \right)$$

where, as said, the function $\varphi(\cdot)$ computes the sums of the elements of each block matrix. This means that

$$y = \left[\begin{array}{c|c} 10 & 6.2 \\ \hline 16 & 9.9 \end{array}\right]$$

that is a 2×2 matrix, entailing a *valid* padding as $2 < 4$.

7.3.3 *Pooling*

The convolutional operation transforms an input matrix into an output matrix of larger, smaller or equal size depending on the employed pudding mode. After convolution, however, building a CNN requires another operational step, known as *pooling* (of size k). Pooling takes as input the output matrix of the convolution and returns another matrix (or even a scalar) of smaller size. The idea of pooling is to form non-overlapping submatrices of size k, i.e., $\mathbf{P}_{k \times k}$, out of the convolutional output matrix, and generate from them a (local) scalar by computing the maximum (*max-pooling*) or the average (*mean-pooling*) of their elements. Figure 7.9 illustrates how the pooling operation works, in the case of max-pooling with a pooling matrix of size 2. We start with matrix $\mathbf{X}$ having size 6, that is transformed into a matrix $\mathbf{y}$ of size 3. Alternatively, one can use mean-pooling that computes the average of the elements for every submatrix.

Compared to mean-pooling, max-pooling is generally a more robust approach in the presence of noisy signals. In fact, the matrix $\mathbf{y}$ resulting from max-pooling will be less sensitive to local changes in the elements of the submatrices than the resulting matrix $\mathbf{y}$ obtained from mean-pooling.

It is worth noting that the pooling step requires no weights' estimation.

Matrix **X**

72	73	82	6	48	18
41	52	26	47	28	16
75	1	69	26	4	18
11	60	57	43	94	83
59	19	16	94	75	32
75	93	97	59	43	90

Max-pooling with matrix $\mathbf{P}_{2x2}$ →

Matrix **y**

73	82	48
75	69	94
93	97	90

Fig. 7.9 Example of max-pooling using a $\mathbf{P}_{2\times2}$ pooling matrix

7.3.4 Full Connectivity

The final step of a CNN is a fully connected neural network, similar to the one described in the previous chapter. This final neural network takes place after producing several convolutional and pooling layers. The fully connected network uses typical activation functions and, in the case of a labeled target variable, it returns as final output the probabilities of each label (using, for example, a softmax activation function).

7.3.5 Multi-channel/Multi-filter CNNs

The initial input of a CNN can be made of more than one single 2D matrix. For example, the entire bundle of existing colors can be obtained by mixing up different degrees of red, green, and blue, through what is known as the RGB color mode. As a consequence, a colored image can be represented by three 2D matrices, where each pixel in each matrix contains different shades of red (for the first matrix), green (for the second matrix), and blue (for the third matrix).

An input composed of c matrices of size $n_1 \times n_2$ is called a *tensor*. The input tensor can be seen as a three-index input matrix $\mathbf{X}_{n_1 \times n_2 \times c}$. When an image is in gray scale, we have that $c = 1$, as the image can be perfectly represented by one 2D matrix containing in each cell (i.e., in each pixel) a number ranging from 0 to 1 (with $0 =$ white, and $1 =$ black) indicating the intensity of the gray scale.

In a related manner, CNNs generally consider more than one convolutional weighting 2D matrix, where every 2D matrix of weights is called *kernel filter*, or more simply *filter*. Also in this case, we can refer to a weighting tensor $\Omega_{m_1 \times m_2 \times d}$ as a three-index matrix.

Figure 7.10 illustrates an example of a multi-channel/multi-filter CNN for colored images recognition in a same pudding mode. The input tensor contains three channels ($c = 3$), one for each baseline color of the RGB mode. This input is convoluted by using five three-dimensional filters (i.e., $d = c = 3$), thus generating as output a tensor of depth five. Each single 2D matrix of the convolutional output is obtained by carrying out a separate convolution for each input channel to then sum up the results from all the channels. Every step, going from the inputs to the outputs, either convolutional or pooling, is known as *feature map*.

This illustrative example can shed light also on the benefits in terms of reduced parametric size of implementing a CNN over a traditional fully connected ANN. As said, such benefit resides in the capacity of CNNs to activate parameter sharing on the one hand, and sparse connectivity on the other hand.

How many parameters does the CNN of Fig. 7.10 need to compute? Considering the filters, we see that there are $m_1 \times m_2 \times c$ weights, plus 5 biases. As pooling layers do not have weights, we have that

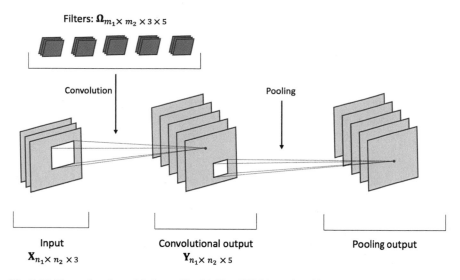

Fig. 7.10 Example of a multi-channel/multi-filter CNN for colored images recognition

$$p_{cnn} = (m_1 \times m_2 \times c \times nf) + nf$$

where nf indicates the number of filters. If, instead of carrying out a convolution, we estimate parameters through a fully connected layer, the number of weights to estimate for obtaining the same number of output units becomes:

$$p_{nn} = (n_1 \times n_2 \times c) \times (n_1 \times n_2 \times nf) = (n_1 \times n_2)^2 \times c \times nf$$

For example, suppose that $m_1 = m_2 = 5$, $n_1 = n_2 = 10$, $c = 3$, and $nf = 5$, than the number of parameters needed to obtain the output layer is 380 for a CNN, and 150,000 for a fully connected ANN, which is around 400 times larger. The reduction in the computational burden made possible by implementing a CNN architecture is thus sizable.

7.3.6 Tuning a CNN by Dropout

As a deep learning architecture, CNN is very likely to be plagued by overfitting. This is due to the large set of parameters that are embedded within this type of neural network. In general, we saw that ANNs can be tuned by regularization methods such as weight decay that uses a Ridge type (or L2) regularization approach. Another way to avoid overfitting is early stopping.

However, when the parametric space is huge, as in the case of images' recognition using CNNs, more efficient tuning alternatives are available. One of these alternatives is the *dropout* algorithm, proposed by Srivastava et al. (2014).

The dropout algorithm entails that, at each training pass, every node can be removed (i.e., dropped out) from the network with probability p (or retained with probability $1 - p$), where p is generally set equal to 0.5. It is clear that, for a network characterized by K nodes (excluding the terminal ones), one has to train 2^K different networks with different compositions of activated neurons. The final network weights are then obtained by averaging the weights resulting from each trained network, thus obtaining a unique final set of weights. From this perspective, the dropout can be seen as an ensemble approach. However, in the testing phase the ultimate predictions are not estimated as an ensemble prediction. Indeed, instead of generating 2^K predictions and taking the average (for numerical outcomes) or the majority vote (for categorical outcomes), the dropout takes predictions from one single network, the one obtained by averaging over the trained networks, by multiplying each weight by the removing probability p. This works properly and reduces sensibly the computational burden.

7.3.7 Application P1: Fitting a CNN Using Keras—The Lenet-5 Architecture

LeNet-5 is a simple CNN featured for the first time by LeCun et al. (1989). Since its appearance, LeNet-5 has become a reference CNN architecture and a foundation for the the development of more complex CNNs. Indeed, many refinements and developments of it have been carried out over the years, thus generating a family of CNNs based on LeNet-5. Although simple in its structure, LeNet-5 enjoys all the essential properties that a CNN should possess, that is, convolutional, pooling, and fully connected layers.

Figure 7.11 shows a graphical representation of the LeNet-5 architecture where, for the sake of simplicity, I consider only 6 filters for the first convolution and 16 filters for the second convolution. The input is a one-channel gray scale 28×28 image with a pudding of 2. We can easily see the sequence of convolutional and pooling layers, ending with two fully connected layers made respectively of 120 and 84 neurons. In this case, the output layer is made of 10 categories (referring, as we will see, to 10 digits).

In what follows, I briefly summarize the LeNet-5 architecture and parameters to then fit it to the MNIST digit dataset using Keras.

LENET-5 ARCHITECTURE AND PARAMETERS:

1. Layer 1—Convolutional layer (Conv1)

    - Input $(n) = 28$, Padding $(p) = 2$, Filter $(m) = 5 \times 5$, Stride $(s) = 1$
    - Convolutional size: $[(n + 2p - m)/s] + 1 = [(28 + 4 - 5)/1] + 1 = 28$
    - N. of filters = 6, Convolutional output: $28 \times 28 \times 6$
    - Activation function: *ReLu*

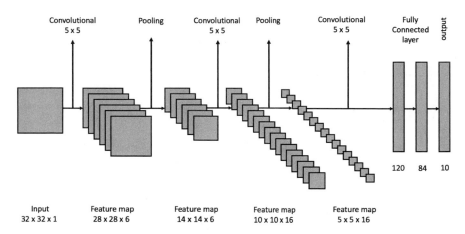

Fig. 7.11 The Lenet-5 convolutional neural network

2. Layer 2—Pooling Layer (Pool1):

   - Input $(n) = 28$, Filter $(m) = 2 \times 2$, Stride $(s) = 2$
   - Pooling size: $[(n + 2p - m)/s] + 1 = [(28 + 4 - 5)/1] + 1 = 14$
   - Pooling outcome: $14 \times 14 \times 6$

3. Layer 3—Convolutional Layer (Conv2):

   - Input $(n) = 14$, Filter $(m) = 5 \times 5$, Stride $(s) = 1$
   - Convolutional size: $[(n + 2p - m)/s] + 1 = ((14 - 5)/1) + 1 = 10$
   - N. of filters = 16, Convolutional output: $10 \times 10 \times 16$
   - Activation function: *ReLu*

4. Layer 4—Pooling Layer (Pool2):

   - Input $(n) = 10$, Filter $(m) = 2 \times 2$, Stride $(s) = 2$
   - Pooling size: $[(n + 2p - m)/s] + 1 = ((10 - 2)/2) + 1 = 5$
   - Pooling outcome: $5 \times 5 \times 16$

5. Layer 5—Fully Connected Layer(Fcl1):

   - N. of weights = 400×120, N. of biases = 120
   - Output: 120×1
   - Activation function: *ReLu*

6. Layer 6—Fully Connected Layer(Fcl2):

   - N. of weights = 120×84, N. of biases = 84
   - Output: 84×1
   - Activation function: *ReLu*

7. Layer 7—Output Layer

   - Activation function: *Softmax*
   - N. of weights = 84×10, N. of biases = 10
   - Output: 10×1

In the code that follows, I provide step-by-step implementation of the previous LeNet-5 architecture in Python using Keras to fit it.

──────────────────────── Python code and output ────────────────────────

```python
# LeNet-5 CNN with Keras

# Clear all
for v in dir(): del globals()[v]

# Import dependencies
import tensorflow as tf
import numpy as np # linear algebra
import pandas as pd # data processing, CSV file I/O (e.g. pd.read_csv)
from keras.models import Sequential
from keras.layers import Conv2D, Dense, MaxPool2D, Dropout, Flatten
from sklearn.model_selection import train_test_split
import matplotlib.pyplot as plt

# Set the working directory
import os
os.chdir('/Users/giocer/Desktop/CCN_application')

# Load train data
df_train = pd.read_csv('mnist_train.csv')

# Put "X_train" and "y_train" as dataframes
X_train = df_train.iloc[:, 1:]
y_train = df_train.iloc[:, 0]

# Put "X_train" and "y_train" as arrays
X_train = np.array(X_train)
y_train = np.array(y_train)

# Load test data
df_test = pd.read_csv('mnist_test.csv')

# Put "X_test" and "y_test" as dataframes
X_test = df_train.iloc[:, 1:]
y_test = df_train.iloc[:, 0]

# Put "X_test" and "y_test" as arrays
X_test = np.array(X_test)
y_test = np.array(y_test)

# Plot the digits
def plot_digits(X, Y):
    for i in range(20):
        plt.subplot(5, 4, i+1)
        plt.tight_layout()
        plt.imshow(X[i].reshape(28, 28), cmap='gray')
        plt.title('Digit:'.format(Y[i]))
        plt.xticks([])
        plt.yticks([])
    plt.show()
plot_digits(X_train, y_train)

# Train-Test Split
X_dev, X_val, y_dev, y_val = train_test_split(X_train, y_train,
                                    test_size=0.03,
                                    shuffle=True,
                                    random_state=2019)

# Make the y a matrix with "columns = class dummies"
T_dev = pd.get_dummies(y_dev).values
T_val = pd.get_dummies(y_val).values

# Reshape to be [samples][width][height][channels]
X_dev = X_dev.reshape(X_dev.shape[0],28,28,1)
X_val = X_val.reshape(X_val.shape[0],28,28,1)
```

```
# Reshape "X_test" for final prediction
X_test = X_test.reshape(X_test.shape[0],28,28,1)

# Define the model
model = Sequential()
model.add(Conv2D(filters=32, kernel_size=(5,5),
                 padding='same',
                 activation='relu',
                 input_shape=(28, 28, 1)))
model.add(MaxPool2D(strides=2))
model.add(Conv2D(filters=64,
                 kernel_size=(5,5),
                 padding='valid',
                 activation='relu'))
model.add(MaxPool2D(strides=2))
model.add(Flatten())
model.add(Dense(120, activation='relu'))
model.add(Dense(84, activation='relu'))
model.add(Dropout(rate=0.5)) # dropout
model.add(Dense(10, activation='softmax'))

# Summarize the model
print(model.summary())

Model: "sequential_2"
_____
Layer (type)                 Output Shape              Param #
=================================================================
conv2d_6 (Conv2D)            (None, 28, 28, 32)        832
_____
max_pooling2d_4 (MaxPooling2 (None, 14, 14, 32)        0
_____
conv2d_7 (Conv2D)            (None, 10, 10, 64)        51264
_____
max_pooling2d_5 (MaxPooling2 (None, 5, 5, 64)          0
_____
flatten_2 (Flatten)          (None, 1600)              0
_____
dense_8 (Dense)              (None, 120)               192120
_____
dense_9 (Dense)              (None, 84)                10164
_____
dropout_2 (Dropout)          (None, 84)                0
_____
dense_10 (Dense)             (None, 10)                850
=================================================================
Total params: 255,230
Trainable params: 255,230
Non-trainable params: 0
_____

# Compile the model
model.compile(loss='categorical_crossentropy',
              metrics=['accuracy'],
              optimizer='adam')

# Fit the model
history=model.fit(X_dev, T_dev,
          batch_size=32,
          epochs=30,
          verbose=1,
          validation_data=(X_val, T_val))

Epoch 1/30
1819/1819 [==============================]
- 56s 31ms/step - loss: 0.2029 - accuracy: 0.9389 - val_loss: 0.0671 - val_
accuracy: 0.9772
```

```
Epoch 2/30
1819/1819 [==============================]
- 61s 33ms/step - loss: 0.0634 - accuracy: 0.9823 - val_loss: 0.0410 - val_
accuracy: 0.9878
...
Epoch 29/30
1819/1819 [==============================]
- 70s 39ms/step - loss: 0.0095 - accuracy: 0.9981 - val_loss: 0.0989 - val_
accuracy: 0.9900
Epoch 30/30
1819/1819 [==============================]
- 68s 37ms/step - loss: 0.0051 - accuracy: 0.9988 - val_loss: 0.1334 - val_
accuracy: 0.9878


# Validate the model
score = model.evaluate(X_val,T_val, batch_size=32)
print(score)
57/57 [==============================]
- 1s 9ms/step - loss: 0.1334 - accuracy: 0.9878
[0.1334291398525238, 0.9877777695655823]


# Graph training and testing results over the epoch
hist = history.history
x_arr = np.arange(len(hist['loss'])) + 1
fig = plt.figure(figsize=(12, 4))
ax = fig.add_subplot(1, 2, 1)
ax.plot(x_arr, hist['loss'], '-o', label='Train loss')
ax.plot(x_arr, hist['val_loss'], '--<', label='Validation loss')
ax.set_xlabel('Epoch', size=15)
ax.set_ylabel('Loss', size=15)
ax.legend(fontsize=15)
ax = fig.add_subplot(1, 2, 2)
ax.plot(x_arr, hist['accuracy'], '-o', label='Train acc.')
ax.plot(x_arr, hist['val_accuracy'], '--<', label='Validation acc.')
ax.legend(fontsize=15)
ax.set_xlabel('Epoch', size=15)
ax.set_ylabel('Accuracy', size=15)
#plt.savefig('figures/15_12.png', dpi=300)
plt.show()


# Make probability predictions

preds=model.predict(X_test)
print(preds.shape)


# Make class predictions
preds = tf.argmax(preds, axis=1)
print(preds.shape)
print(preds)
df=pd.DataFrame(preds)
print(df.head)
0       5
1       0
2       4
3       1
4       9
  ..
59995  8
59996  3
59997  5
59998  6
59999  8


[60000 rows x 1 columns]
```

Fig. 7.12 Graphical representation of the digits in the MNIST dataset

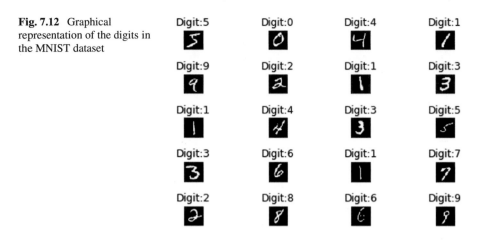

After clearing the Python environment, importing the dependencies, and setting the working directory, I load the MNIST training and testing datasets (in CSV format), both as dataframes and arrays. Using the training dataset, I then plot the digits to visualize the images contained in the dataset (see Fig. 7.12).

Using the `train_test_split` function, I generate the (actual) training features and outcome ([X_dev,y_dev]) and the validation features and outcome ([X_val, y_val]) choosing a size of 3% for the validation sample. In the next step, using the `pd.get_dummies()` function, the outcome variable is transformed into a matrix with columns equal to the class dummies. Because we have 10 digits, the matrix has 10 columns each containing a binary (0/1) variable taking 1 when the class corresponds to the specific digit. We do this both for y_dev and y_val, as Keras requires to supply a multi-class outcome in this form. Similarly, we need to reshape the feature matrices, X_dev and X_val, to transform it into tensors, where four dimensions are considered: observations, width, height, and channels (that, in this case, is equal to one as images are in grey scale). We also reshape X_test.

The inputs are ready for Keras to read them. We can define our model by invoking the `model = Sequential()` statement. Each convolutiuonal layer is set up using `model.add(Conv2D())`, whereas each pooling layer by setting up `model.add(MaxPool2D())`. The arguments of these functions are self-explanatory. After initializing `model.add(Flatten())`, we can start to insert the fully connected layers using the `model.add(Dense())` function, which includes also a "dropout" function to carry out dropout regularization. We finally summarize the model to obtain a compact description of our CNN architecture. At the end of the summary, we can observe that our CNN requires to train 255,230 parameters.

We go ahead by compiling our model using a `categorical_crossentropy` loss function and the Adam (adaptive moment estimation) optimizer, which is a variant of the traditional stochastic gradient descent. We then fit the model to the training data, using a batch size of 32 and 30 epochs. As validation data, we consider [X_val,y_val]. We finally evaluate the out-of-sample performance of our CNN by

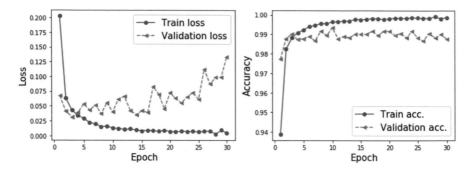

Fig. 7.13 Patterns of the training and testing *loss* and *accuracy* as function and the epochs for the Lenet-5 convolutional neural network estimated using Keras

printing its "score" thus discovering that the testing loss is minimum at a value of 0.13, and the best testing accuracy reached is equal to 0.987. Finally, we graph the patterns of the training and testing *loss* and *accuracy* as function of the epochs, and make probability and class predictions over the original (not so far used) testing dataset (i.e., X_test) (Fig. 7.13).

7.4 Recurrent Neural Networks

Recurrent Neural Networks (RNNs) are special ANN architectures used to model and predict *sequential data*. As argued at the beginning of this chapter, sequential data are data characterized by some degree of *sequential ordering*. Typical examples are text streams, audio clips, video clips, and time-series data. In all these examples, in fact, information comes with a sequential ordering meaning that every instance, coming logically immediately after previous instances, records some of the information brought by previous instances, thus yielding a cumulative information process. Generally, the presence of sequential ordering helps prediction as data are no longer independent as in non-sequential data.

Compared to feedforward neural networks (including also CNNs), RNNs exhibit a recursive temporal dynamic behavior that can take place either at hidden layer level and/or at outcome layer level. The most popular recursive structure is however designed at hidden layer level. Figure 7.14 compares traditional feedforward neural networks with an RNN.

A feedforward network entails that all the relevant information transits from the input (x) to the hidden layer (z), and then from the hidden layer to the output (y). It takes the form of a one-way directed graph where no loops take place. Differently, an RNN allows for loops. More specifically, the hidden layer not only receives information from the input, but also from the hidden layer of the previous time/step.

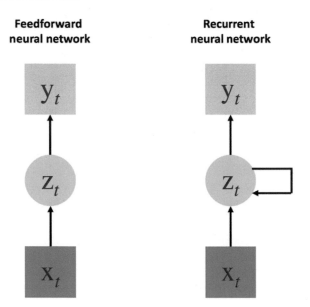

Fig. 7.14 Feedforward versus recurrent neural network architectures

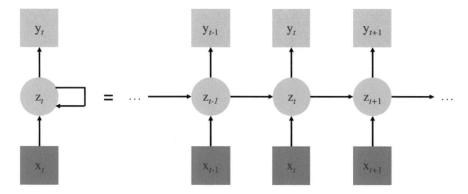

Fig. 7.15 The *unfolded* structure of a recurrent neural network (RNN)

In this regard, the current hidden layer saves information flowing from the past layers and this allows for memorizing past events (*memory storing*).

To better appreciate how information flows through an RNN, it is convenient to describe its *unfolded* architecture showed in Fig. 7.15.

From this figure, it is easy to see the recursive nature of an RNN implying that, at each time, every hidden layer obtains two distinct signals, one from the current input (x_t), and one from the previous hidden layer (z_{t-1}). This process takes place from time 0 (initial time) to time T (final time).

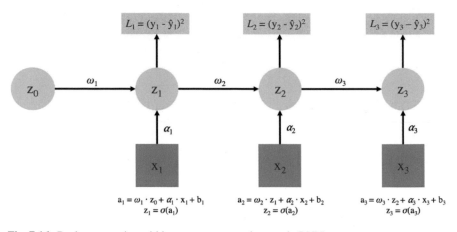

Fig. 7.16 Back-propagation within a recurrent neural network (RNN)

7.4.1 Back-Propagation for an RNN

In theory, fitting an RNN is not different from fitting a standard ANN. Indeed, also in this case, we can use back-propagation. Unfortunately, especially in the presence of long sequences, back-propagation does not work well with RNNs because of the so-called vanishing/exploding gradient phenomenon.

To appreciate the nature of the vanishing/exploding gradient phenomenon, Fig. 7.16 unfolds a 3-time RNN ($t = 1, 2, 3$) where at each time a loss function L_t is computed, with $\widehat{y}_t$ indicating the RNN prediction and y_t the actual outcome; with ω_t, we indicate the weight transporting information from the hidden layer z_{t-1} to the hidden layer z_t, while α_t is the weight bringing to z_t the information stored in the observed input x_t; finally, a_t indicates the pre-activation function, and $\sigma(\cdot)$ the sigmoid activation function, which eventually produces z_t.

As in any other ANN, the aim is that of minimizing the overall loss function L over the vector of parameters ω and α. Here, we are particularly interested in what occurs when we try to find the optimal ω. First, the overall loss function is the sum of the time-specific loss functions, that is

$$L = \sum_{t=1}^{T} L_t \tag{7.7}$$

Minimizing the whole loss requires to minimize the time specific losses $L_t, t = 1, \ldots, T$. As we saw in the previous chapter, this minimization is made possible through the gradient descent updating algorithm of the parameters, entailing that

$$\omega_t := \omega_t + \gamma \frac{\partial L_t}{\partial \omega_t} \tag{7.8}$$

Consider now the loss function at time 3, namely, L_3 and the parameter ω_1. By applying the derivative chain rule, we have that

$$\frac{\partial L_3}{\partial \omega_1} = \frac{\partial L}{\partial z_3} \cdot \frac{\partial z_3}{\partial a_3} \cdot \frac{\partial a_3}{\partial z_2} \cdot \frac{\partial z_2}{\partial a_2} \cdot \frac{\partial a_2}{\partial z_1} \cdot \frac{\partial z_1}{\partial a_1} \cdot \frac{\partial a_1}{\partial \omega_1}. \tag{7.9}$$

By collecting similar terms, we have

$$\frac{\partial L_3}{\partial \omega_1} = \frac{\partial L}{\partial z_3} \cdot [\frac{\partial z_3}{\partial a_3} \cdot \frac{\partial z_2}{\partial a_2} \cdot \frac{\partial z_1}{\partial a_1}] \cdot [\frac{\partial a_3}{\partial z_2} \cdot \frac{\partial a_2}{\partial z_1}] \cdot \frac{\partial a_1}{\partial \omega_1} \tag{7.10}$$

This last equation can be written as

$$\frac{\partial L_3}{\partial \omega_1} = \frac{\partial L}{\partial z_3} \cdot \frac{\partial a_1}{\partial \omega_1} \cdot \prod_{i=1}^{3} \frac{\partial z_i}{\partial a_i} \cdot \prod_{i=1}^{2} \frac{\partial a_{i+1}}{\partial z_i}. \tag{7.11}$$

Now, by considering only the second production operator, we have that

$$\prod_{i=1}^{2} \frac{\partial a_{i+1}}{\partial z_i} = \omega_2 \cdot \omega_3 \tag{7.12}$$

Suppose to have T times and that $\omega_t = \omega$. Then, we have

$$\prod_{i=1}^{T-1} \frac{\partial a_{i+1}}{\partial z_i} = \omega^{T-1} \tag{7.13}$$

If T is sufficiently large, this equation either entails that (i) if $|\omega| < 1$ a vanishing gradient takes place; or (ii) if $|\omega| > 1$ an exploding gradient takes place. Only when $|\omega| = 1$ an RNN with a long-term dependency works properly.

7.4.2 Long Short-Term Memory Networks

In the presence of long-term dependencies (i.e., long sequences) RNNs come across the vanishing/exploding gradient problem. In practice, this limits the use of RNNs in contexts characterized by long sequences, such as high-frequency time-series, or long text speech recognition.

To cope with this issue, Hochreiter and Schmidhuber (1997) proposed the so-called Long Short-Term Memory networks (or LSTMs), a special class of RNNs able to deal with long-term dependencies without incurring the vanishing/exploding gradient problem (Gers et al., 2000). LSTMs are today's most popular models for sequence prediction in deep learning and have been subject to many extensions.

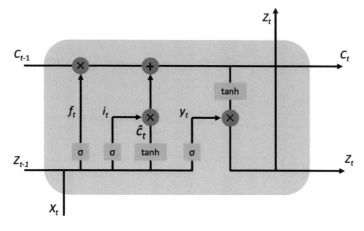

Fig. 7.17 Information flow within a Long Short-Term Memory network (LSTM)

What is the logic of LSTMs, and why do they circumvent the vanishing/exploding gradient problem? To begin with, think about human behavior. To take the decision about whether or not to do something, we rely on predicting certain states of the world. For example, we may decide to take or not an umbrella to go out by predicting whether or not it will rain, based on external temperature and the presence of some clouds. To make this decision (i.e., to make the underlying prediction), we generally do not consider all the previous states of the environment, but we do rely on a few recent previous observations of it. We may say that storing in memory too much information, not only would be costly but also detrimental for prediction, as too old information is expected to be useless for today's prediction purposes, only adding on undesirable noise.

Retaining only information that really matters for prediction is thus more efficient and adequate if one wants to increase prediction accuracy. Indeed, the vanishing/exploding gradient issue takes place exactly because of a too-long memory storing of past states of the world, some having no role for prediction purposes, except accumulating noisy signals. Therefore, retaining only the information that really matters for good prediction, by discarding too long memory sequences, would be a suitable approach for an RNN to fit the data properly without incurring gradient calculation problems.

An LSTM does it. The trick is in putting at work a special object, the so-called *memory cell*. The memory cell is like a box containing the necessary information to predict our target in a proper way. As a bin, the cell can be (partially) emptied of useless information, and filled in only with the information really useful for prediction.

In what follows, I describe two representations of a typical LSTM, one graphical and one mathematical in the form of a system of equations.[1] Figure 7.17 illustrates

---

[1] This part relies on Christopher Olah's "Understanding LSTM Networks"—colah's blog: http://colah.github.io/posts/2015-08-Understanding-LSTMs.

the LSTM architecture and shows how the information flows within it. We can immediately see how, in an LSTM, information is added or removed by means of the so-called *gates*, elementary sub-neural networks regulating the flux of information. To properly describe an LSTM architecture, it is easier to start from the initial inputs, namely, the observable input, X_t, and the lagged hidden input, Z_{t-1}, which comes from a preceding time/step. I describe separately the different components of the LSTM as follows:

1. **Forget gate**. The forget gate takes as inputs X_t and Z_{t-1} and, through a sigmoid activation function, generates as output f_t, a value ranging from 0 to 1. The memory cell value, C_{t-1}, storing information coming from the previous time/step, is thus multiplied by f_t. If $f_t = 1$, all the previous information contained in the cell will flow through the current prediction architecture, whereas if $f_t = 0$ all the previous information is discarded. Intermediate values of f_t define the percentage of past information to retain in the current state. Observe that the symbol "$\times$" indicates a vector multiplication (here, however, we can think of scalar values for the sake of exposition). Mathematically, the forget gate equation is equal to

$$f_t = \sigma_g(\omega_f X_t + \gamma_f Z_{t-1} + b_f).$$

2. **Input gate**. The input gate takes as inputs X_t and Z_{t-1} and, through a sigmoid activation function, generates as output i_t, a value ranging from 0 to 1. This value is used to decide the percentage of information to consider for the candidate new memory cell $\widehat{C}_t$. The input gate equation is

$$i_t = \sigma_g(\omega_i X_t + \gamma_i Z_{t-1} + b_i)$$

3. **Memory cell update**. A candidate new memory cell, $\widehat{C}_t$, is generated using again as inputs X_t and Z_{t-1} through a hyperbolic tangent activation function. It means that $\widehat{C}_t$ ranges from -1 and $+1$. Its equation is

$$\tilde{C}_t = \sigma_c(\omega_c X_t + \gamma_c Z_{t-1} + b_c)$$

The output of the input gate, i_t, sets how much one decides to update the old information contained in C_{t-1}. It means that the new updated memory cell has this equation:

$$C_t = f_t \circ C_{t-1} + i_t \circ \tilde{C}_t$$

4. **Output gate**. The output gate takes as inputs X_t and Z_{t-1} and, through a sigmoid activation function, generates the output y_t as a value ranging from 0 to 1. Its equation is

$$y_t = \sigma_g(\omega_o X_t + \gamma_o Z_{t-1} + b_o)$$

5. **Hidden unit update**. Finally, the new hidden unit Z_t will be generated through this equation:

$$Z_t = y_t \circ \sigma_h(C_t).$$

where Z_t is equal to a hyperbolic tangent transformation of the updated memory cell, scaled by the value of the output (taking, as said, a value between 0 and 1).

We can summarize the entire LSTM as a system of equations as follows:

$$
\begin{aligned}
f_t &= \sigma_g(\omega_f X_t + \gamma_f Z_{t-1} + b_f) \\
i_t &= \sigma_g(\omega_i X_t + \gamma_i Z_{t-1} + b_i) \\
y_t &= \sigma_g(\omega_o X_t + \gamma_o Z_{t-1} + b_o) \\
\tilde{C}_t &= \sigma_c(\omega_c X_t + \gamma_c Z_{t-1} + b_c) \\
C_t &= f_t \circ C_{t-1} + i_t \circ \tilde{C}_t \\
Z_t &= y_t \circ \sigma_h(C_t)
\end{aligned}
\tag{7.14}
$$

This system can be trained over a training dataset and evaluated over a testing dataset as any other learning architecture. It is worth stressing that in none of the equations Z_t depends directly on Z_{t-1}. This prevents the occurrence of the vanishing/exploding gradient, but does not compromise a suitable estimation of the deep parameters.

7.4.3 Application P2: Univariate Forecasting Using an LSTM Network

In this application, I consider a dataset called `hardware` containing quarterly sales data from a huge U.S. regional distributor of building products. This company has various product lines but in this application—based on past experience and using LSTM—I want to forecast *dimensional lumber sales* (i.e., variable `dim`).[2]

To forecast future dimensional lumber sales by an LSTM network, I use the Keras platform. An important aspect when implementing LSTM with Keras is the shape of the inputs one has to provide. Based on Tensorflow, in fact, the main inputs of the Keras LSTM fitting functions are *tensors*. Basically, a tensor is a 3-index (or 3D) matrix. Specifically, when applying LSTMs, the required input is a tensor with three dimensions: (1) *observations*, (2) *time-steps*, and (3) *features*. Figure 7.18 shows two different tensors, where the one on the left panel is a tensor of shape $(1 \times 1 \times 1)$, while the one on the right panel is a tensor of shape $(1 \times 1 \times 2)$. Each single square of this drawing stores one single value (a number).

---

[2] The Python code of this application draws on "Time Series Prediction with LSTM Recurrent Neural Networks in Python with Keras" by Jason Brownlee. Website: https://machinelearningmastery.com/time-series-prediction-lstm-recurrent-neural-networks-python-keras.

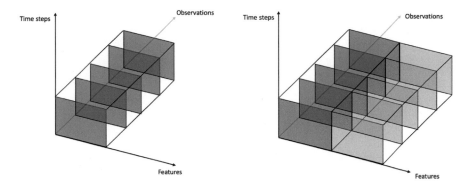

Fig. 7.18 Structure of a *Tensor* used as input in a Long Short-Term Memory (LSTM) networks. A tensor is a 3D array with three axes: *observations, time-steps, features*

Table 7.1 Example of a univariate time-series dataset with three lags

	Y_{t-3}	Y_{t-2}	Y_{t-1}	Y_t
1	.	.	.	32
2	.	.	32	100
3	.	32	100	25
4	32	100	25	9
5	100	25	9	29
6	25	9	29	62
7	9	29	62	90
8	29	62	90	74
9	62	90	74	0
10	90	74	0	5

Generally, however, statistical datasets are provided as 2D matrices, with shape *observations × features*. In the time-series case, features also contain lagged dependent variables. For example, Table 7.1 displays a dataset storing a univariate time-series with three lags.

The shape of this data matrix is 10×4. We can reshape the information contained in this data matrix into a tensor with a shape $10 \times 3 \times 1$, meaning that we have 10 observation times, three lags of the dependent variable, and 1 feature (namely, the variable Y). Observe that also the order matters. For Keras to read correctly the tensor we will provide, the initial 2D dataset to transform into a tensor must have the largest lagged variable appearing as the first variable, and the subsequent lags appearing afterward, with the dependent variable at time t placed in the last position as in Table 7.1.

We are now ready to go through the Python code to fit our LSTM. The whole code is set out below.

```
─────────────────────────── Python code and output ───────────────────────────
# Time-series univariate forecasting using LSTM

# Clear the environment
for v in dir(): del globals()[v]

# Import dependencies
import numpy
import matplotlib.pyplot as plt
from pandas import read_csv
import math
from keras.models import Sequential
from keras.layers import Dense
from keras.layers import LSTM
from sklearn.preprocessing import MinMaxScaler
from sklearn.metrics import mean_squared_error
import os

# Set random seed for reproducibility
numpy.random.seed(5)

# Change the current working directory
os.chdir('/Users/cerulli/Dropbox/Giovanni/Book_Stata_ML_GCerulli/deep/sjlogs')

# Generate dataX and dataY starting from the intial dataset (= 1D-array)
def create_dataset(dataset,lag=1):
dataX, dataY = [], []
for i in range(len(dataset)-lag-1):
a = dataset[i:(i+lag), 0]
dataX.append(a)
dataY.append(dataset[i + lag, 0])
return numpy.array(dataX), numpy.array(dataY)

# Load the intial dataset as dataframe
df = read_csv('hardware.csv', usecols=[1], engine='python')

# Transform "df" into an array of floats
dataset = df.values
dataset = dataset.astype('float32')

# Plot the time-series actual values
plt.plot(dataset,
         label="Actual")
plt.xlabel('Time')
plt.ylabel('Sales of dimensional lumber')
plt.legend()
plt.show()

# Normalize [0:1] the dataset
scaler = MinMaxScaler(feature_range=(0, 1))
dataset = scaler.fit_transform(dataset)

# Split the data into train and test sets
train_size = int(len(dataset) * 0.70)
test_size = len(dataset) - train_size
train, test = dataset[0:train_size,:], dataset[train_size:len(dataset),:]

# Set the lag
lag = 3

# Create the train and test datasets for y and X
trainX, trainY = create_dataset(train, lag)
testX, testY = create_dataset(test, lag)

# Look at the 2D-shape of these datasets
print(trainX.shape)
(59, 3)
```

```
print(trainY.shape)
(59,)
print(testX.shape)
(24, 3)
print(testY.shape)
(24,)

# Reshape the previous inputs in a 3D-array: [observations, time-steps,
 features]
trainX = numpy.reshape(trainX, (trainX.shape[0], trainX.shape[1], 1))
testX = numpy.reshape(testX, (testX.shape[0], testX.shape[1], 1))

# Look at the 3D-shape of trainX and testX
print(trainX.shape)
print(testX.shape)

# Create and fit the LSTM
model = Sequential()
model.add(LSTM(4, input_shape=(lag, 1)))
model.add(Dense(1))
model.compile(loss='mean_squared_error', optimizer='adam')
model.fit(trainX, trainY, epochs=100, batch_size=1, verbose=2)

Epoch 1/100
59/59 - 1s - loss: 0.1240 - 1s/epoch - 18ms/step
Epoch 2/100
59/59 - 0s - loss: 0.0637 - 53ms/epoch - 896us/step
Epoch 3/100
59/59 - 0s - loss: 0.0368 - 52ms/epoch - 883us/step
...
Epoch 98/100
59/59 - 0s - loss: 4.6825e-04 - 51ms/epoch - 868us/step
Epoch 99/100
59/59 - 0s - loss: 3.7587e-04 - 51ms/epoch - 869us/step
Epoch 100/100
59/59 - 0s - loss: 3.9733e-04 - 51ms/epoch - 857us/step

# Make train and test predictions
trainPredict = model.predict(trainX)
testPredict = model.predict(testX)

# Invert predictions to get original data scale
trainPredict = scaler.inverse_transform(trainPredict)
trainY = scaler.inverse_transform([trainY])
testPredict = scaler.inverse_transform(testPredict)
testY = scaler.inverse_transform([testY])

# Compute the root mean squared error (RMSE)
trainScore = math.sqrt(mean_squared_error(trainY[0], trainPredict[:,0]))
print('Train Score: %.2f RMSE' % (trainScore))
testScore = math.sqrt(mean_squared_error(testY[0], testPredict[:,0]))
print('Test Score: %.2f RMSE' % (testScore))

Train Score: 4.00 RMSE
Test Score: 14.30 RMSE

# Plot only training results
plt.plot(trainY[0],label="actual")
plt.plot(trainPredict[:,0], label="forecast")
plt.legend(loc="upper left")
plt.show()

# Plot only testing results
plt.plot(testY[0],label="actual")
plt.plot(testPredict[:,0], label="forecast")
plt.legend(loc="upper left")
plt.show()
```

```
# Plot training and testing results
Y = numpy.append(trainY[0],testY[0])
Y_hat = numpy.append(trainPredict[:,0],testPredict[:,0])
plt.plot(Y,label="actual")
plt.plot(Y_hat, label="forecast")
plt.legend(loc="upper left")
plt.axvline(trainY[0].shape[0], color='k', linestyle='--')
plt.show()
```

After clearing the environment, importing the dependencies, setting the random seed for reproducibility, and changing the current working directory, we define a function `create_dataset` taking as inputs the initial dataset (a 1D vector of data representing the variable Y_t) and the number of lags of Y_t we want to generate. As outputs, this function produces two datasets, `dataX` and `dataY`, where `dataX` contains all the lags of Y_t ordered as in Table 7.1.

Then, we go on by loading the initial dataset as a data frame, transforming it into an array of floats. Thus, we plot the time-series actual values, just for visualization purposes (not reported). For Keras to fit the LSTM network properly, we have to normalize the data to vary between 0 and 1. After this step, we form the train (70% of observations) and test (30% of observations) sets, and set the number of lags equal to three.

We go on by creating the train and test datasets both for Y_t (dependent variable) and X (the lagged variables of Y_t) using the `create_dataset` function. We show that the output datasets `trainX` and `testX` are 2D arrays. As said above, because Keras requires these arrays to be 3D, we reshape them using the `numpy.reshape` function to become [*observations, time-steps, features*]-shaped.

We have now all the ingredients to create and fit the LSTM network using the proper Keras functions. Focusing on this specific chunk of the code, we have that

```
model = Sequential()
model.add(LSTM(4, input_shape=(lag, 1)))
model.add(Dense(1))
model.compile(loss='mean_squared_error', optimizer='adam')
model.fit(trainX, trainY, epochs=100, batch_size=1, verbose=2)
```

After invoking the `model = Sequential()` initialization, we fit an LSTM network with 4 LSTM units; the `input_shpe()` option takes as values the number of lags (equal to 3, in this case) and the number of features (equal to 1, in this case). Then, we compile the model and fit it using 100 epochs and a batch size of 1.

Subsequently, we make train and test predictions, invert predictions to get the original data scale, and compute the root mean squared error (RMSE). This latter result shows some degree of overfitting of this network, with a training error of 4 against a larger test error of 14.30.

Finally, we plot training and testing results and generate the plot visible in Fig. 7.19, where in-sample and out-of-sample prediction patterns are jointly plotted and compared. As expected by looking at the previous test and train RMSEs, while the in-sample fit is very good, the out-of-sample fit proves to be not perfect with forecast values largely smaller than the actual ones. Considering however that the

Fig. 7.19 Long Short-Term Memory networks (LSTM) univariate in-sample and out-of-sample forecasting. Dataset: `hardware`. Variable: `dim`: *dimensional lumber sales*. Number of lags: 3

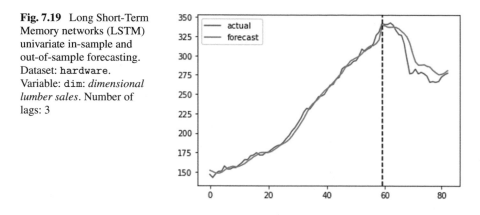

series exhibited an unusual decrease in that part of the sample, we can be sufficiently satisfied with our forecasting result.

7.4.4   Application P3: Multivariate Forecasting Using an LSTM Network

In this application, we apply LSTM forecasting to a multivariate setting. For this purpose, we use German quarterly macroeconomic data on income, consumption, and investment obtained from Lütkepohl (2010). We are interested in forecasting consumption at time t (variable `cons`) based on consumption, investment (`inv`), and income (`inc`) at times $t - 1$ and $t - 2$, thus assuming a number of lags equal to 2. Of course, the number of lags can be properly tuned.

In a multivariate context, prediction can be more accurate as one introduces further predictive information compared to a univariate setting. Indeed, if the added features (and their lags) have substantial explaining power on the target variable, this has the potential to increase forecasting ability.

As in the univariate case, also for the multivariate application the initial dataset must be provided with a proper ordering of the variables. Table 7.2 shows the case of a multivariate setting with two features (X_1 and X_2) and two lags ($t - 1$ and $t - 2$). The initial dataset first shows the largest lag in the first positions, and finally ends up with the dependent variable at time t.

The entire code to apply a multivariate LSTM forecast of consumption as a function on income and investment is showed below.[3]

---

[3] The Python code of this application draws on "Multivariate Time Series Forecasting with LSTMs in Keras" by Jason Brownlee. Website: https://machinelearningmastery.com/multivariate-time-series-forecasting-lstms-keras.

Table 7.2 Example of a multivariate time-series dataset with two features and two lags

Time	Y_{t-2}	$X_{1,t-2}$	$X_{2,t-2}$	Y_{t-1}	$X_{1,t-1}$	$X_{2,t-1}$	Y_t
1	.	.	.	.	.	.	36
2	.	.	.	36	72	16	84
3	36	72	16	84	56	72	67
4	84	56	72	67	17	19	32
5	67	17	19	32	86	42	72
6	32	86	42	72	89	52	60
7	72	89	52	60	20	19	31
8	60	20	19	31	97	68	12
9	31	97	68	12	51	92	30
10	12	51	92	30	94	80	50

```
────────────────────────── Python code and output ──────────────────────────
# Time-series multivariate forecasting using LSTM

# Clear the environment
for v in dir(): del globals()[v]

# Import dependencies
from math import sqrt
from numpy import concatenate
from matplotlib import pyplot
from pandas import read_csv
from sklearn.preprocessing import MinMaxScaler
from sklearn.metrics import mean_squared_error , mean_absolute_error
from keras.models import Sequential
from keras.layers import Dense
from keras.layers import LSTM
import matplotlib.pyplot as plt
import numpy
import os

# Change the current working directory
os.chdir('/Users/cerulli/Dropbox/Giovanni/Book_Stata_ML_GCerulli/deep/sjlogs')

# Allow for perfect reproducibility of results
from numpy.random import seed
seed(5)
import tensorflow
tensorflow.random.set_seed(4)

# Load dataset
dataset = read_csv('lutkepohl2.csv', header=0, index_col=0)
print(dataset)

       cons_2  inv_2  inc_2  cons_1  inv_1  inc_1  cons
time
3         415    180    451     421    179    465   434
4         421    179    465     434    185    485   448
5         434    185    485     448    192    493   459
6         448    192    493     459    211    509   458
7         459    211    509     458    202    520   479
        ...    ...    ...     ...    ...    ...   ...
88       2164    860   2545    2206    870   2580  2225
89       2206    870   2580    2225    830   2620  2235
90       2225    830   2620    2235    801   2639  2237
91       2235    801   2639    2237    824   2618  2250
92       2237    824   2618    2250    831   2628  2271
```

```
[90 rows x 7 columns]

# Form an array from the dataframe
values = dataset.values

# Ensure all data is float
values = values.astype('float32')

# Normalize features
scaler = MinMaxScaler(feature_range=(0, 1))
scaled = scaler.fit_transform(values)

# Split into train and test sets
values=scaled
n_train = 70
train = values[:n_train, :]
test = values[n_train:, :]

# Split into input and outputs (NOTE: y = last column)
train_X, train_y = train[:, :-1], train[:, -1]
test_X, test_y = test[:, :-1], test[:, -1]

# Save the feature matrices as 2D arrays
train_X_2d = train_X
test_X_2d = test_X

# Reshape input to be 3D [observations, timesteps, features]
n_features = 3
n_lags = 2
train_X = train_X.reshape((train_X.shape[0], n_lags, n_features))
test_X = test_X.reshape((test_X.shape[0], n_lags, n_features))

# Design network
model = Sequential()
model.add(LSTM(10, input_shape=(n_lags, n_features)))
model.add(Dense(1))
model.compile(loss='mae', optimizer='adam')

# Fit network
lstm = model.fit(train_X, train_y,
                 epochs=60,
                 batch_size=10,
                 validation_data=(test_X, test_y),
                 verbose=2,
                 shuffle=False)

Epoch 1/60
7/7 - 1s - loss: 0.2890 - val_loss: 0.9276 - 1s/epoch - 210ms/step
Epoch 2/60
7/7 - 0s - loss: 0.2645 - val_loss: 0.8855 - 23ms/epoch - 3ms/step
Epoch 3/60
7/7 - 0s - loss: 0.2422 - val_loss: 0.8442 - 22ms/epoch - 3ms/step
...
Epoch 58/60
7/7 - 0s - loss: 0.0171 - val_loss: 0.0528 - 21ms/epoch - 3ms/step
Epoch 59/60
7/7 - 0s - loss: 0.0169 - val_loss: 0.0535 - 21ms/epoch - 3ms/step
Epoch 60/60
7/7 - 0s - loss: 0.0167 - val_loss: 0.0544 - 24ms/epoch - 3ms/step

# Plot history
pyplot.plot(lstm.history['loss'], '--', label='train loss')
pyplot.plot(lstm.history['val_loss'], label='test loss')
pyplot.legend()
pyplot.show()

# Make train prediction
```

```
yhat_train = model.predict(train_X)

# Make "train_X" a 2D array again
train_X = train_X_2d

# Make test prediction
yhat_test = model.predict(test_X)

# Make "test_X" a 2D array
test_X = test_X_2d

# Invert scaling for train forecast
inv_yhat_train = concatenate((yhat_train, train_X[:,:]), axis=1)
inv_yhat_train = scaler.inverse_transform(inv_yhat_train)
inv_yhat_train = inv_yhat_train[:,0]

# Invert scaling for test forecast
inv_yhat_test = concatenate((yhat_test, test_X[:,:]), axis=1)
inv_yhat_test = scaler.inverse_transform(inv_yhat_test)
inv_yhat_test = inv_yhat_test[:,0]

# Invert scaling for train actual
train_y = train_y.reshape((len(train_y), 1))
inv_y_train = concatenate((train_y, train_X[:, :]), axis=1)
inv_y_train = scaler.inverse_transform(inv_y_train)
inv_y_train = inv_y_train[:,0]

# Invert scaling for test actual
test_y = test_y.reshape((len(test_y), 1))
inv_y_test = concatenate((test_y, test_X[:, :]), axis=1)
inv_y_test = scaler.inverse_transform(inv_y_test)
inv_y_test = inv_y_test[:,0]

# Calculate train and test RMSE
rmse_train = sqrt(mean_squared_error(inv_y_train, inv_yhat_train))
print('Train RMSE: %.3f' % rmse_train)
rmse_test = sqrt(mean_squared_error(inv_y_test, inv_yhat_test))
print('Test RMSE: %.3f' % rmse_test)

Train RMSE: 40.404
Test RMSE: 114.739

# Calculate train and test MAE
mae_train = sqrt(mean_absolute_error(inv_y_train, inv_yhat_train))
print('Train MAE: %.3f' % mae_train)
mae_test = sqrt(mean_absolute_error(inv_y_test, inv_yhat_test))
print('Test MAE: %.3f' % mae_test)

Train MAE: 5.448
Test MAE: 9.958

# Plot test forecast
plt.plot(inv_y_test,label="actual")
plt.plot(inv_yhat_test, label="forecast")
plt.legend(loc="upper left")
plt.show()

# Plot train and test forecast together
Y = numpy.append(inv_y_train,inv_y_test)
Y_hat = numpy.append(inv_yhat_train,inv_yhat_test)
plt.plot(Y,label="actual")
plt.plot(Y_hat, label="forecast")
plt.legend(loc="upper left")
plt.axvline(n_train, color='k', linestyle='--')
plt.show()
```

This code is similar to the one of the univariate case, but presents a few differences.

Similarly, we set out by clearing the environment, importing dependencies, changing the current working directory, and allowing for perfect reproducibility of results. After that, we load the `lutkepohl2` dataset that we transform into a 2D array. The dataset has a form similar to the template in Table 7.2. For example, the variable `inv_2` is investment at time $t - 2$, and so on.

After ensuring that all data is float, and normalizing the features to be between 0 and 1, we split the data into a train set of size 70 observations and a test set of size 20 observations, for a total of 90 time observations.

In tune with the ordering of the data, we use numpy slicing syntax to form the feature matrix (`train_X` and `test_X`) and the target vector (`train_y` and `test_y`). As said, since Keras requires tensors as main inputs, we reshape `train_X` and `test_X` to become 3D arrays of shape [*observations*, *time-steps*, *features*].

We have now all the ingredients to design and fit our multivariate LSTM network. We thus run this bunch of code:

```
# Design network
model = Sequential()
model.add(LSTM(10, input_shape=(n_lags, n_features)))
model.add(Dense(1))
model.compile(loss='mae', optimizer='adam')

# Fit network
lstm = model.fit(train_X, train_y,
                epochs=60,
                batch_size=10,
                validation_data=(test_X, test_y),
                verbose=2,
                shuffle=False)
```

We fit an LSTM network with 10 cells, 2 lags, and 3 features (`inv`, `inc`, and `cons`) using the mean absolute error (MAE) as loss function. Also, we consider 60 epochs with a batch size of 10. We can plot the fitting history to discover at which epoch the algorithm obtains the smallest test loss, with the result visible in Fig. 7.20, where we see that the minimum is reached at the epoch 38.

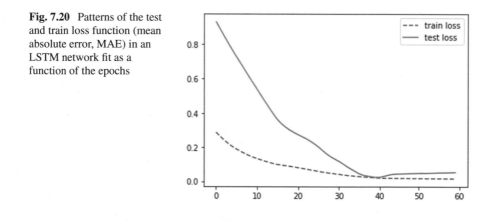

Fig. 7.20 Patterns of the test and train loss function (mean absolute error, MAE) in an LSTM network fit as a function of the epochs

Fig. 7.21 Patterns of the actual and forecast (both in-sample and out-of-sample) time-series values for a multivariate LSTM network fit using as loss function the mean absolute error (MAE). Dataset: `lutkepohl2`. Target variable: `cons` (consumption)

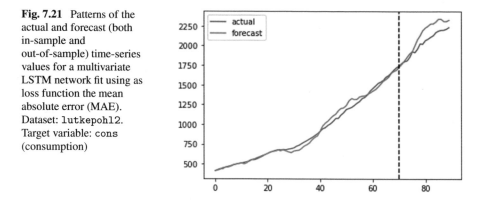

We are now ready to make train and test predictions. However, we need to denormalize the scale of the values of the predictions to make them equal to the original scale. To do this, we use the `train_X` and `test_X` as 2D arrays.

Having done that, we can easily compute train and test RMSE and MAE. Focusing on MAE, our fit obtains a train MAE of around 5.5%, and a test MAE of around 10%, two times larger but still acceptable.

Finally, Fig. 7.21 plots the in-sample (train) and out-of-sample (test) forecasts to visualize our ultimate results. We can see that the out-of-sample fit is decisively worse than the in-sample one, in tune with the previous results on train and test MAE.

7.4.5 Application P4: Text Generation with an LSTM Network

Natural language processing (NLP) refers to a sub-field of artificial intelligence (AI) dealing with giving computers the capacity to process, analyze, understand, and generate text and/or spoken words as human beings do. There are many different types of NLP applications and developments, but one I find particularly fascinating is using NLP to generate new text starting from a given sentence. In this case, we should rely on a generative NLP model, although for this purpose we still can be within the perimeter of supervised Machine Learning.

In this application, I would like to generate plausible text sequences starting from a given seed-sentence based on a *training* operated on a specific document. The used document is Oscar Wilde's celebrated book "The picture of Dorian Gray", freely available at the Project Gutenberg's website (https://www.gutenberg.org) where this and many other books whose copyright is expired are available as text files.

Conceptually, we want to let our LSTM network to learn a mapping between a given sequence of characters (X-input) and the subsequent character (y-output). Put simply, we generate a dataset from a given text where X is a sequence of characters

Table 7.3 Text sequence data representation. This dataset is obtained by generating sequences of *length* equal to 6 and *pace* equal to 1 starting from the initial text: "Gio does good"

Sequence	X	y
1	Gio do	e
2	io doe	s
3	o does	
4	does	g
5	does g	o
6	oes go	o
7	es goo	d

with pre-fixed length, and *y* is the subsequent character. In what follows, I provide an illustrative example to understand how to generate a dataset from a text.

Suppose that our text is made of only this single sentence: "Gio does good". Also, suppose that we consider character sequences of *length* equal to 6 and *pace* equal to 1. Under these assumptions, we can easily build a dataset as the one set out in Table 7.3.

The dataset presents 7 instances (single text sequences), 1 feature (X), and 1 target variable (y). If we transform the characters into integer numbers, the problem becomes one of classifying a new integer by predicting it based on 6 previous integers.

The input data X can be represented as a 2D array with shape [*observations*, *sequence-length*] as visible in Table 7.4. As seen, however, Keras requires the X-input to be a 3D array (a tensor) of shape [*observations*, *sequence-length*, *features*], thus a reshape of the 2D array into a 3D array must be carried out. In this case, the X-input tensor will have shape [*observations*=7, *sequence-length*=6, *features*=1].

Table 7.4 Text sequence data representation. The input data X of Table 7.3 can be represented as a 2D array with shape [*observations*, *sequence-length*]

		Sequence-length					
		1	2	3	4	5	6
Observations	1	G	i	o		d	o
	2	i	o		d	o	e
	3	o	d	o	e	s	
	4		d	o	e	s	
	5	d	o	e	s		g
	6	o	e	s		g	o
	7	e	s		g	o	o

As for the output data, i.e., our target variable y, Keras requests to provide it as a matrix of categorical dummies. For example, if y contains the characters $[a,b,c]$, we build from y the following matrix with columns $[D_a, D_b, D_c]$:

	y	D_a	D_b	D_c
1	a	1	0	0
2	b	0	1	0
3	c	0	0	1
4	a	1	0	0
5	b	0	1	0

We have now all the elements to process the text and fit an LSTM for predicting the next character after a series of characters. This code is showed below.[4]

Python code and output

```
# Fitting an LSTM for "text generation" - Part 1

# Clear the environment
for v in dir(): del globals()[v]

# Small LSTM Network to Generate Text
import numpy
from keras.models import Sequential
from keras.layers import Dense
from keras.layers import Dropout
from keras.layers import LSTM
from keras.callbacks import ModelCheckpoint
from keras.utils import np_utils
import sys
import os

# Change the current working directory
os.chdir('/Users/cerulli/Dropbox/Giovanni/Book_Stata_ML_GCerulli/deep/sjlogs')

# Load the ascii text and covert to lowercase
filename = "ThePictureOfDorianGray.txt"
raw_text = open(filename, 'r', encoding='utf-8').read()
raw_text = raw_text.lower()

# Using "set()" generate unique characters from the text and sort them
chars = sorted(list(set(raw_text)))
print(chars)

# Create a dictonary mapping a unique char into a unique integer
char_to_int = dict((c, i) for i, c in enumerate(chars))
print(char_to_int)

# Summarize the properties of the text
n_chars = len(raw_text)
n_vocab = len(chars)
print("Total Characters: ", n_chars)
print("Total Vocab: ", n_vocab)

# Form the input ('dataX') and target ('dataY') encoded as integers
seq_length = 100
dataX = []
```

---

[4] The Python code of this application draws on: "Text Generation With LSTM Recurrent Neural Networks in Python with Keras" by Jason Brownlee. Website: https://machinelearningmastery.com/text-generation-lstm-recurrent-neural-networks-python-keras.

```
dataY = []
for i in range(0, n_chars - seq_length, 1):
seq_in = raw_text[i:i + seq_length]
seq_out = raw_text[i + seq_length]
dataX.append([char_to_int[char] for char in seq_in])
dataY.append(char_to_int[seq_out])
n_patterns = len(dataX)
print("Total Patterns: ", n_patterns)

# Reshape ´dataX´ as tensor to be [observations, seq-length, features]
X = numpy.reshape(dataX, (n_patterns, seq_length, 1))

# Normalize X between [0;1] dividing by the size of the vocabulary
X = X / float(n_vocab)

# Tranfrom the output variable into a matrix of category-dummies
y = np_utils.to_categorical(dataY)

# Define the LSTM model
model = Sequential()
n_lags = X.shape[1]
n_features = X.shape[2]
model.add(LSTM(256, input_shape=(n_lags, n_features)))
model.add(Dropout(0.2))
model.add(Dense(y.shape[1], activation=´softmax´))
model.compile(loss=´categorical_crossentropy´, optimizer=´adam´)

# Define the checkpoint
filepath="weights-improvement-epoch:02d-loss:.4f.hdf5"
checkpoint = ModelCheckpoint(filepath, monitor=´loss´, verbose=1, save_best_
only= True, mode=´min´)
callbacks_list = [checkpoint]

# Fit the LSTM model
model.fit(X, y, epochs=50, batch_size=64, callbacks=callbacks_list)
```

After running some standard steps, the code shows how to load our ascii text (ThePictureOfDorianGray.txt), and convert all the characters to be lowercase. At this point, the entire text is stored in Python with name raw_text. Using the set() function, we then generate unique characters from the text and sort them. Thus, chars is the list of unique and sorted characters.

We go on by creating a Python dictionary mapping a unique character into a unique integer, thus making a one-to-one correspondence between characters and integer numbers. Then, we summarize the properties of the text discovering that the total number of characters present in the book is 448,543, while the total vocabulary is made of 66 characters.

The code goes on by forming two lists, both encoded as integers: (1) the input feature (dataX) and the target outcome (dataY). We thus generate a data representation similar to the one in Table 7.3. For this purpose, we first set the sequence length (seq_length) equal to 100, to then loop over the number of characters by: (1) creating the single instance (seq_in; seq_out); (2) appending the single instances one after another to form a data structure similar to that of Table 7.3. This data structure has 448,443 instances, exactly 100 fewer instances than the total number of characters.

Now, we have to reshape dataX for this list to become a tensor of shape [*observations, sequence-length, features*]. For this purpose, we use, as usual, the numpy.reshape() function. Also, we normalize the numbers contained in X to range between 0 and 1, as Keras requires.

The target outcome list dataY needs some manipulation. As said, it is necessary to encode it in order to transform it into a matrix of categorical dummies. For this purpose, we use the function np_utils.to_categorical().

We have now all the elements to correctly define our LSTM network. We consider one layer made of 256 LSTM cells, and a dropout probability of 0.2. Given the multiclass nature of *y*, we consider as a loss function the *categorical cross-entropy*.

Before fitting the network, however, we define a *checkpoint* for storing the weights that minimize the loss function along the fitting iteration. We first define a filepath, and then we use the ModelCheckpoint() function, taking filepath as its main argument. Notice that save_best_only=True allows us to save only the best weights within the generated file.

We can fit the model, using 50 epochs, and a batch size of 64 observations. Also, we provide the callbacks option to store in a file the best weights, that we will use to generate new test sequences.

The final part of this application aims at generating the new text. The code below—a continuation of the previous code—carries it out.

```
———————————————————————— Python code and output ————————————————————————
# Fitting an LSTM for "text generation" - Part 2

# Generating new text with the fitted LSTM Network

# Load the network weights with smallest loss value
filename = "weights-improvement-01-3.2338.hdf5"
model.load_weights(filename)
model.compile(loss='categorical_crossentropy', optimizer='adam')

# Convert the integers back to characters
int_to_char = dict((i, c) for i, c in enumerate(chars))

# Pick-up a random sequence as seed sequence (and call it "pattern")
start = numpy.random.randint(0, len(dataX)-1)
pattern = dataX[start]
print("Seed:")
print("\"", ''.join([int_to_char[value] for value in pattern]), "\"")

# Generate 1000 characters starting from our seed sequence
for i in range(1000):
x = numpy.reshape(pattern, (1, len(pattern), 1))
x = x / float(n_vocab)
prediction = model.predict(x, verbose=0)
index = numpy.argmax(prediction)
result = int_to_char[index]
seq_in = [int_to_char[value] for value in pattern]
sys.stdout.write(result)
pattern.append(index)
pattern = pattern[1:len(pattern)]
print("Done.")
```

First, we load the network weights with the smallest loss value, and compile the model again. Then, we convert the integers back to characters using the same loop seen above. We go on by picking up a random sequence (that we call `pattern`) as our seed, and printing it. We use it as the basis for generating the new text.

The most important chunk of code is the one that generates 1,000 characters starting from our seed sequence. Here, we set out by reshaping `pattern` to become a tensor (i.e., a 3D array) so that we can obtain predictions from it. The generated tensor is named x. Next, in order to generate the new text, the code proceeds as follows:

1. Given x, we yield the predicted character that we store into the variable `index` as integer, and into the variable `result` as character.
2. Using the function `sys.stdout.write()`, we print `result` (notice that we do not use `print()` in this case as `print()` returns the result in a new line while we want the characters to appear one after another as allowed by `sys.stdout.write()`).
3. We append `index` at the end of the current `pattern`, and then we update `pattern` by deleting the first character in `pattern` (the selection of `pattern`, in fact, starts from 1 which is the second element of `pattern`).
4. After updating `pattern`, we can run through the second iteration of the loop, print another character as predicted by the network, and so forth until iteration 1,000 is reached.

The ultimate result of this procedure is as follows:

```
Seed:
" hat you had sent it down to
selby, and that it had got mislaid or stolen on the way. you never got i "
t would be a gertlrm tfat to hev farlen an axouasso theng in the sore
io the sore in the sorm and she was a menter of the siree of she was
oo the sorm and saed the man who had been sore that the sore whs
had been sore that he had sooething the saale tfat her oans on the
sorm and she was ao alt oanner and the saale and taek of his oon
things that he had been so the wirlont po the world of the sorm
and she was a mont of the sorm and she was a menter of the siree
of she was oo the sorm and saed the fare of the sorm and saed
the fare of the sorm and saed the fareen and the saie oo the
sorm and saed the fare of the sorm and saed the fare of the
sorm and saed the fare of the sorm and saed the sore oo the
sorm and saed the fare of the sorm and saed the fare of the
sorm and saed the sore oo the sorm and saed the fare of the
sorm and saed the fare of the sorm and saed the sore oo the
sorm and saed the fare of the sorm and saed the fare of the
sorm and saed the sore oo the sorm and saed the fare
Done.
```

The result is interesting, but not perfectly fitting our expectations. In order to better the new text generation, some further steps are needed, including: predicting a smaller number of characters (1,000 may be too much); eliminating punctuation from the original text; increasing the number of training epochs, even to many hundreds; tuning some parameters, such as the batch size; adding a larger number of layers and LSTM memory units within layers; including a dropout option, possibly tuning the dropout probability. All these changes can be computationally costly, but are likely to produce a better readability of the new text thereby generated.

7.5 Conclusion

Deep learning is today at the forefront of new developments in artificial intelligence. In many application contexts, such as imaging recognition and natural language processing, deep learning is contributing to increasingly blurring the boundary between human intelligence and machine intelligence, although the extent of its success depends on the domain of application. For example, in chatbot development—computer programs designed to allow conversation between humans and machines—deep learning has so far proved to be less successful than in other artificial intelligence domains.

It goes without saying that the field of deep learning is large, including also unsupervised deep learning methods such as *autoencoders*, and is constantly expanding. This chapter has provided an introduction to the issue by providing Python applications of both CNNs and RNNs that can be easily extended to similar or even more complex settings.

References

Gers, F. A., Schmidhuber, J., & Cummins, F. (2000). Learning to forget: Continual prediction with LSTM. *Neural Computation, 12*(10), 2451–2471.

Hochreiter, S., & Schmidhuber, J. (1997). Long short-term memory. *Neural computation* (Vol. 9(8), pp. 1735–1780). Publisher: MIT Press.

LeCun, Y., Boser, B. E., Denker, J. S., Henderson, D., Howard, R. E., Hubbard, W. E., & Jackel, L. D. (1989). Backpropagation applied to handwritten zip code recognition. *Neural Computation, 1*(4), 541–551.

Lütkepohl, H. (2010). *New introduction to multiple time series analysis* (1st ed.). Berlin, Heidelberg, New York: Springer.

Nikparvar, B., & Thill, J.-C. (2021). Machine learning of spatial data. *ISPRS International Journal of Geo-Information, 10*(9), 600. Number: 9.

Chapter 8
Sentiment Analysis

8.1 Introduction

This chapter presents the theory and practical applications in Stata, R, and Python of the so-called *sentiment analysis* (SA). As a subfield of natural language processing (NLP), the SA main purpose is to produce a predicting mapping between human textual expressions (generally stored in documents such as emails, web pages, electronic documents, etc.) and specific human feelings. Nowadays, many applied fields of application—such as marketing, advertisement, politics, social psychology, business, and economics—make daily use of SA to target potential new consumers, classify documents, predict people's preferences, or forecast macroeconomic dynamics.

In a nutshell, SA can be defined as a methodology integrating text mining and statistical learning for sentiment prediction. As such, it is a powerful approach to improving predictive analytics, whose popularity has tremendously increased over the years.

After presenting the SA definition and logic, this chapter discusses the three fundamental steps needed to correctly carry out SA, that is, feature engineering, statistical learning, and quality assessment. The chapter goes on by presenting *text vectorization* through the "bag of words" algorithm, which is key to allow for transforming textual documents into datasets. We propose three software applications: (1) classifying the topic of newswires based on their text (Stata); (2) detecting positive/negative sentiment toward movies based on people's reviews (R); building an email "spam" detector (Python). Some conclusions end the chapter.

G. Cerulli, *Fundamentals of Supervised Machine Learning*, Statistics and Computing,
https://doi.org/10.1007/978-3-031-41337-7_8

8.2   Sentiment Analysis: Definition and Logic

As part of artificial intelligence (AI), sentiment analysis (SA) is a natural language processing (NLP) technique using text mining (TM) to generate predicting mappings between human textual expressions and human associated feelings (Medhat et al., 2014; Zhang et al., 2018).

In practical implementations, SA can be seen of a data processing technique that allows to identify and extract subjective information, opinions, and emotions from different textual documents. The main purpose of sentiment analysis is to generate a sort of *polarity* that people can exhibit with regard to a specific topic (such as the war, the climate change, or the pandemic) through the prediction of class labels going from, let's say, "completely positive" to "completely negative", passing through intermediate values which reflect the neutrality that an individual can assume with respect to the subject.

Initially developed in strategic marketing (for example, to support the purchase of specific goods), sentiment analysis has found several applications in many other areas of study, such as politics (to gauge, for example, left-wing or right-wing preferences), social psychology (to measure, for example, the emotional reactions of people to specific social phenomena), as well as to economics for forecasting in advance the trend of some macroeconomic variables, such as consumption, investment, or inflation.

SA makes large use of text mining (TM) (Zong et al., 2022). TM is a technique that uses natural language processing (NLP) to turn the free, unstructured text of documents—i.e., databases such as web pages, newspaper articles, e-mails, and comments on social media—into structured and/or normalized data. In other words, TM allows computers to read texts of different nature to provide specific intelligent tasks. Among the main purposes of TM we can consider: identifying, within texts, the main thematic groups; classifying documents into predefined categories; discovering hidden associations (links between topics, authors, as well as temporal trends); extracting specific information (names of individuals, names of companies, etc.); training search engines; extracting concepts for creating ontologies (Majumdar, 2021; Silge & Robinson, 2017).

Carrying out sentiment analysis applications entails three fundamental steps: (1) feature engineering (through some TM techniques); (2) statistical learning, and (3) quality assessment:

- the feature engineering phase concerns all that is needed to produce a *vectorized representation* of the document, that is, a representation of the text as a standard dataset, with N rows and p variables. This phase is made of sub-phases, such as the identification and selection of the relevant terms, the stemming and/or lemmatization procedures, the weighting, the definition of the stop-words, the features' dimensionality reduction, and possibly the integration with any meta-information;
- the Machine Learning phase applies specific Machine Learning algorithms to the feature engineered document with the objective of achieving unsupervised clustering algorithm for thematic grouping, or supervised algorithm for automatic

classification. It constructs the mapping between the textual expressions and the sentiments;

- the quality evaluation phase consists of the calculation of efficacy (or accuracy) indicators related to the outputs obtained in the second phase, as well as the interpretation of the results thus obtained.

In what follows, we will focus on these three steps to then provide SA applications in Stata, R and Python.

8.2.1   Step 1. Textual Feature Engineering

Textual feature engineering is the process of generating features from texts. A popular way to do this process is the so called "bag of words", which is based on the *n-gram* algorithm. The n-gram algorithm is straightforward to apply, as it generates features by counting the number of times certain words appears in the text. The uni-gram is the simplest (but highly used) case in which one counts the number of times a single word (or token) appear in the text.

To better grasp the functioning of the n-gram algorithm, Table 8.1 provides a simple example of how to generate features form three different texts using a uni-gram approach. For each word, we form a variable whose values indicate the number of times the single word appears in the specific text. For example, the word "also" appear one time both in text 2 and 3, while does not appear at all in text 1. In contrast, the word "football" appears two times only in text 2. Observe that the first variable, n_token, counts the number of words in every text. We can notice that some words have not been used to generate variables. These words—called *stopwords*—are common words generally appearing in a large set of different texts, such as "I", "is", "and", or "to". Because they are commonly used, these words do not add any relevant information to characterize the specific text, and thus they can be excluded from generating variables.

In a 2-gram setting, the algorithm builds a variable for any pair of words obtained in each text. Based on the previous texts, Table 8.2 shows an example on how to generate features from texts using a bi-gram algorithm (single-token variables are not showed for the sake of simplicity).

As we can see, the one generated by the bag of words algorithm is a huge and sparse matrix. To reduce the size of this matrix, thus making computation more affordable, two NPL procedures are generally used in applications: *stemming* (mainly used for sentiment analysis) and *lemmatization* (mainly used for chatbots).

The logic behind stemming and lemmatization is however similar. Many words are semantically similar as, for example, "play", "playing", and "played"; thus, building a unique "token" representative of the three words might be more convenient than considering three distinct words.

The stemming algorithm adopts the strategy of cutting off the end or the beginning of the words with the aim of retaining only a meaningful *root* common to all

Table 8.1 Example of how to generate features from texts using a uni-gram algorithm

Id	Text	n_token	t_also	t_enjoy	t_football	t_gio	t_guitar	t_like	t_name	t_play	t_playing	t_research	t_sometimes	t_watch
1	My name is Gio and I like playing guitar	9	0	0	0	1	1	1	1	0	1	0	0	0
2	Also, I like to play football and watch football	9	1	0	2	0	0	1	0	1	0	0	0	1
3	Sometimes, I also enjoy doing research	6	1	1	0	0	0	0	0	0	0	1	1	0

Table 8.2 Example of how to generate features form texts using a bi-gram algorithm. Single-token variables are not showed for simplicity

Id	Text	n_token	t_also_ enjoy	t_also_ like	t_enjoy_ research	t_football_ watch	t_gio_ like	t_like_ play	t_like_ playing	t_name_ gio	t_play_ football	t_playing_ guitar	t_sometimes_ also	t_watch_ football
1	My name is Gio and I like playing guitar	9	0	0	0	0	1	0	1	1	0	1	0	0
2	Also, I like to play football and watch football	9	0	1	0	1	0	1	0	0	1	0	0	1
3	Sometimes, I also enjoy doing research	6	1	0	1	0	0	0	0	0	0	0	1	0

the inflected words. This strategy can lead to generate words whose meaning in the baseline language (English, for example) is completely nonsensical. This is however not a great problem when the generated variables are used only for predictive purposes, as it happens for the sentiment analysis. The quality of the stemming may vary depending on the languages, as some languages can benefit more than others (Hollink et al., 2004). Among the many stemmers provided in the literature, the Porter stemmer is among the most popular (Porter, 1980).

Unlike stemming, lemmatization looks for meaningful words (the so-called lemmas) best representing a certain set of words that are semantically related to these lemmas. It takes into account the morphological structure of the words. For this purpose, lemmatization makes use of detailed dictionaries which the algorithm can use to create a mapping between the inflected words and the main lemmas. Lemmatization is thus less data-driven than stemming and incorporates external information for "reading" the semantic context within which the words are located (Balakrishnan & Lloyd-Yemoh, 2014).

8.2.2 Step 2. Statistical Learning

In sentiment analysis, we aim at generating a mapping between the texts and a label variable that refers to the texts. This label variable is what we call here the *sentiment*. For example, consider a series of movie reviews, characterized by a large number of words used by some reviewers to comment on a movie. Every reviewer's comment can be either positive, negative, or neutral, thus expressing the point of view (i.e., the sentiment, or feeling) of the reviewer with regard to the considered movie.

The mapping we aim to learn is obtained by transforming a series of texts into a series of variables that form the training dataset, where the target variable (i.e., the sentiment) is known in advance (for example, someone could have collected this information for us). As stressed several times in this book, there are a plethora of statistical learning methods suitable to learn this mapping, and then predicting sentiments when they are unknown (that is, on unlabeled data).

Statistical learning methods are most of the time necessary in this case, as textual mapping—based on the "bag of words" approach—entails a high-dimensional statistical setting, where the number of features is generally much larger than the number of texts. Traditional statistical models, such as the logit, probit, or multinomial regressions, fail to classify texts in this case, as they are unable to deal with high-dimensional settings. In this case, we have to rely on supervised Machine Learning models such as random forests, support vector machines, or boosting that are suitable to be applied in high dimensional data settings.

8.2.3 Step 3. Quality Assessment

Quality assessment refers to the ability of statistical learning models to accurately predict the sentiment. Generally, our sentiment target is a multivalued variable, most of the time taking only two values ("positive" against "negative", "happy" against "unhappy", "good" against "bad", and so forth). We can use all the goodness-of-fit measures and procedures that we have listed for classification in Chap. 2. Generally, when running sentiment analysis, many learners are fitted and compared in terms of accuracy, where accuracy is typically measured by the ROC. In what follows, we will provide a series of examples using Python, Stata, and R.

8.3 Applications

8.3.1 Application S1: Classifying the Topic of Newswires Based on Their Text

In this application, we use the `reuters` dataset on news from Reuters.[1] The dataset collects 2,000 newswires for 147 countries, on 120 topics. We aim at classifying this news based on their text. For this purpose, we use the Stata user-written command `ngram` (Schonlau et al., 2017), implementing the "bag of words" algorithm (i.e., the *vectorization* of the words contained in the texts). We run `ngram` over the `title` variable, using the options `threshold(3)` and `stemmer`. The option `threshold(3)` set to 3 is the minimum number of occurrences required across all observations to count a particular n-gram. The larger this threshold (set, by default, equal to 5), the smaller the number of variables created, as words with too few occurrences will not be used to generate related variables. The option `stemmer` applies Porter's stemming algorithm to collapse words having the same root. After applying a unigram (which is the default, that is, option `degree(1)`) "bag of words", we obtain a series of derived features all starting with the suffix "t_", that we can use as predictors in a random forest classification of the topic "trade". In other words, we want to know how good is the random forest in correctly classifying news when the main topic of the news is a trade-related subject. The below Stata code carries out this analysis.

---

[1] Data are freely available at the UCI Machine Learning archive http://archive.ics.uci.edu/ml/datasets/Reuters-21578+Text+Categorization+Collection.

```
────────────────────────────────── Stata code and output ──────────────────
.* Classyfing the topic of newswires based on their text
.
. * Import data
. sysuse "reuters", clear
.
. * Use the stata command "ngram" for "bag of word" text vectorization
.
. * Set the "locale" (language)
. set locale_functions en_US
.
.
. * Extract unigrams with stemming
. ngram title, thresh(3) stem
Removing stopwords specified in stopwords_en.txt
stemming in ´en´
.
.
. * Compute the prediction accuracy in classifying a topic of the document
. * using a "random forest" classifier
.
. * Form the target and the features
. global y "to_trade" // topic = "trade"
.
. global X "t_*"
.
.
. * Split the dataset into training, validation, and testing samples
. splitsample, gen(sample) split(.6 .2 .2) rseed(1010)
.
.
. preserve
.
. * Remove the testing units (sample=3)
. drop if sample==3
(400 observations deleted)
.
. replace sample=0 if sample==2
(400 real changes made)
.
. label define slabel 1 training 0 testing
.
. label values sample slabel
.
.
. * Find the optimal parameters "numvars" and "depth" of the random forest classifier
. gridsearch rforest $y $X , ///
> type(class) iter(500) seed(1) par1name(numvars) par1list(3 6 9 12) ///
> par2name(depth) par2list(0 3 6 9 12) criterion(accuracy) method(trainvar sample)
counter 1 1
counter 1 2
Warning: Predicted values are all identical or missing (standard deviation=0)
counter 1 3
Warning: Predicted values are all identical or missing (standard deviation=0)
counter 1 4
Warning: Predicted values are all identical or missing (standard deviation=0)
counter 1 5
Warning: Predicted values are all identical or missing (standard deviation=0)
counter 2 1
counter 2 2
Warning: Predicted values are all identical or missing (standard deviation=0)
counter 2 3
Warning: Predicted values are all identical or missing (standard deviation=0)
counter 2 4
Warning: Predicted values are all identical or missing (standard deviation=0)
counter 2 5
```

```
Warning: Predicted values are all identical or missing (standard deviation=0)
counter 3 1
counter 3 2
Warning: Predicted values are all identical or missing (standard deviation=0)
counter 3 3
Warning: Predicted values are all identical or missing (standard deviation=0)
counter 3 4
Warning: Predicted values are all identical or missing (standard deviation=0)
counter 3 5
Warning: Predicted values are all identical or missing (standard deviation=0)
counter 4 1
counter 4 2
Warning:
Predicted values are all identical or missing (standard deviation=0)
counter 4 3
Warning: Predicted values are all identical or missing (standard deviation=0)
counter 4 4
Warning: Predicted values are all identical or missing (standard deviation=0)
counter 4 5
Warning: Predicted values are all identical or missing (standard deviation=0)
(20 real changes made)
```

	accuracy	numvars	depth	seconds
1.	.975	3	0	18

```
. * Explore the grid search results
. matrix list r(gridsearch)

r(gridsearch)[20,4]
        numvars      depth   accuracy      seconds
   r1          3          0  .97500002       18.389
   r2          3          3  .97500002    2.9809999
   r3          3          6  .97500002    3.4530001
   r4          3          9  .97500002    4.0830002
   r5          3         12  .97500002    4.5929999
   r6          6          0  .97500002    17.667999
   r7          6          3  .97500002    3.2030001
   r8          6          6  .97500002        4.079
   r9          6          9  .97500002    4.9899998
  r10          6         12 .97500002    5.6589999
  r11          9          0  .97500002    18.188999
  r12          9          3  .97500002        3.569
  r13          9          6  .97500002        4.704
  r14          9          9  .97500002        5.921
  r15          9         12  .97500002    7.0110002
  r16         12          0  .97500002    19.122999
  r17         12          3  .97500002        3.888
  r18         12          6  .97500002    5.4180002
  r19         12          9  .97500002    6.6620002
  r20         12         12  .97500002    8.0839996
. restore


. * Fit "rforest" only on the training dataset at optimal parameters
. rforest $y $X if sample==1 , type(class) iterations(500) numvars(3) depth(0) seed(10)


. * Obtain predictions
. predict y_hat

. * Compute the accuracy over "sample-3" (the pure out-of-sample)
. preserve

. keep if sample==3
```

```
(1,600 observations deleted)

. gen accuracy=($y==y_hat)

.
. * Desplay the testing MCE
. qui sum accuracy

. di "The testing ACCURACY is " r(mean)
The testing ACCURACY is .9775

. restore
```

Let's comment on the Stata code and output. After importing the dataset, we use the command `ngram` for text vectorization. We set the "locale" for the English language, as stemming depends on the specific language of the texts. We thus extract unigrams with stemming. We have all the ingredients to compute the prediction accuracy in classifying the topic "trade" of the document. For this purpose, we form the target and the features and split the dataset into training, validation, and testing samples.

We apply the random forest classifier using the Stata commands `gridsearch` and `rforest`. The `gridsearch` command carries out cross-validation (CV) for finding the optimal values for the hyper-parameters of the random forest classifier that are `numvars` (number of splitting variables at each split), and `depth` (baseline depth of the trees). We use a one-fold CV, where the same training and validation sets are used for each value of the grid of the hyper-parameters. For CV, therefore, we remove the testing units (that is, we exclude observation with `sample=3`).

After running `gridsearch` with `rforest`, we explore the results and find that the optimal values for the hyper-parameters are `numvars=3` and `depth=0`. Then, we fit `rforest` only over the training dataset at the optimal parameters, thus obtaining the predictions. We conclude the application by computing the accuracy over the testing dataset (`sample=3`), thus getting a pure out-of-sample accuracy of 97.75%, a rather good accuracy level. We can conclude that our algorithm is able to predict whether the title of a news regards the topic "trade" with an average accuracy of 97.75%.

8.3.2   Application R1: Sentiment Analysis of IMDB Movie Reviews

In this application, we run sentiment analysis using the popular IMDB dataset, containing a large set of highly polar movie reviews, using R (Naldi, 2019). Every review can be either "positive" or "negative", thus allowing for carrying out a binary sentiment classification (Maas et al., 2011).[2]

---

[2] The dataset can be downloaded at this link: https://www.kaggle.com/code/lakshmi25npathi/sentiment-analysis-of-imdb-movie-reviews. More dataset information are available here: http://ai.stanford.edu/~amaas/data/sentiment/.

The original dataset contains 25,000 training observations, but in this application we only use 5,000 reviews. We split the exercise in two parts: (1) first, we show how to generate *word vectorization* from the texts of the reviews, by also cleaning the texts; (2) second, we apply a sentiment analysis by classifying the reviews using three popular classifiers: gradient boosting machine (GBM), neural network (NN), and random forests (RF). Our main purpose is to set which is the best classifier, and for this task we first plot the ROC curve for each classifier, and then compare their area under the curve (AUC) as our leading goodness-of-fit measure. Observe that we optimally fine-tune the hyper-parameters of each classifier using K-fold cross-validation. The classifier obtaining the largest AUC will win this competition.

We implement sentiment analysis in R by generating the word vectorization from the texts of the reviews. This leads to construct the so-called *document term matrix* (DTM), the sparse matrix whose columns contain the frequency of the reviews' words. The following R code produces this matrix and save it as a Stata dataset.

———————————————————— R code and output ————————————————————

```
# Sentiment analysis: construction of the Document Term Matrix

# Clear the environment
> rm(list = ls())

# Set the current directory
> setwd("/Users/cerulli/Dropbox/Giovanni/Book_Stata_ML_GCerulli/sentiment/SA_in_R")

# Load needed libraries
> library("tm")
> library("stringr")
> library("SnowballC")
> library("textstem")
> library("devtools")
> library("foreign")

# Read the data
> Sample <- read.csv("Test.csv")
> ls(Sample)
[1] "label" "text"

# Select only the "text" and form the VCorpus
> DB <- VCorpus(VectorSource(Sample$text))

# Form a dataframe containing the label
> Label <- as.data.frame(Sample$label)

# Rename the label variable as "label"
> names(Label)[1] <- "label"

# Clean the texts
> toSpace <- content_transformer(function(x, pattern) {return (gsub(pattern, "", x))})
> DB <- tm_map(DB, toSpace, "-")
> DB <- tm_map(DB, toSpace, "\texteuro")
> DB <- tm_map(DB, toSpace, ":")
> DB <- tm_map(DB, toSpace, "\u0096")
> DB <- tm_map(DB, removePunctuation)
> DB <- tm_map(DB, toSpace, "'")
> DB <- tm_map(DB, toSpace, "'")
> DB <- tm_map(DB, toSpace, "-")
> DB <- tm_map(DB, content_transformer(tolower))
```

```
> DB <- tm_map(DB, removeNumbers)
> DB <- tm_map(DB, removeWords, stopwords("english"))  # remove english stopwords
> DB <- tm_map(DB, stripWhitespace)
> DB <- tm_map(DB,stemDocument)   # stemming

> # Create the Document Term Matrix
> dtmr <-DocumentTermMatrix(DB,
+                         control=list(wordLengths=c(1, Inf),
+                                      removePunctuation = TRUE,
+                                      removeNumbers= TRUE,
+                                      stopwords = TRUE,
+                                      bounds = list(global = c(500,Inf))))

# Transform the DTM into a data frame
> adtm.df<-as.data.frame(as.matrix(dtmr))
> ls(adtm.df)
  [1] "aaron"           "abandon"         "abbott"          "abc"
  [5] "abil"            "abl"             "aboard"          "abomin"
  [9] "abort"           "abound"          "aboutbr"         "abraham"
  ...

# Write the sparse matrix and the label into a Stata dataset
> write.dta(cbind.data.frame(Label,adtm.df),"/Users/cerulli/Dropbox/Giovanni/Book_Stata_ML_GCerulli/
sentiment/SA_in_R/mydata2.dta")
```

Let's comment on the code and results. After cleaning the R environment, and setting the current directory, we load the needed libraries, and in particular the tm library, specialized in carrying out text mining. Then, we read the data contained in the CSV file Test.csv. This represents our sample, made of two variables, the binary sentiment variable (with 1 = "positive", and 0 = "negative") called label, and the variable containing in each cell a different review, called text. We select only the variable text, and form the vectorized corpus using the function VCorpus(VectorSource()).

We also form a dataframe containing the label, and rename the label variable as "label". In order to remove any symbol that is not informative to generate prediction, we use the tm_map() function, eliminating symbols not informative for prediction, or even weird characters. This procedure cleans the texts. Finally, using tm_map(DB,stemDocument), we use a proper stemmer for the texts.

We have now the main ingredients to apply the function DocumentTermMatrix() that creates the document term matrix (DTM). We use some standard options of this function, including:

- wordLengths=c(1, Inf): this option discards words shorter than the minimum word length (1, in this case) or longer than the maximum word length (whatever length, in this case);
- removePunctuation=TRUE: when this option is set to TRUE (as in this case), it removes all punctuation characters;
- removeNumbers=TRUE: when this option is set to TRUE (as in this case), it removes all numbers from the set of words;
- stopwords=TRUE: when this option is set to TRUE (as in this case), it removes stopwords;

- bounds = list(global = c(500,Inf)): this option indicates to discard words that appear in less documents than the lower bound (500, in our case), or in more documents than the upper bound (infinity).

After this step, we transform the DTM into a data frame and save the sparse matrix and the label as a Stata dataset called mydata.dta. If we look at this dataset, we can observe that there are 5,000 observations (i.e., reviews), and 147 variables, each containing the frequency of the word in the single review. Actually, the number of total words is much larger, around 5,000. For illustrative purposes (mainly due to computational time), however, we consider here only 147 words obtained when we set the lower bound in the option bounds equal to 500. Smaller values of this lower bound would increase the number of retained words, by possibly increasing the goodness-of-fit at expense of a greater computational burden.

We are now ready to apply our classifiers. The following code provides the R script and the results.

―――――――――――――――――――――――― R code and output ――――――――――――――――――――――――

```
# Sentiment Analysis: plot the ROC curve for several models

# Set the directory
> setwd("/Users/cerulli/Dropbox/Giovanni/Book_Stata_ML_GCerulli/sentiment/SA_in_R")

# Clean the R environment
> remove(list = ls())

# Load the dependencies
> library(haven)
> library(plyr)
> library(dplyr)      # for data manipulation
> library(caret)      # for model-building
> library(purrr)      # for functional programming (map)
> library(pROC)       # for AUC calculations
> library(PRROC)      # for Precision-Recall curve calculations
> library(gbm)        # for gbm importance
> library(tree)       # for tree classification
> library(nnet)       # for neural network classification
> library(foreign)

# Load the full dataset
> data <- read.dta("mydata2.dta")

# Form the target variable "y" (i.e., the sentiment)
> y <- as.data.frame(data[,1])

# Generate the random index "train_index" for train and test sets
> set.seed(999)
> train_index = sample(1:nrow(data), nrow(data)/2)

# Generate the "data.train" dataset
> data.train=data[train_index,]
> table(data.train$label)

   0    1
1259 1241

# Generate the "data.test" dataset
> data.test=data[-train_index,]
> table(data.test$label)
```

```
   0    1
1236 1264


# Generate the "y_train"
> y_train=y[train_index,]

# Generate the "y_test"
> y_test=y[-train_index,]

# Use the CARET package to fit the models using K-fold cross-validation

# Step 1. Define the "trainControl()" function
> ctrl <- trainControl(method = "repeatedcv",
+                       number = 10,
+                       repeats = 1,
+                       summaryFunction = twoClassSummary,
+                       classProbs = TRUE)

# Step 2. Transform the train and test target variable into a factor variable
# and rename the value so that: 1=YES, 2=NO

# For train
> data.train$label <- as.factor(data.train$label)
> levels(data.train$label) <- c("NO", "YES")

# For test
> data.test$label <- as.factor(data.test$label)
> levels(data.test$label) <- c("NO", "YES")

# Step 3. Fit 3 learners (or models):
#         1. GBM (Gradient Boosting Machine)
#         2. RF  (Random Forest)
#         3. NN  (Neural Network)

# GBM
> GBM_fit<- train(label ~ .,
+                 data = data.train,
+                 method = "gbm",
+                 verbose = FALSE,
+                 metric = "ROC",
+                 trControl = ctrl)

# RF
> RF_fit <- train(label ~ .,
+                 data = data.train,
+                 method = "rf",
+                 verbose = FALSE,
+                 metric = "ROC",
+                 trControl = ctrl)

# NN
> NN_fit <- train(label ~ .,
+                 data = data.train,
+                 method = "nnet",
+                 verbose = FALSE,
+                 metric = "ROC",
+                 trControl = ctrl)


# Define function "test_roc" to obtain the ROC data
> test_roc <- function(model, data) {
+   roc(data$label,
+       predict(model, data, type = "prob")[, "YES"])
+ }

# Apply the "test_roc" function to every model of "model_list"
# and store the results into the list "model_list_roc"
```

```
> model_list_roc <- model_list %>%
+   map(test_roc, data = data.test)

# Print the AUC for each model
> model_list_roc %>%
+   map(auc)

$GBM
Area under the curve: 0.8177

$RF
Area under the curve: 0.8153

$NN
Area under the curve: 0.8111

# Define an empty list "results_list_roc"
> results_list_roc <- list(NA)

# Initialize a counter named "num_mod" equal to 1
> num_mod <- 1

# For each model, put into the list "results_list_roc" 3 dataframes
# containing "tpf" and "fpr"
> for(the_roc in model_list_roc){
+   results_list_roc[[num_mod]] <-
+     data_frame(tpr = the_roc$sensitivities,
+                fpr = 1 - the_roc$specificities,
+                model = names(model_list)[num_mod])
+   num_mod <- num_mod + 1
+ }

# Transform the list "results_list_roc" into a unique dataframe "results_df_roc"
> results_df_roc <- bind_rows(results_list_roc)
> print(results_df_roc)
# A tibble: 4,907  3
      tpr   fpr model
    <dbl> <dbl> <chr>
 1  1     1     GBM
 2  1     0.999 GBM
 3  1     0.998 GBM
 4  1     0.998 GBM
 5  1     0.997 GBM
 6  0.999 0.997 GBM
 7  0.999 0.996 GBM
 8  0.999 0.995 GBM
 9  0.999 0.994 GBM
10  0.999 0.994 GBM
#  with 4,897 more rows

# Plot the ROC curve for all 3 models
> custom_col <- c("#000000", "#009E73", "#0072B2")
> ggplot(aes(x = fpr,  y = tpr, group = model), data = results_df_roc) +
+   geom_line(aes(color = model), size = 1) +
+   scale_color_manual(values = custom_col) +
+   geom_abline(intercept = 0, slope = 1, color = "gray", size = 1) +
+   theme_bw(base_size = 18)
```

Let's comment on the code and its outputs. We start by setting the current directory, cleaning the R environment, and loading the dependencies. Then, we load the full dataset mydata.dta, and form the target variable y (i.e., the sentiment). Also, we generate the random index train_index for indexing the train and test sets,

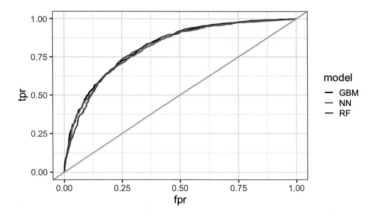

Fig. 8.1 Sentiment analysis of the IMDB movie reviews: plot of the ROC curve for the GBM (Gradient Boosting Machine), RF (Random Forests), and NN (Neural Network)

thus generating the data.train and data.test datasets. Also, we generate the y_train, and y_test, as our targets.

We go on by using the CARET package to fit the models using repeated K-fold cross-validation (notice that only one repetition is used in this illustrative example). We proceed here in three steps: (1) we define the trainControl() function as usual with CARET; (2) we transform the train and test target variables into factor variables and rename their value so that: 1=YES, 2=NO; (3) we fit three classifiers: GBM (Gradient Boosting Machine), RF (Random Forests), and NN (Neural Network). We choose as goodness-of-fit metric the ROC for every classifier.

To obtain the ROC data and draw the ROC curve, we define function test_roc, to then apply it to every fitted model, storing the results into the list model_list_roc. We can now print the area under the curve (AUC) for each model, obtaining one value for each learner: 0.8177 (for GBM), 0.8153 (for RF), and 0.8111 (for NN). The winner is GBM, obtaining the largest AUC, but the difference in the AUC is so small that we can consider the three approaches as equivalent.

We want now to provide a graphical representation of the ROC curves. For this purpose, we define an empty list results_list_roc, and initialize a counter named num_mod equal to 1. For each learner, we put into the list results_list_roc three data frames containing the tpf and fpr. Then, we transform the list results_list_roc into a unique dataframe called results_df_roc, and print it.

Finally, using the ggplot() function, we plot the three ROC curves in one single graph visible in Fig. 8.1.

8.3.3   Application P1: Building a "Spam" Detector in Python

As sentiment analysis application in Python, this application proposes to build a "spam" detector (Rathee et al., 2018). We consider the spam.csv dataset, containing 5,572 messages tagged according to whether they are "ham" (good messages), or "spam" (bad messages). Each line of the dataset contains two columns: v1, which contains the label (either "ham" or "spam"), and v2, which contains the raw text. This dataset can be freely downloaded at this page: https://www.kaggle.com/datasets/uciml/sms-spam-collection-dataset. Below we display the Python code and output.

```
──────────────────────────────── Python code and output ────────────────────────────────
>>> # Sentiment Analysis with Python: building a "spam" detector

>>> # Import basic dependencies
... import pandas as pd
>>> import os
>>>
>>> # Set the current directory
...
os.chdir('/Users/cerulli/Dropbox/Giovanni/Corsi_Stata_by_G.Cerulli/Course_Stata_Machine_Learning/M
> y_ML_Course/Sentiment_Analysis/SA_Python')
>>>
>>> # Import the dataset "spam.csv" as Pandas dataset
... df = pd.read_csv('spam.csv',encoding='Windows-1252')
>>>
>>> # Drop the last 3 columns
... df.drop(columns=['Unnamed: 2','Unnamed: 3','Unnamed: 4'],inplace=True)
>>>
>>> # Renaming the two left columns
...
df.rename(columns={'v1':'tag','v2':'text'},inplace=True)
>>>
>>> # Describe the data
... df.info() <class 'pandas.core.frame.DataFrame'> RangeIndex: 5572
entries, 0 to 5571 Data columns (total 2 columns): tag      5572
non-null object text     5572 non-null object dtypes: object(2)
memory usage: 87.2+ KB
>>> df.head()
      tag                                               text
0   ham  Go until jurong point, crazy.. Available only ... 1    ham
Ok lar... Joking wif u oni... 2  spam  Free entry in 2 a wkly comp
to win FA Cup fina... 3   ham  U dun say so early hor... U c already
then say... 4   ham  Nah I don't think he goes to usf, he lives
aro...
>>>
>>> # Look at the number of "ham" and "spam"
... print(len(df[df.tag=="ham"])) 4825
>>> print(len(df[df.tag=="spam"]))
747
>>>
>>> # Form the features X ("text")
... X = df['text']
>>>
>>> # Form the target y ("tag")
... y = df['tag']
>>>
>>> # Set a random seed
... from numpy import random, array
>>> random.seed(0)
>>>
>>> # The "CountVectorizer()" class allows to vectorize the training dataset
... # Option "ngram_range=(1,2)" means that we are considering both
```

```
1-gram and 2-gram
>>> # Load the "CountVectorizer()"
... from sklearn.feature_extraction.text import CountVectorizer
>>>
>>> # Initialize "CountVectorizer()"
... vect=CountVectorizer(ngram_range=(1,2), ...
stop_words="english", ...                              max_df=0.7) ... ...
# Generate the "document term matrix" X using the "fit_transform()"
function
>>> X=vect.fit_transform(X)
>>>
>>> # Split features and target into a train and test set
... from sklearn.model_selection import train_test_split
>>> X_train, X_test, y_train, y_test = train_test_split(X,y,test_size=0.3)
>>>
>>> # Import the Naive Bayes classifier
... from sklearn.naive_bayes import BernoulliNB
>>>
>>> # Initialize the model
... model = BernoulliNB()
>>>
>>> # Fit the model over the train set
... model.fit(X_train, y_train) BernoulliNB()
>>>
>>> # Predict over the test data
... pred_test = model.predict(X_test)
>>>
>>> # Import "accuracy_score"
... from sklearn.metrics import accuracy_score
>>>
>>> # Compute the accuracy
... accuracy = accuracy_score(y_test, pred_test)
>>>
>>> # Print the accuracy
... print(accuracy) 0.9168660287081339
>>>
>>> # Predict the tag of an untagged message
... msg_new=array(['Hi dear, are you available for a pizza
tonight?'])
>>> msg_new=vect.transform(msg_new)
>>> print(model.predict(msg_new))
['ham']
```

Let's comment on the code and results. We set out by importing basic dependencies, setting the current directory, and importing the dataset spam.csv as a Pandas data frame. Then, we drop the last 3 columns (useless for our purposes), rename the two left columns as tag and text, and describe the data. We can observe that the number of observations is, as expected, 5,572 with the number of spams equal to 747, and that of hams equal to 4,825. After this inspection, we form the input dataset X (containing only the variable text), and the target y (containing only the variable tag). Also, we set a random seed for replication purposes.

The most important step of this application concerns the transformation of the input test (variable X) into a document term matrix (DTM), as we did in Stata and R in the previous sections. In Python, for this purpose, we use the Scikit-learn module CountVectorizer(), which allows to "vectorize" the training dataset, thus generating a proper DTM.

The module CountVectorizer() takes as input a corpus (texts) that, in our case, is contained in the data frame X, plus a series of options determining the final form of the bag-of-words sparse matrix. Here we use only a few but relevant options: the option `ngram_range=(1,2)` allows for considering both 1-gram and 2-gram; the option `stop_words="english"` considers a pre-fixed vocabulary eliminating all the stop words of the English language; the option `max_df=0.7` permits to eliminate all the words that appear in more than 70% of documents, as these words are probably unable to discriminate between a ham or a spam, given that they are highly frequent within the messages. After initializing `CountVectorizer()` with these options, we generate the DTM named (again) X, using the `fit_transform()` function. Then, we split features and target into a train set and a test set using `train_test_split()`, and import the Naïve Bayes classifier through the module `BernoulliNB()`.

We fit the model over the train set and predict over the test data, thus obtaining an accuracy of around 92%. Just as an illustrative example, we predict the tag of an untagged message. The nature of the message considered is likely to be a ham. This is confirmed by the predicted label.

8.4 Conclusions

The aim of this chapter was to provide the reader with a primer on the theory and practice of sentiment analysis. Sentiment analysis has a large reach of applications, spanning from marketing, business, and economics to sociology and social psychology, and it is one of the most frequently used techniques within the field of natural language processing (NLP).

We may rethink sentiment analysis as a method to integrate text mining and Machine Learning, with the aim of giving human language (stored in textual documents) a predictive power for carrying out various tasks, including document classification, human preferences prediction, and so on.

Of course, the information processing needed to transform texts into viable features (a process called in the literature as *feature engineering*) can require more laborious steps than those outlined in the applications offered by this chapter, but the underlying working philosophy is the one herein developed.

References

Balakrishnan, V., & Lloyd-Yemoh, E. (2014). *Stemming and lemmatization: A comparison of retrieval performances*, Seoul, Korea (pp. 174–179). https://eprints.um.edu.my/13423/.

Hollink, V., Kamps, J., Monz, C., & de Rijke, M. (2004). Monolingual document retrieval for European languages. *Information Retrieval, 7*(1), 33–52.

Maas, A. L., Daly, R. E., Pham, P. T., Huang, D., Ng, A. Y., & Potts, C. (2011). Learning word vectors for sentiment analysis. In *Proceedings of the 49th Annual Meeting of the Association for Computational Linguistics: Human Language Technologies* (pp. 142–150). Portland, Oregon, USA: Association for Computational Linguistics. http://www.aclweb.org/anthology/P11-1015.

Majumdar, P. (2021). *Learn Emotion Analysis with R: Perform Sentiment Assessments, Extract Emotions, and Learn NLP Techniques Using R and Shiny.* BPB Publications.

Medhat, W., Hassan, A., & Korashy, H. (2014). Sentiment analysis algorithms and applications: A survey. *Ain Shams Engineering Journal, 5*(4), 1093–1113.

Naldi, M. (2019). A review of sentiment computation methods with R packages. ArXiv:1901.08319 [cs].

Porter, M. (1980). An algorithm for suffix stripping. *Program* (Vol. 14(3), pp. 130–137). Publisher: MCB UP Ltd.

Rathee, N., Joshi, N., & Kaur, J. (2018). Sentiment analysis using machine learning techniques on Python. In *2018 Second International Conference on Intelligent Computing and Control Systems (ICICCS)* (pp. 779–785).

Schonlau, M., Guenther, N., & Sucholutsky, I. (2017). Text mining with n-gram variables. *The Stata Journal, 17*(4), 866–881. Publisher: SAGE Publications. https://doi.org/10.1177/1536867X1801700406.

Silge, J., & Robinson, D. (2017). *Text mining with R: A tidy approach.* Beijing, Boston: Oreilly & Associates Inc.

Zhang, L., Wang, S., & Liu, B. (2018). Deep learning for sentiment analysis: A survey. *WIREs Data Mining and Knowledge Discovery, 8*(4), e1253. https://onlinelibrary.wiley.com/doi/abs/10.1002/widm.1253.

Zong, C., Xia, R., & Zhang, J. (2022). *Text data mining* (1st ed.). Springer-Nature New York Inc.

Author Index

Index

© The Editor(s) (if applicable) and The Author(s), under exclusive licence to Springer   389
Nature Switzerland AG 2023
G. Cerulli, *Fundamentals of Supervised Machine Learning*, Statistics and Computing,
https://doi.org/10.1007/978-3-031-41337-7